Origins of Semiosis

Approaches to Semiotics
116

Mouton de Gruyter
Berlin · New York

Origins of Semiosis

Sign Evolution in Nature and Culture

Edited by

Winfried Nöth

Mouton de Gruyter
Berlin · New York 1994

Mouton de Gruyter (formerly Mouton, The Hague)
is a Division of Walter de Gruyter & Co., Berlin.

♾ Printed on acid-free paper which falls within the guidelines of the ANSI to ensure permanence and durability.

Library of Congress Cataloging-in-Publication-Data

Origins of semiosis : sign evolution in nature and culture / edited by Winfried Nöth.
p. cm. – (Approaches to semiotics ; 116)
Includes bibliographical references (p.) and index.
ISBN 3-11-014196-5 (acid-free paper) : 1. Semiotics. 2. Evolution. 3. Language and languages – Origin. 4. Communication – Philosophy. I. Nöth, Winfried. II. Series.
P99.075 1994
302.2–dc20 94-21517
CIP

Die Deutsche Bibliothek – Cataloging-in-Publication-Data

Origins of semiosis : sign evolution in nature and culture / ed. by Winfried Nöth. – Berlin ; New York : Mouton de Gruyter, 1994
(Approaches to semiotics ; 116)
ISBN 3-11-014196-5
NE: Nöth, Winfried [Hrsg.]; GT

"Structure as idyll" Copyright of photos: Roland Posner
Arbeitsstelle für Semiotik, Technische Universität Berlin 1994.
Reproductions: Eckhard Muser, Berlin.

Printing: Gerike GmbH, Berlin. – Binding: Lüderitz & Bauer, Berlin. Printed in Germany.

For Walter A. Koch
From his students, colleagues, and friends

Contents

Introduction

Winfried Nöth

The life of signs in the minds of their users is an infinite process. As Peirce has taught us, every act of semiosis creates new signs in a dialogic process of interpretation, taking place not only between different interpreters, but also within the interpreter's own semiotic self.[1] Such processes of semiosis can only be "interrupted", but never really be "ended" (Peirce, CP 5.284). Can there be an origin to this never ending process, or does semiosis begin with the evolution of our cosmic universe (cf. Anderson 1989)?

The question of our origins and its semiotic representation was a topic once hotly debated by our orthodox forefathers, who asked:[2] "Should Adam be depicted with or without a navel?" Drawn with a navel, Adam could not have been the first of humankind. Drawn without a navel, he would not have been a complete man. In modern semiotics, Adam's navel has been the so-called semiotic threshold, the dividing line between the semiotic and the nonsemiotic universe (cf. Nöth 1990: 81-83). Some modern semioticians, guided by the paradigm of language, have pleaded for drawing Adam without a navel by setting up the semiotic threshold at a level as high as the human mind. The authors of the present volume vote for Adam with a navel and are resolved to pursue their search for the origins of semiosis further back in the evolution of life.

Semiogenesis, the evolution and development of semiosis since its origins, is a broad field of research. According to Koch (1982b: 18), it "covers the interrelated geneses of such phenomena as animal communication, computer systems, oral language, written language, theater, film, gestural systems, philosophy or religion as languages, the language of the drums and flags, pheromones or logic as sign systems, DNA as genetic 'code', painting etc. . . ., an array of apparently heterogeneous and yet somehow isomorphic systems." There can be no search for *one* origin of a field as broad as this one. Instead, a plurality of origins and a plurality of evolutionary and developmental paths from primitive semiosis to the most complex and highly evolved forms of present-day semiosis will have to be pursued.

1. Semiosis and semiotics

Semiosis, according to Peirce, is the "intelligent, or triadic action" of a sign (CP 5.472). The triad consists of the sign, as a signifying stimulus, the object represented by the sign, and the "significate outcome" of the sign in the mind of its interpreter (CP 5.473). Semiosis is thus the process in which a sign has a cognitive effect on its interpreter (cf. CP 5.484). Semiotics is then "the doctrine of the essential nature and fundamental varieties of possible semiosis" (CP 5.488).

The scope of semiotics has been defined in a narrower and in a broader sense (cf. Nöth 1990: 81-83). The narrower definition has been advocated in the tradition of Saussure's semiolinguistics, or better linguosemiotics, which considers language as the sole key to the world of semiosis. Only language gives structure to our perceptions of the world. Since "nothing is distinct before the appearance of language", the nonsemiotic world is "a vague, uncharted nebula" (Saussure 1916: 111-112). Such and other anthropocentric views of semiosis have obstructed the view of semioticians for an evolutionary perspective of semiosis.

The evolution of semiosis can only be studied in the framework of a semiotics transcending the limitations of anthroposemiotics. Such a perspective is inherent in Peirce's view of the "entire universe" being "perfused with signs, if not composed exclusively of signs" (CP 5.488, fn.). One of the broadest variants of this pansemiotic view of semiosis is the one advocated by Koch (1986a: 54), who defines semiotics as a metadiscipline which "enables us to discover traces of semiosis . . . in everything: from original atoms through extravagant thoughts". In contrast to this partly pansemiotic view of the universe, which discovers traces of semiosis as early as the origins of cosmos, Peirce's semiotic view of the universe requires the presence of an interpreting mind as an essential criterion of the beginning of semiosis, since "nothing is a sign, unless it is interpreted as a sign" (CP 2.308). On the other hand, semiosic interpretation, "intelligence", and mind, according to Peirce, are not categories restricted to *Homo loquens*. Semiosis begins with the evolution of life.

2. Biosemiosis and the semiotic threshold

In the evolution of our universe, biogenesis evolves parallel to semiogenesis. As Sebeok (1986: 15) points out, cosmic evolution before biogenesis was mere informational evolution. With biogenesis, by contrast, we reach the semiotic threshold at which signs begin to emerge. According to Sebeok (1986: 15),

> semiosis is what distinguishes all that is animate from the lifeless. Before semiosis, there was information. According to the standard (big bang) theory of cosmic genesis, the universe began totally devoid of information, but it quickly evolved out of that initial state as a measure of the multiform. The essence of information is change, the prerequisite of semiosis is life.

To establish the semiotic threshold at an evolutionary point as low as the origin of life has not remained undisputed by semioticians (cf. 1.). Much of the debate about the distinction between signs and non-signs, however, has been of a terminological nature, concentrating on such questions as whether signals or indices are "already" or "not yet" signs or whether signs must be intentional and necessarily intraspecific (cf., e.g., Schult 1992). Semioticity, however, is a matter of degree, and there is a hierarchy of criteria by which lower forms of semiosis can be distinguished from higher ones (cf. Nöth 1990: 169). A hierarchy of criteria to determine several lower thresholds in the biological evolution of semiosis has been set up by Bentele (1984: 138ff.). Further thresholds leading to ever higher levels of semiosis in human communication have been distinguished by Posner (1992). Koch (1986a: 54), by contrast, emphasizes not the dividing line, but the continuity between the non-semiotic and the semiotic world, when he proposes his Evolutionary Cultural Semiotics as a project to "embark upon the reconstruction of an objective, ultimately indivisible, flow of reality, of which the evolution of matter, of life, of consciousness, of culture, and of signs are but special points of emergence".

Nevertheless, in spite of all evolutionary continuity, the emergence of signs from the world of mere physical processes remains to be determined. Peirce draws this dividing line between dyadic and triadic interactions (see also Nesher 1990: 2, 4). His semiotic threshold is the one which leads from dyads of interactions between physical causes and effects to triadic interactions in which an organism interprets (forms an *interpretant* of) its environment, the signifying stimulus (*representamen*), relative to a goal (the *object*) which is distinct from the environmental stimulus.

Not every stimulus-response interaction involves semiosis. Consider the event of a stone dropping by chance (stimulus) on an organism that reacts with death to the ensuing injury. This interaction exemplifies dyadic, and not triadic, interaction. Semiosis begins with responses that are goal-directed (Short 1986: 114), but requires neither consciousness nor intentionality (unless intentionality is defined as the pursuing of a goal). Metabolic "reactions" of an organism to environmental stimuli are goal-directed actions. Organisms select, and hence evaluate, environmental energy or matter *for the purpose*[3] of their own survival, while at the same time rejecting other environmental stimuli as unsuitable. This process exemplifies semiotic thirdness at the lowest threshold of semiosis. Peirce even saw the action of a "mind" and psychology involved at this level. Furthermore he postulated that the agent need not be an individual but can be a whole species in its process of evolution:

> Mind has its universal mode of action, namely by final causation. The microscopist looks to see whether the motions of a little creature show any purpose. If so, there is mind there. Passing from the little to the large, natural selection is the theory of how forms come to be adaptive, that is, to be governed by a *quasi* purpose. It suggests a machinery of efficiency to bring about the end . . . But the being governed by a purpose or other final cause is the very essence of the psychical phenomenon, in general. (Peirce CP 1.269)

3. Framework for the evolution of semiosis

A holistic evolutionary world model to determine the emergence of semiosis within a continuum of discrete stages, beginning with cosmogenesis and continuing to ever higher systems of sign use, has been developed by W.A. Koch in his writings since 1973.[4] According to this model, the world of semiosis emerges from the origins of cosmos in the following successive stages (Koch 1986a: 14-20; 1986b: 12; 1987a: 2; 1991c: 217):

(1) *Cosmogenesis* or physicogenesis (Koch 1987b: 67): beginning 15 billion years ago with the physical expansion of the universe at the moment of the big bang (Koch 1986c: 93-114).
(2) *Galactogenesis*, galactic evolution: 12.5 billion years.
(3) *Geogenesis*, the origin of earth: 4.5 billion years.
(4) *Biogenesis*, the appearance of life on earth (from bacteria to protozoa): from 4 to 1.7 billion years ago.
(5) *Sociogenesis*, the origins of social organization: 500 million years ago.
(6) *Semiogenesis* [in the narrower sense], the origins of human culture: beginning with *Homo habilis* 2 million years ago.

(7) *Glottogenesis*, origins of language: 50,000 years ago.
(8) *Eikonogenesis*, origins of pictorial representation: 30,000 years ago.
(9) *Graphogenesis*, the invention of writing: 5,000 years ago.
(10) *Typographogenesis*, the invention of printing: 1400 A.D.
(11) *Mediogenesis*, the diffusion of television: 1950 A.D.

In contrast to the terminology adopted in this introduction and elsewhere in this volume, semiogenesis as the sixth evolutionary level of Koch's model does not refer to the semiotic threshold at which semiosis begins, but to a higher prototype of semiosis, the one in which artefacts begin to be used as sign vehicles in anthroposemiotics (Koch 1986c: 126). The earliest beginnings of semiosis are at the lower levels of biogenesis and sociogenesis where they emerge in the form of biosemiosis, zoosemiosis, and sociosemiosis. The genetically preceding levels, from cosmo- to geogenesis, provide the material basis of semiosis, but the sphere of the material world also evinces structural isomorphies with the world of semiosis. These isomorphies in the evolution from matter to semiosis are indicators of evolutionary continuity and coherence and justify the assumption of an indivisible evolutionary whole with traces of semiosis since the very origin of cosmos (cf. Koch 1987b: 54, 71).

Besides its chronological dimension, evolution has a variety of aspectual facets for whose study Koch (esp. 1992) provides an elaborate conceptual framework. In contrast to scholars who emphasize the differences between *evolution* and individual *development*, Koch emphasizes the similarities between *genesis* in *phylogeny*, the evolution of the species, and genesis in *ontogeny*, the process of individual maturation. Following Ernst Haeckel's biogenetic rule, Koch (1992: 172, 182) formulates a general principle of *echogenesis* according to which phylogenetic processes tend to be recapitulated ("echoed") in the processes of ontogeneses and other geneses.

Genesis, evolution at large, is furthermore distinguished from *metagenesis*, the discovery and re-experiencing of genesis in human consciousness, e.g., in religious *mythogenesis*, philosophical theories of *cosmogenesis*, or *epistemogenesis*, the "epistemic reconstruction of the entire objective-subjective world" (cf. Koch 1986a: 10, 17; 1991a). The relation between both is typically one of chronological reversal: metagenesis tends to mirror genesis by ascribing historical primacy to phenomena which evolved later in the course of evolutionary history. For example, whereas *Homo sapiens* comes last in genetic evolution, the human I is imagined to be first in mythogenesis and epistemogenesis.

Besides ontogenesis, Koch (1990: 23-33; 1992: 181-186) distinguishes several other variants of developmental evolution, including *actogenesis*, the process of production and reception of signs in actual semiosis, and *eugenesis*, the processes of regressive simplification as well as progressive complexification of semiotic systems in their historical development. The latter mode of genesis is related to two more general evolutionary principles, *anagenesis*, the complexification of matter, and *catagenesis*, the decomplexification of matter.

4. Preview

Koch's holistic model provides a useful framework for the contributions to the present volume. *Origins of Semiosis* considers six evolutionary levels, biogenesis (Part I), anthroposemiotic, sociogenesis, and cultural semiogenesis (Part II), glottogenesis (Part III), eikonogenesis, and graphogenesis (Part IV), but there are also excursus on cosmogenesis (Nöth) and typographogenesis (Wienold).

Part I covers general aspects of the theory of evolution and focuses on the level of biogenesis (cf. Koch 1982a, 1984, 1985, 1986b, 1992) including evolutionary aspects of zoosemiotics. In contrast to the present introduction, which seeks to *differentiate* between various evolutionary levels, Myrdene Anderson opens the volume with a paper on the ways and means of *transition* between the stages of evolution. Udo L. Figge, exploring the interaction of the organism with its biosemiotic umwelt, reveals the evolutionary primacy of receptive in contrast to productive semiosis and concludes that the latter has only developed as an epiphenomenon of nonsemiotic processes. Winfried Nöth investigates the role of opposition in biosemiosis and at other evolutionary levels from cosmogenesis to glottogenesis. Signe and Holger Preuschoft examine the differences between the communicative potential of the four major channels of nonverbal communication in primate semiosis with reference to the evolutionary paths that lead to anthroposemiosis.

Part II deals with anthroposemiotic sociogenesis and cultural semiogenesis (cf. Koch 1982a, b, 1983, 1984, 1986a, 1986c, 1987a, b; Koch, ed. 1982, 1989a, b, c, d, 1990a). The sociobiologist Peter Meyer opens with a paper on the survival value of certainty in species specific communication and on the function of emotional semiosis in the phylogeny of human communication. The literary semiotician Peter Grzybek follows with an investigation of the anthroposemiotic sphere of mesocosm and its mythological, cognitive, and epistemogenetic place between microcosm and macrocosm. The potential of visual indexical semiosis in hominid and early human reading of animal tracks is the topic of the paper by the anthropologist Gordon W. Hewes. The following ethological contribution by Paul Leyhausen and Jörg Pfleiderer presents the empirical results of an investigation into the gestural and facial features which constitute the male intimidation display as a human innate releasing schema. Both phylogenetic and ontogenetic perspectives are pursued by Susan Vogel in her investigation of the sociosemiotic and cultural functions of play in mother-child interaction. The final paper on infant semiosis is by the child psychologist and psychobiologist Colwyn Trevarthen gives a comprehensive semiotic account of the nature and stages of protosymbolic communication from the first days of semiotic ontogeny until the emergence of speech.

Part III deals with glottogenesis in its broadest sense (cf. Koch 1970, 1974b, 1991b; Koch, ed. 1990b), covering language phylogeny and ontogeny, attempts at linguistic reconstruction of prehistorical languages, and neurological actogenesis. Robin Allott introduces the broader semiotic framework of his motor theory of language origins which states that human language has evolved from earlier neurological programs for the control of body movements and that speech production programs have become cross-modally connected in the human brain with those earlier programs of nonverbal behavior. Bernhard Bichakjian interprets language evolution as a Darwinian process of *eugenesis* (Koch 1992: 184), which has favored the historical selection of ever more advantageous linguistic features as these develop in early language acquisition (paedomorphosis). The Russian linguist Vitaly Shevoroshkin dares to pursue the historical reconstruction of language not only as far as Nostratic, the assumed proto-language of the Indo-European, Afro-Asiatic, Uralo-Dravidian, and Altaic languages, but even further back via Austric-Australian, Sino-Caucasian, Niger-Nilotic, and Southern-African Khoisan to a Proto-World language (see also Koch 1991b: 35-36) spoken some 90,000 years ago. Roland Harweg investigates the genesis of language referring to time, demonstrates that event-related "material" time references are primary as opposed to clock- or calendar-related "formal" time references, which have evolved from primitive reifications of the solar and lunar motions, and relates these findings to Koch's thesis of metagenesis as the chronological reversal of genesis. Yoshihiko Ikegami explores the primacy of the agent (human source) in relation to the role of the sentient (human goal) in syntactic structure, relates this observation to the bodily basis for semiosis in action and sensation, and reveals cultural differences in the encoding of these categories in the form of narrative personae. The last contribution to part III, Helmut Schnelle's paper on "Language and Brain", is concerned with the neurological actogenesis (Koch 1990: 29, 1992: 183) of language. With his insights into the neurobiological, neuroanatomical, and neuro-logical foundations of language and cognition, Schnelle also provides a frame of reference for other papers of this volume dealing with neurosemiotic topics (e.g., Bouissac, Meyer).

Part IV deals with phylogenetic and ontogenetic aspects of eikonogenesis and graphogenesis (cf. Koch 1981, 1982a, 1983, 1984, 1985, 1986a, b, c, 1989, 1993a, b; Koch, ed. 1982, 1989a, b, c, d, 1990b). In an interchapter, Roland Posner presents a contribution on "Structure as Idyll" which constitutes a link between the chapters on natural and cultural genesis: inspired by Koch's structuralist visions of homologies between nature and culture, Posner wants to give an iconopoetic illustration of the principle of meaning attribution, the tendency of humans to attribute meaning, under certain conditions, even to the phenomena of nature. In his exemplification of this principle, Posner invites the reader to explore his or her semantic reactions at the perception of a sequence of scenes from nature, which appear first in black and white and then in color.

Part IV opens with the paper by the paleo-ethnologist Emmanuel Anati on constants in 40,000 years of art which develops seventeen theses on the basic patterns and archetypes of rock art from twenty major sites around the world. Paul Bouissac evolves the thesis that there are two evolutionary roots of spatial and aesthetic semiosis, an older deictic (indexical) root, which humans share with other vertebrates, and a more recent nondeictic root, which only humans evince in representational or modeling activities. Both phylogenetic and ontogenetic aspects of eikono- and graphogenesis are the topic of Martin Krampen's paper dealing with the development of pictorial and graphemic representation in children as compared with nonhuman primates. Aspects of the origins of writing in pictorial representations are discussed in Richard W. Janney and Horst Arndt's investigation of the semiotic potential of writing in comparison to pictorial representation. The final contribution by Götz Wienold explores text semiotic functions of the ancient text type of inscription and places it in the context of the evolution of the alphabet from its earliest beginnings to 20th century typography.

The all-comprehensive scope set by the topic of *Origins of Semiosis* necessarily encompasses a great thematic diversity. The disciplinary horizon of the contributors to this volume is also highly diverse: it extends from anthropology to zoology and from linguistics to visual aesthetics. Unity in this diversity may be provided by the transdisciplinary framework of semiotics which guides the authors' explorations in the evolution of semiosis, even though the framework of this volume is by no means one of a unifying school of specific semiotic theory. All of the contributors to *Origins of Semiosis* have previously participated in one or several of Walter A. Koch's inspiring conferences in Evolutionary Cultural Semiotics (e.g., Koch, ed. 1989a, b, c, d, 1990a, b, c) and have known him as a colleague, friend, "indefatigable teacher, writer, editor, organizer, visionary, and seminal figure in semiotics", a "rock in the stream" of semiosis, as Myrdene Anderson (1990: 137) has so aptly described him. The diversity offered by the perspectives of *Origins of Semiosis* find an additional unity in the authors' dedication of this volume to Walter A. Koch.[5]

Notes

1. For references to Peirce's theory of semiosis as an unlimited process, see Nöth (1990: 43).
2. For details see Kirchner (1903: 18-19).
3. From the perspective of a biologist, Thorpe (1978: 36-76) defines the origins of life as the appearance of purposiveness.
4. For Koch's writings in evolutionary and developmental semiotics, see Koch (1973a, 1974b, 1981, 1982a, b, 1984, 1985, 1986a, b, c, 1987a, b, 1989, 1990, 1991a, b, c, 1992, 1993a, b).
   The evolution of European semiotics since the 1960s from its roots in structural linguistics to ever expanding horizons is mirrored in the development of Koch's own writings. Beginning with diachronic structural research at the level of the phoneme (Koch 1970 [first ed. 1963]), his research program progressively expanded to higher levels of linguistic analysis, via the morpheme to the texteme (Koch 1969, 1971a, 1973, 1974a). With his studies in linguistic and semiotic poetics, Koch (1966, 1981, 1983, 1984) transcended the limits of the language system. The framework of language itself is transcended in Koch's (1971, 1973a, 1974b) influential studies in general and applied semiotics. The next step of this expansive development, the one to evolutionary semiotics (see above), is both a step beyond semiotics in its synchronic and anthropocentric tradition and a step leading back to the origins of Koch's own research in diachronic linguistics, the field linking up with the prehistory of modern semiotics in historical linguistics.
5. The editor is greatly indebted to Susan Vogel for invaluable editorial assistance from the point of view of the semiotically competent native

speaker. Thanks are also due to technical and editorial help by Renate Förster-Fernández, Guido Ipsen, Anne Kroesche, and Jan Siebert.

References

Anderson, Myrdene
1989 "Cosmological semiotics", in: W.A. Koch (ed.) 1989b,21-30.
1990 "Decocting Koch", *Semiotica* 82: 137-151.
Anderson, Myrdene – Floyd Merrell (eds.)
1991 *On semiotic modeling*. Berlin: Mouton de Gruyter.
Ballmer, Thomas T. (ed.)
1985 *Linguistics and dynamics*. Berlin: de Gruyter.
Bentele, Günter
1984 *Zeichen und Entwicklung*. Tübingen: Narr.
Deely, John, Brooke Williams, Felicia E. Kruse (eds.)
1986 *Frontiers in semiotics*. Bloomington: Indiana University Press.
Jorna, René, Barend van Heusden, Roland Posner (eds.)
1992 *Signs, search, and communication: Semiotic aspects of artificial intelligence*. Berlin: de Gruyter.
Kirchner, Josef
1903 *Die Darstellung des ersten Menschenpaares*. Stuttgart: Enke.
Koch, Walter A.
1966 *Recurrence and a three-modal approach to poetry*. The Hague: Mouton.
1969 *Vom Morphem zum Textem*. Hildesheim: Olms.
1970 *Zur Theorie des Lautwandels*. (2nd edition.) Hildesheim: Olms.
1971a *Taxologie des Englischen*. München: Fink.
1971b *Varia semiotica*. Hildesheim: Olms.
1973a "Einleitung", in: W.A. Koch (ed.), XI-LV.
1973b *Das Textem*. Hildesheim: Olms.
1974a "Tendenzen der Linguistik", in: W.A. Koch (ed.), 190-311.
1974b "Semiotik und Sprachgenese", in: W.A. Koch (ed.), 312-346.
1981 "Evolution des Kreativen", in: H. Schnelle (ed.), 158-173.
1982a "The biogenesis of language, literature, and art," in: W.A. Koch (ed.), 447-465.
1982b "Semiogenesis", in: W.A. Koch (ed.), 15-104.
1983 *Poetry and science*. Tübingen: Narr.
1984 "Art: Biogenesis and semiogenesis", *Semiotica* 49: 283-304.

1985 "Tension and suspense: On the biogenesis and the semiogenesis of the detective novel, soccer, and art", in: T.T. Ballmer (ed.), 279-321.

1986a *Evolutionary cultural semiotics*, trans. by Susan Vogel. Bochum: Brockmeyer.

1986b *Genes vs. memes*. Bochum: Brockmeyer.

1986c *Philosophie der Philologie und Semiotik*. Bochum: Brockmeyer.

1987a *Hodos and kosmos*. Bochum: Brockmeyer.

1987b "A plea for 'evolutionary cultural semiotics'", in: W.A. Koch – A. Eschbach (eds.), 53-131.

1989 *The well of tears*. Bochum: Brockmeyer.

1990 "The naturalness of language", in: W.A. Koch (ed.) 1990c: 2-42.

1991a *Gott und die Welt*, Bochum: Brockmeyer.

1991b *Language in the upper pleistocene*. Bochum: Brockmeyer.

1991c "Worlds and signs: Explicate and implicate order", in: M. Anderson – F. Merrell (eds.), 211-246.

1992 "Ecogenesis and echogenesis", in: T.A. Sebeok – J. Umiker-Sebeok (eds.), 171-211.

1993a *The biology of literature*. Bochum: Brockmeyer.

1993b *The roots of literature*. Bochum: Brockmeyer.

Koch, Walter A. (ed.)

1973-1974 *Perspektiven der Linguistik*. Vol. 1-2. Stuttgart: Kröner.

1982 *Semiogenesis*. Frankfurt/Main: Lang.

1989a *The nature of culture*. Bochum: Brockmeyer.

1989b *Das Ganze und seine Teile – The whole and its parts*. Bochum: Brockmeyer.

1989c *Culture and semiotics*. Bochum: Brockmeyer.

1989d *Evolution of culture*. Bochum: Brockmeyer.

1990a *Aspekte einer Kultursemiotik*. Bochum: Brockmeyer.

1990b *Geneses of language*. Bochum: Brockmeyer.

1990c *Natürlichkeit der Sprache und der Kultur*. Bochum: Brockmeyer.

Koch, Walter A. – Achim Eschbach (eds.)

1987 *A plea for cultural semiotics*. Bochum: Brockmeyer.

Nesher, Dan

1990 "Understanding sign semiosis as cognitition and as self-conscious process", *Semiotica* 79: 1-49.

Nöth, Winfried

1990 *Handbook of semiotics*. Bloomington: Indiana University Press.

Peirce, Charles Sanders
1931-1958 *Collected papers of Charles Sanders Peirce*, vols. 1-6 ed. Charles Hartshorne and Peter Weiss, vols. 7-8 ed. Arthur W. Burks. Cambridge, Mass.: Harvard University Press.

Posner, Roland
1992 "Believing, causing, intending: The basis for a hierarchy of sign concepts", in: René J. Jorna et al. (eds.), 215-270.

Saussure, Ferdinand de
1916 [1969] *Course in general linguistics*, trans. by Wade Baskin. New York: McGraw Hill.

Schnelle, Helmut (ed.)
1981 *Sprache und Gehirn*. Frankfurt: Suhrkamp.

Schult, Joachim
1992 "Species, signs, and intentionality", in: Thomas A. Sebeok – Jean Umiker-Sebeok (eds.), 317-332.

Sebeok, Thomas A.
1986 *I think I am a verb*. New York: Plenum.

Sebeok, Thomas A. – Jean Umiker-Sebeok, (eds.)
1992 *Biosemiotics*. Berlin: Mouton de Gruyter.

Short, Thomas L.
1986 "Life among the legisigns", in: John Deely et al. (eds.), 105-119.

Thorpe, W.H.
1978 *Purpose in a world of chance: A biologist's view*. Oxford: University Press.

PART I

Evolution and biosemiosis

Breathing life into signs: Ways and means of semiosic transition and transformation

Myrdene Anderson

1. Sleight of sign

Taking off from the assumption that in Indo-European languages we habitually tend to "thingify" signs (even life), I aim to dynamicize things, pushing and pulling them into open-ended temporal, processual, even if unmannered, flows. Indeed, what follows reveals an on-going tug-of-peace between the comfort of digital categories and the disconcerting resolution via interpolated analogue processes. Even the distinction of digital and analogue dissolves into dichotomies or continuities, depending on one's perspective and one's purpose.

The perspective and the purpose here build on the metaphor of "breathing". Breathing reveals diphasic, complementary processes of "in" and "out". Yet there are tangible moments of "both" and "neither" – at least more tangible than those conditions in the flipping of an electric light switch. Breathing also suggests a living process, something shared in the circadian and metabolic rhythms of all known and all imagined life forms. Life forms have inspired a discourse speculating on evolutionary origins and on developmental births and deaths, and that discourse of evolution and development will be tapped as well.

To address the origin(s) of semiosis entails suspending – which is tantamount to suppressing – all awareness of our limited knowledge of the perhaps unknowable. We must suspend as well the shortcomings and flaws in our evident knowledge of the presumed knowable, that knowable sitting still long enough to be amenable to our habits of understanding. In the spirit of provisionality of constructions, I specify "origins" as well as "origin" to underline the possibility, even likelihood, of multiple sources, multiple perspectives, or just muddied ones. The important distinction between knowledge and understanding also merits serious but separate treatment.

Any of this knowledge *and* knowing, and their metonymic complements of lacking knowledge *and* unknowing *and* unknowability, if not suppressed would otherwise soberly encumber our exercise. In fact, were we unwilling to make this congenial gesture, our interrogation of origin(s) would be rendered ridiculous, wrong-headed, futile, and further, even pathologic.

It must be admitted, however, that to suppress either knowledge *or* awareness of lack of knowledge *or* provisionality of knowability, also entails costs on the debit side of the ledger – albeit a forever open-ended and dedicatedly *un*balanced ledger. Do we, in pinpointing "origins" of semiosis, guarantee enlightenment as to the contemporary shape or habits of semiosis? Not necessarily, I argue. Will contemporary processes of semiosis and of our semiosic inquiries enlighten their origins? Not necessarily, again, but contemporary habits of inquiry provide the watershed for our constructions, reconstructions, and deconstructions. Are we able truly to conceive of the origin(s) of time, of space, of matter/energy, of information, that is to say, of semiosis? Not unless our linguistic ability (indeed, proclivity) – to make sentences of statements in response to sentences of inquiry – can convince us that we comfortably conceive of the doubly unmarked (but not therefore marked) "before before" or "beyond beyond". In our relativistic (including linguistic and cultural) universe, no longer can we conjure up a "god's-eye view".

All this notwithstanding, however, semioticians do explore such paths of knowing and unknowing as a matter of course, in pursuit of the sign and of signing. To tackle the origin(s) of the sign – whether that be consonant with the origin of semiosis or/and the origin of communication – therefore appears a noble enterprise, particularly when Walter A. Koch has laid so much of the groundwork, for example, in two of his books of 1986, *Philosophie der Philologie und Semiotik – Literatur und Welt: Versuche zur Interdisziplinarität der Philologie*, and *Evolutionary Cultural Semiotics: Essays on the Foundation and Institutionalization of Integrated Cultural Studies*. What I will do as a footnote, or a headnote, is to expand on a simple sketch of how systems breathe. Their breathing *is* semiosis, and it commences with the hiccough signaling a discontinuity in a developmental or an evolutionary stream of signs. Following that emergence or origin, there may follow other "sighs" and startles, of various shapes and prognoses (in developmental trajectories), and of uneasy tentative post-hoc patterns (in evolutionary trajectories).

2. Development and evolution

In pursuit of semiosis itself, one looks for emergence, submergence, diversification, convergence, subversion, collapse, extinction, yes – but one is also alert for the overdetermined and underdetermined seams in this fabric. In addition, one calls into question those very seams that appear to partition the units of ordinary analysis, although the seams (boundaries between coordinate categories and edges between non-coordinate ones) are the active sites for semiosis (Anderson 1989a, b, c; Anderson et al. 1984).

Here I will content myself with two genres of emergence, or self-organization, termed development and evolution (Dyson 1985; Karger 1986; Salthe 1993; Salthe and Anderson 1989; Anderson 1990a, b). Development and evolution apply to all semiosic processes, not just the biological ones associated with these terms. Hence, biological development and biological evolution constitute ordinary examples of these generic processes of self-organization.

In biology, development progressively actualizes organisms (tokens) in their ontogenies, and evolution erratically precipitates forms (types), such as species, in their phylogenies. Ontogenies and phylogenies mutually construct each other in a kinky, semiosic flow. Biological entities also realize themselves in a different score when viewed energetically (ingestion maintaining organismic growth and "development" in a larger spatial field of ecology) or informatically (copulation contributing to differential replication and species continuity or discontinuity, perhaps radical evolutionary discontinuity, in a temporal field of genealogy).

The terminology of biology provides handholds, sometimes slippery ones, for the apprehension of the dynamics of other open systems. Biological origins cycle through developmental births (inceptions for finite, relatively closed loops or stageal spirals) and through evolutionary speciation (threaded discontinuities in utterly open emergent processes). In other fields, the deployment of the terms "development" and "evolution" has been less consistent and often not disambiguated at all, the terms used interchangeably or only one of them selected for all dynamical processes involving continuity and change. In the corporate world, "research and development" maps relatively well onto "evolution and development". "Economic development" is viewed deterministically, that is, in predictable and indeed developmental stages, although many of the dynamics are not accessible to any apriori assumptions. Marx's evolution was decidedly developmental, in that historical and social time had a shape and direction. Historians tend to find developmental "rises and falls" in epochs, although occasionally it is posited that the person makes history (evolutionarily) rather than history making the person (developmentally). We can follow the gyrations of development and evolution in the unfoldings of tragedy (suspenseful) and comedy (surprising).

A major feature of developmental beginnings and ends is their relative predictability. Development signals itself through stages, and these stages can be gingerly anticipated, albeit fostering suspense in the observer. Evolution in contrast fosters surprise. Evolution has no shape, except that which can be imposed after the fact. Indeed, after the fact evolution will appear developmental, even though we cannot predict its shape into the future.

One reason that past evolution appears developmental has to do with our habits of speech. As languaging creatures, humans (including scientists) tell stories, and our stories are temporal, linear, with beginnings, middles, and ends. We have difficulty portraying in speech simultaneity or contingency. All the possibilities inherent in every instant of an evolutionary process (and to a much lesser extent, every instant of developmental processes too), remain vague if not utterly opaque. All we have in our reconstructions are stories of the temporal flow of events. We are seldom confident as to the mix of "chance and necessity" in the events we do uncover.

Anthropologists concern themselves with the origins of any number of sign processes. One category of origin would be the evolutionary emergence of a string of biologically ancestral species within the realms of animals, vertebrates, mammals, primates, and humans, each through accidents of punctuated equilibria. But the narrative of the prehistory of these forms will often smack of development – from small to large, from shorter-lived to longer-lived, from presumably simpler to presumably more complex in form and/or in social organization, Whenever so-called evolution has a shape, as indeed it had for Darwin and later for Teilhard de Chardin, the process resembles the tamer development than the wild evolution.

In every instance, however, the temporal and spatial scale of the figured phenomenon and of the observer – including the focus, perspective, and agenda of the observer – all *participate* in the mode of self-organization. Further, the motivation of the semiotician will never reduce to mere classification. Little will be gained, and much may be sacrificed, if we ponder processes with the intent of attaching a label. In fact, virtually all processes may be explored through both lenses, as both developmental and evolutionary.

3. Transition and transformation

Both development and evolution describe irreversible change, whether gradual (but *not* consequently slow), continuous transitions of degree, or punctuated (but *not* consequently rapid), discrete, transformations of kind. When organismic change is transition-like, or perceived to the observer as gradual, we call it "growth", over against transformation-like organismic stages, called "development". Each of these falls into the genre of metadevelopment as "growth and development". Although growth appears gradual, it occurs in spurts linked to the rest of development. The overall developmental process generates individual tokens of things, and is slightly deterministic in that it manifests temporally-sequenced stages, capped by equifinality. That equifinality in an organismic trope is "death" (Anderson 1991; Biddick 1993).

The developmental stages – of an organism or of a dynasty or of a scientific paradigm – are often placed in a syntagmatic narrative having a beginning, middle, and end. What comes *before* the beginning, in terms of an origin, is often off the discursive map. In fact, what comes before the beginning, is not the sigh of development but the hiccough of evolution. Evolution precipitates new twists, new genres, surprises that could not have been anticipated.

When formal change is transition-like, we are still able to find the continuous threads of change, expressed as statistical modifications of degree rather than of kind. Transition-like change in a *token* may conveniently be understood as developmental. When formal change in a *token* is transformation-like, a modification of kind within the token, we similarly recognize developmental stages. But when formal change in a *type* (or in a token at a bifurcation point) is transformation-like, we confront the evolutionary abduction of a newly created kind and/or process. Basically, the evolutionary process generating the phylogeny of kinds of things is open-ended, nondeterministic, and teleonomic (rather than narrowly teleologic).

Origins cycle through evolutionary phylogenies at the junctures of new sign types, and through developmental ontogenies at the junctures of new sign tokens. Seeking origins in either type of self-organization process entails probing time. Sometimes our inquiries become nonsensical, when the question of "what/how/why before" comes down to "before before". Time itself collapses, as does space, when we push through the edges of our epistemological system.

At the present, semioticians (e.g., Merrell 1994a, b; Gould 1994) tend to insert analogue continua between all conventional digital nodes in time as well as in space and in substance. In part, this might be a reaction to the overdose of digitalities, dichotomies, and classifications generally during the long reign of structuralistic, semiological praxis. While continua may refresh us, and open up those mysterious gaps left unexplored in an earlier semiotic, the present trend may itself become overdetermined and sterile. For one thing, one can not ask questions of origins at all within the emerging paradigm.

4. Signs as futures

Signs reveal as well as conceal the ripples from their kinky, convoluted past, where structures and functions, patterns and relationships, flip-flop with mutation-like abandon (Bohm 1980; Koch 1991). Signs also invite – without determining – their own futures. They spill toward fulfillment, but remain perpetually incomplete, cycling in a dynamic and thereby life-like, execution-driven movement. One could also map this sinuous flow onto Peircean categories, as Nathan Houser (1991) and Floyd Merrell (1991a, 1991b, 1994a, 1994b) among others, already have done with respect to Peirce's ten species of sign.

Because signs flow, by necessity and by accident, they incessantly become themselves (developmentally) and sporadically become other signs (evolutionarily). In so doing, they generate by-products which themselves, indeterminately, contribute to the overall shape of the process. Of course, to say "by-products" may mislead, for it draws our awareness back dangerously close to "products" and to "function". In this just-so linear rationalization of "by-products" and "products" we err in privileging an exclusively developmental mode of thinking, which in larger doses leads to bald teleology.

Indeed, friction and waste are also by-products or unintended consequences of semiotic processes involving, as they do, information and energy. While "by-product" suggests an unintentionality potentially redeemable by *positively* recoverable (or tamed) outcomes, "waste" carries an overwhelmingly *negative* connotation. "Noise" likewise suggests a friction-like drag on the parent system, interfering with its own integrity. Here I wish to neutralize to some degree these connotations. Under certain very normal conditions entailing open, irreversible, nondeterministic systems, "waste" and "noise" can transcend chaos and erupt into order (Prigogine 1984), rendering an instance of "origin".

The metaphor of "waste" to describe the inevitable by-productions of unbridled semiosic flows may still require some laundering before we can use it non- or pre-judgmentally. One potential in the semantic field of "waste" lies in its liminality to the figured system. First of all, "waste" may be on its way into or out of consciousness, out of or into nothingness. Secondly, "waste" may reinforce the figured system or may bootstrap itself from the ground to the figure itself. These behaviors suggest the movement of a system – including either the marked figure or both the figure and its ground – into transition states having unequivocal genealogical connections with antecedent states.

Aside from these integral roles for "waste", and regardless of its degree of liminality or degree of circumscription within the system, "waste" may provide the pivot swinging the system with (or without) waste, or the waste itself, into a transformed orbit which can only be described as evolutionary. It is the evolutionary which generates surprise, and more unequivocal origins. But once set into flow, into respiration, the sign proceeds to develop, precipitating suspense, eventually to spin into originary eddies again.

Semiosic systems appear both profligate and stingy. In their flow of habitual breathing, they hurtle toward new developmental stages and fresh evolutionary states, each opening up generous genres of "origin". At the same time, in the same breath, as it were, semiosic systems succumb to rationalization measures whereby some processes are economized and others eliminated and still others utterly redefined. These are parsimonious genres of "origin", with even the semiosic systems selected out (given dysfunction or by accident) leaving residues for fresh invention.

A favorite image suggesting infinite and incessant continuity-cum-change, transition-cum-transformation, has been the Möbius strip. In such a contorted but still linear universe, the only origin would be hopping onto the strip. I have noted that some semioticians appear to be pushing continuity to its limits as an antidote to the structural slavery to paradigms, taxonomies, typologies, keys, lists, and other dimensional systems of classification.

The intuitive appeal of creation and dissolution, origin and collapse, growth and decay, cannot be sated by a compulsively analogue semiotic, however. Even though each phenomenon can be wrung of digitality, and of difference, by the imposition of interpolation, we are also deprived of any cultural and linguistic and cognitive discontinuities, including origins. A somatic trope such as breathing and a visual image such as an hourglass may prove instructive. Both privilege time, and as such both will be friendly to contemporary moves toward continuity.

Breathing, as has been pointed out, represents all the pervasive diphasic processes in the universe – rhythms with phases of rest and activity, for instance. Is the originary breath inhaling fuels or exhaling wastes, or perhaps their monolithic unity? There are moments of juncture when the system is neither at rest nor in action, or/and both at rest and in action, just as in a stream of vocal sounds in language, each phoneme bleeds into the following one even as it anticipates that following sound.

The being of breathing merges with the becoming of breathing, at the "now" of the waist of the hourglass. Any real hourglass hangs in developmental space, with an evidently finite amount of sand above the waist and an evidently finite space for collection of time-processed sand beneath it. Clearly, the actual hourglass is a human tool-toy provisioned by the human hand and human mind, and fashioned by a human creator, and more subject to entropy than negentropy. Nonetheless, on pondering the hourglass as a catalyst for understanding, we fixate on the behavior of time, not of sand or of hands or of humans. Thus integrating time and the provisional, provisioning extrinsic elements means that the sand above and below the waist/waste of "now" will be infinite, not finite, rendering the hourglass into a vitally *open* model.

Our mind's eye might sketch in a myriad of Möbius strips connecting the grains of sand with futures, hanging above, with the grains of sand with pasts, deposited below. The futures have evolutionary potentials to precipitate into circumscribed developmental pasts. Each grain and each movement possesses a tangle of antecedents and relationships and origins, and an even messier tangle of futures. The Möbius strips become more unruly as they come to resemble the double helixes of DNA, with the complication that the bonding sites can swivel in all directions and not just link one pair of spirals. Synchronic, structural models of the micro-organic appear like spaghetti finger-painting, and our model must acknowledge both time and explicit nonlinearities. The Möbius strips/multiple helixes bifurcate and coalesce with abandon, with different perspectives affording different views to an observer – each perspective facilitating *and* obstructing some useful, some useless, *and* some misleading illusions.

On closer monitoring, the entropic exhale of the dropping grains becomes syncopated, regularly irregular, irregularly regular – more than suspenseful and sometimes surprising, not unlike heartbeats. Many grains and momenta will be negentropically scooped up as the system inhales, creating fresh semiosic watersheds, bringing new futures, new origins, new pasts, new nows, into view.

References

Anderson, Myrdene
1989a "Cosmological semiotics: Rhymes and reasons in the cosmos", in: Walter A. Koch (ed.) 1989b, 21-30.
1989b "Dialectics of emergence and hierarchization in nature-cum-culture", in: Walter A. Koch (ed.) 1989c, 442-455.
1989c "Paradigms of evolution and paradigms of culture", in: Walter A. Koch 1989a, 1-8.
1990a "Biologie und Semiotik", in: Walter A. Koch (ed.) 1990c, 373-392.
1990b "Biology and semiotics", in: Walter A. Koch (ed.) 1990b, 254-281.
1990c "Culture: Evolution or development?" in: Walter A. Koch (ed.) 1990a, 129-140.
1991 "The conduct of artificial life: A quest for artificial death?" in: M. Anderson – F. Merrell (eds.), 195-209.

Anderson, Myrdene – Floyd Merrell (eds.)
1991 *On semiotic modeling*. Berlin: Mouton de Gruyter.

Anderson, Myrdene – John Deely – Martin Krampen – Joseph Ransdell – Thomas A. Sebeok – Thure von Uexküll
1984 "A semiotic perspective on the sciences: Steps toward a new paradigm", *Semiotica* 51: 7-47.

Biddick, Kathleen
1993 "Stranded histories: Feminist allegories of artificial life", *Research in Philosophy and Technology* 13: 165-182.

Bohm, David
1980 *Wholeness and the implicate order*. London: Routledge.

Corrington, Robert (ed.)
1994 *Semiotics 1993*. Lanham: University Press of America.

Dyson, Freeman
1985 *Origins of life*. Cambridge: University Press.

Gould, Stephen Jay
1994 "The persistently flat earth", *Natural History* 103.3: 12-19.

Houser, Nathan
1991 "A Peircean classification of models", in: M. Anderson – F. Merrell, 431-439.

Karger, Angelika H.
1986 *Zeichen und Evolution: Theoretische Grundlagen und Anwendungen semio-morphogenetischer Prozesse*. Köln: Pahl-Rugenstein.

Koch, Walter A.
1986a *Evolutionary cultural semiotics*, trans. by Susan Vogel. Bochum: Brockmeyer.

1986b *Philosophie der Philologie und Semiotik.* Bochum: Brockmeyer.

1991 "Worlds and signs: Explicate and implicate order", in: M. Anderson – F. Merrell, 211-246.

Koch, Walter A. (ed.)

1989a *Evolution of culture.* Bochum: Brockmeyer.

1989b *Das Ganze und seine Teile - The whole and its parts.* Bochum: Brockmeyer.

1989c *The nature of culture.* Bochum: Brockmeyer.

1990a *Aspekte einer Kultursemiotik.* Bochum: Brockmeyer.

1990b *Semiotics in the individual sciences.* Bochum: Brockmeyer.

1990c *Semiotik in den Einzelwissenschaften.* Bochum: Brockmeyer.

Merrell, Floyd

1991a *Signs becoming signs: Our perfusive, pervasive universe.* Bloomington: Indiana University Press.

1991b *Unthinking thinking: Jorge Luis Borges: Mathematics, and the new physics.* West Lafayette, Ind.: Purdue University Press.

1994a "The knowledge sin(drome)", in: R. Corrington [forthcoming].

1994b *Semiosis in the postmodern age.* West Lafayette, Ind.: Purdue University Press [forthcoming].

Prewitt, Terry – John Deely – Karen Haworth (eds.)

1989 *Semiotics 1988.* Lanham: University Press of America.

Prigogine, Ilya, – Isabelle Stengers

1984 *Order out of chaos.* New York: Bantam.

Salthe, Stanley N.

1993 *Development and evolution: Complexity and change in biology.* Cambridge Mass.: MIT Press.

Salthe, Stanley N. – Myrdene Anderson

1989 "Modeling self-organization", in: T. Prewitt – J. Deely – K. Haworth (eds.), 14-23.

Semiotic principles and systems: Biological foundations of semiotics

Udo L. Figge

1. Introduction

Semiotic behavior of animals (including humans) cannot be explained on the basis of only one semiotic principle, e.g., the principle of communicative needs. On the contrary, it is important to distinguish two semiotic principles: one principle which underlies receptive semiotic behavior and deals with the further processing of information extracted from sensory input, and a second principle which underlies productive semiotic behavior, using some existing effector organ to express inner states that would otherwise remain imperceptible.

Semiotic behavior is based on semiotic systems. A semiotic system establishes relationships between certain inner states of an organism that normally are inaccessible and imperceptible from outside and its periphery, making those states accessible according to the first semiotic principle or perceptible according to the second principle. A partial semiotic system realizes either the first or the second semiotic principle. A complete semiotic system is a combined realization of both principles in one and the same individual.

2. Semiotic principles

The world is made up of matter and energy. Consequently, an organism can only exchange matter and energy with its environment. Matter taken in by organisms is, for the most part, directly transmitted in the process of metabolism. Energy, however, can be assimilated only in plants and photobacteria by means of photosynthesis. In animals, energy coming from outside either fades away (e.g., energy transmitted by a blow) or stimulates organs sensitive to energy, i.e., sense organs, thus generating *information*. In animals, information can thus result from a specific processing of energy. There are, however, also sense organs that transform matter into information, namely, the chemical senses. All such information, whether going back to energy or to matter, contributes to the concept an organism forms of its environment.

Now, it can happen that an organism does not use newly acquired information as such, directly adding a new element to its representation of the world, but rather goes on processing it. This kind of further processing results in a state inside the organism that is quite dissimilar from the original information. Thus, a bee stimulated by energy to form the percept of a certain color may transform this percept into the concept of nectar. As nectar is almost colorless, this concept has nothing to do with the original color percept (which goes back to petals). This example illustrates the following first semiotic principle:

> *First semiotic principle: The subsequent processing of energy or matter by an organism results in a particular inner state.*

"Subsequent processing" of this sort is something like a diversion of the usual flow. That is why I like to call it "oblique".

One of the effects of metabolism is the excretion of matter. Moreover, metabolism enables organisms to expend energy so that they can act upon their environment. Now, certain inner states of organisms which, in principle, are not perceptible to their environment are coupled with peripheral organs in a way that certain forms of expending energy or excreting matter function as manifestations of these inner states. A typical example of such an inner state is mating motivation. It is a genuine inner state, insofar as it normally passes unnoticed by the environment. Animals, however, have learned to use various peripheral organs in order to manifest mating motivation, e.g., glands produce smelling substances (matter) or vibrating organs produce sound waves (energy). These are examples of the following second semiotic principle:

> *Second semiotic principle: Effector organs are used to manifest inner states that are otherwise imperceptible.*

"Manifesting" in this context means "making perceptible in an indirect manner".

The first semiotic principle already proves advantageous whenever obliquely triggered inner states allow a better adaptation of an organism's behavior to its environment. It makes possible an oblique gain in environmental information leading to an enrichment of the organism's conception of the world. Thus, bees must have improved in foraging behavior since the moment they became enabled, by the first semiotic principle, to transform obliquely a color percept into the representation of a nearby nectar source. However, this principle assumes a new quality, as it were, when the energy or the matter taken in by an organism is emitted by another organism of the same species according to the second semiotic principle. In this case, the organism has the possibility of arranging its interaction with organisms of its own species in such a way that its fitness will increase. This is clearly the case when the inner state triggered off obliquely is a motivation to mate with an individual of the same species which has manifested its readiness to mate.

3. Semiotic systems

An organic system is separated from its environment by its outer integument. Nevertheless, there are influences which the system exerts on its environment and which the environment exerts on the system. Accordingly, the system can assume states that are almost autonomous with regard to the environment as well as states due to environmental influences or aiming at an effect on the environment. An example of an autonomous state is the hormone level of an organism. Hormones have to do with the self-regulation of the organism. An effect of the environment on an organism can be the perception that an enemy is approaching. Such a perception can call forth a defensive reaction, e.g., the secretion of poison. The states underlying perception and secretion are examples of states an organic system can assume in connection with interactions between the system itself and its environment. Now, there are certain inner states which are autonomous with regard to the environment, but whose interaction with the environment would be advantageous to the organism if such interaction were possible. It is from such inner states that the *development of semiotic systems* (semiogenesis) originates.

A prominent example of such an inner state is mating motivation. Without doubt, it would be to the advantage of an organism if its mating motivation led to copulation immediately. In order to copulate, however, an individual needs a partner. Now, if partners are not at hand at the right time or are at hand, but not in a mating mood, copulation becomes impossible, and the mating motivation fades away without any effect. This produces a dilemma: an inner state that should have an immediate effect on the environment runs the risk of remaining autonomous because the effect is not attainable. Of course, there are direct solutions to this problem. Many species live in narrow territorial communities (at least temporarily, for instance sea lions). In certain species of insects, rudimentation of wings guarantees a strong limitation of mobility in the individuals of one or the other sex. In a certain deep-sea anglerfish and in a certain species of worms (*Bonellia*), evolution went still further: the males are extremely stunted and live fixed on the females (Vogel – Angermann 1984: 169). A common solution to the dilemma, however, is a semiotic one. By producing a signal on its surface, the organism makes the autonomous state perceptible in an indirect manner, according to the second semiotic principle, and thus deautonomizes it. This is always done by means of the usual processes of influencing the environment, i.e., by expending energy or by excreting matter.

In the case of bisexual reproduction, there must be another organism able to take in and then process this energy or matter in such a way that it also achieves mating motivation. Ultimately, the manifestation of an autonomous inner state can only be advantageous if there are other organisms processing the corresponding perceptions according to the first semiotic principle, i.e., processing them obliquely, as signals, and thus achieving the corresponding inner state.

The production and the reception of mating signals are an example of what I call a semiotic system. In general, a *semiotic system* is always a component of an organic system. Its central function is to establish specific relationships between the inside of this organic system and its periphery. As mentioned above, a semiotic system presupposes that inside the total system there are inner states that can neither be perceived nor manipulated from outside although the perceptibility or manipulability of these states would be advantageous to the total system. Furthermore, it presupposes that the total system has exits and/or entrances at its surface. The exits are mostly peripheral organs that can be moved somehow. The entrances are always organs of perception.[1] There are, in principle, two kinds of processes a semiotic system enables an organic system to execute:

(1) a specific processing of perceptions and an eventual subsequent modification of inner states,
(2) the manifestation of inner states that are not perceptible to the environment by means of peripheral organs.

A perception can be processed either directly or obliquely, i.e., according to the first semiotic principle. Only in this second case is it processed as a signal. The manifestation of an inner state always results in the production of a signal, according to the second semiotic principle.

4. Kinds of semiotic systems

Semiotic systems can be divided into complete and partial ones. For a biological species, being endowed with a *complete* semiotic system means that each of its normal adult individuals is enabled both to manifest certain inner states by producing certain signals and to perceive, and obliquely process, such signals in a way that may result in the modification of the corresponding inner states. A *partial* semiotic system enables an individual either to produce or to perceive and obliquely process certain signals, but not both. Receptive partial semiotic systems can be differentiated according to the source of the signals. Signals can originate from individuals of the receiver's own species or from some other source in its environment.

An example of a partial semiotic system with signals having their origin in the environment is the capability of the honeybee to extract oblique knowledge about the position of a nectar source from the perception of the shape, the color, and the smell of flowers. Most systems connected with mating motivation as an inner state are examples of partial semiotic systems with intraspecific signal flow. They are partial systems insofar as the individuals of one sex are only able to produce these signals, whereas individuals of the other sex are only able to perceive and process them. An example of a complete semiotic system is, of course, human language. Nevertheless, with a view to making some remarks about the evolution of complete semiotic systems in general, I will now give a sketch of two other such systems.

One of these systems has been observed in certain species of collective amoebae (cf. Shimomura et al. 1982, Bonner 1983, Vogel – Angermann 1984: 184-185). These amoebae are primitive protozoan unicellular organisms whose life cycle includes two very different phases, an individual and a collective one. During their individual phase, they feed on other unicellular organisms, namely on bacteria. When a feeding place is grazed out, they aggregate into a multicellular slime mold with spores containing single amoebae. These spores can be spread by insects or by water, burst at a warm moist place and then set free their passengers. This is the beginning of another individual phase. Now, just before the collective phase sets in, several physiological processes occur in the individual amoebae. On the one hand, these processes result in the synthesis and the excretion of different substances, especially a pheromone called acrasin. On the other hand, they bring about the insertion of protein molecules in the outer cell membrane, which function as chemoreceptors. The collective phase begins with a rhythmical excretion of acrasin as a manifestation of an inner state that can be characterized as a kind of aggregation motivation. The absorption of acrasin by other individuals triggers off a similar inner state. Of course, one could ask whether the production and reception of a certain substance forms only a relatively simple causal chain without any semiotic characteristics. The system, however, presents a notable peculiarity. Although the chemical processes that constitute it are clearly cellular-metabolic processes, they do not contribute to the maintenance of cellular functions. On the contrary, the formation of a slime mold is tantamount to the death of a part of the population. The creation of the physiological bases of the processes and the processes themselves clearly serve to manifest and to trigger a kind of motivation. Thus, they must be considered as components of a semiotic system.

The other complete semiotic system is the wagging dance of the honeybee (cf. Lindauer 1961, Frisch 1965, Figge 1986). Normally, a bee performs a wagging dance inside the hive on the vertical surface of a comb when it comes back from a successful flight to a nectar source. Essentially, these dances are characterized by three different kinds of variations:

(1) variations in the number of wagging runs per unit of time in connection with variations in the speed of the dance,
(2) variations in the angle between the direction of the wagging run and a vertical line,
(3) variations in the vivacity and the absolute duration of the dance.

These variations result from

(1) variations in the energy consumed by the bee in its flight to the nectar source, mostly in covering the distance, sometimes also in surmounting head wind or differences in altitude,
(2) variations in the direction of the nectar source (the line between the hive and the source, possible roundabouts due to obstacles being neglected),
(3) variations in the quality of the nectar sources.

The following relationships have been observed between these two sets of variations:

(1) The more energy the bee has been forced to spend on its flight, the more slowly it dances, and, consequently, the less wagging runs it executes per unit of time. Thus, under normal wind and terrain conditions, ten wagging runs per 15 seconds correspond to a flight of

100 meters, six runs to 500 meters, four runs to 1,000 meters, and one run to 10 kilometers, which is the maximum. When the distance is less than 100 meters the bee performs its dance without wagging.

(2) When the nectar source from which the bee comes is situated exactly in the direction of the current position of the sun (its azimuth), the wagging run is oriented vertically upwards, in a direction opposite to gravity. When the position of the source and the position of the sun form, e.g., an angle of 15° to the right or to the left, the wagging line inclines by 15° to the right or to the left. Thus, a positive heliotaxis is rendered by a negative geotaxis.

(3) The more persistently and vehemently the bee dances, the higher it estimates the quality of the nectar source.

Through their dance movements, bees manifest knowledge about the distance, the direction, and the quality of nectar sources (or other places vitally important to them). The wagging runs of these dances are accompanied by noises which the bees produce, most probably by means of rhythmical innervations of their flight musculature (Michelsen et al. 1986). The other bees perceive these manifestations partly acoustically and partly by means of tactile perception, following the dancer and touching it with their antennae. An oblique processing of these perceptions allows them to form knowledge very similar to that of the dancer.

5. Evolution of semiotic systems

In an evolutionary perspective, the productive component of the chemical semiotic system of the collective amoebae can be regarded as a by-product of normal cellular metabolism. The excretion of chemical substances is a common cellular process, especially in "stress" situations, such as those caused by scarcity of food. In one of the many species of slime molds, *Polysphondylium violaceum*, the acrasin is a peptide called "glorin" by its discoverers. Peptides are substances which originate from the decomposition of protein compounds and which are therefore found in every cell. In species of the genus *Dictyostelium* the acrasin is cyclical adenosin monophosphate (cAMP), which plays a central part in cellular processes. In particular, it is synthesized in great quantities during a certain phase of cell division (Vogel – Angermann 1984: 39) and, in particular, it contributes to the differentiation of cells during the formation of a slime mold. As to the receptive component, it is by no means uncommon for protein molecules to develop on the surface of the cell membrane and for some of them to function as receptors for chemical substances. Nor is it uncommon for cells to react chemotactically to such substances. Thus, it can be concluded that this component originally evolved as a partial semiotic system allowing to process obliquely substances taken in and by that means making chemotaxis possible. This means that the two components of the system did not co-evolve as such. Rather, they evolved separately, the first one as a by-product of metabolism and the second one as a partial chemotactic system. Their meeting was due to an evolutionary accident that came about the moment the substances taken in by the chemotactic system happened to be those given out by individuals of the same species. Only then did the excretion of acrasin turn into the manifestation of an inner state, namely, aggregative motivation, according to the second semiotic principle.

In discussing the phylogeny of the honeybee dance, Frisch (1965: 323-330) argues that its elements can be traced back to nonsemiotic movements. Normally, such a dance is performed between the return from a successful flight and the start for another flight. This suggests indeed that the discharge of emotions and of surplus energy and intentional movements should be regarded as precursors to the elements of the honeybee dance. Thus, the productive component of the system can be considered a by-product of other activities.

Now, in order to turn movements into the indication of a direction, a distance, and a quality, the bee did not have to change its behavior at all. Instead, it had to make other bees understand its wagging runs as such an indication. Recruiting works only if the recruits are aware that they are to be recruited. Since, however, it is difficult to imagine how a bee could have achieved something like that, I think it more plausible to assume that the bee dance as a semiotic system originated with the followers rather than the dancers themselves. In this interpretation, the nonsemiotic precursor is the attention most animals pay to the behavior of their conspecifics, and bees, too, are attentive animals. Under certain circumstances, it can be advantageous for an individual to guess the actual causes that are behind the emotions and motivations of other individuals. In the case of the bees, guessing from where the dancer comes and where it goes helps the follower to avoid an expensive search on its own account. So it is most probable that the bee dance as a semiotic phenomenon originated, according to the first semiotic principle, as the capacity to process certain perceptions obliquely. The semiotization of the dance itself can be explained as a process of underscoring movements in order to make them more clearly perceptible or, in more technical terms, as a ritualization of these movements. Ritualization is a way of following the second semiotic principle, using emphatic movements for producing signals and, by that means, manifesting inner states.

In many respects, the semiotic system of the collective amoebae and that of the honeybee are quite different from each other. Above all, they differ in their sensory modalities (chemical vs. tactile-acoustic) and in the inner states they are connected with (motivation vs. knowledge). Therefore, it is completely out of the question that there could be any evolutionary relationship between the two. They are not homologous. They can, however, be considered *analogous* because they are governed by the same semiotic principles.

There is still another common feature. In both cases, the receptive and the productive component *evolved independently* of one another and were then *joined by an evolutionary accident*. In the amoebae as well as in the bees, the development of the receptive component came about as an effect of the first semiotic principle. The individuals became enabled to process certain perceptions obliquely and, as a result, to adapt certain previously unadaptable inner states to states of their environment. In the case of the amoebae, these inner states were probably motivations to reach certain places (chemotaxis). In the case of the bees, it was more than that, namely, a kind of knowledge about the position of such places. In the amoebae as well as in the bees, the productive component was initially nothing but a capacity to exert an influence on the environment in a way that corresponded by chance to the evolving receptive component (excretion of chemical substances and performance of movements). Later on, this capacity was used, according to the second semiotic principle, to manifest inner states that corresponded to the states connected with the receptive component. Thus, these components did not co-evolve in the strict sense of the word. Rather, they evolved towards the same point (pace Lindauer 1990: 247-248). Though their meeting at this point was not inevitable, it was not an extremely improbable event. The reason is that in a given species, the number of classes of inner states, the number of effectory modalities, and the number of sensory modalities are relatively small, so that there is a good chance that a combination of a certain kind of inner state and a certain effectory modality will correspond to a combination of the same kind of inner state and an equivalent sensory modality, if such combinations evolve.

6. Conclusions

The following conclusions can be drawn from the foregoing evolutionary reflections on the two complete semiotic systems sketched above.

(1) The different semiotic systems observed in the animal kingdom are normally analogous rather than homologous insofar as they realize the same semiotic principles.

(2) The first semiotic principle - a sort of looking behind perception - is self-sufficient. Receptive semiotic systems can exist without a productive counterpart. Moreover, such systems already develop as semiotic systems. Productive semiotic systems, by contrast, develop as by-products[2] of nonsemiotic processes or activities, as epiphenomena of nonsemiotic phenomena. In order to achieve semiotic status, according to the second semiotic principle, they must combine with a receptive system. A prerequisite of such a combination is that the sensory modality of the receptive system matches the effectory modality of the productive system and that both systems are connected with the same class of inner states. These conditions are not difficult to fulfill, since in a given species, the number of effector modalities, the number of sensory modalities, and the number of classes of inner states are normally relatively limited. Thus, although the meeting of a receptive and a productive system is always due to an evolutionary accident, it is never a highly improbable event.

(3) Often, the combination of a receptive and a productive semiotic system results in the pairing of two partial semiotic systems, as in the case of most sexual signaling systems, which are semiotically dimorphic. However, it can develop into a complete semiotic system, enabling all normal adult individuals of the species to show both productive and receptive semiotic behavior, as in the case of the honeybee dance or of human language.

Communicative needs are often regarded as a uniform principle underlying semiotic behavior.[3] However, as argued above, such a uniform principle does not exist. Communication, especially in the sense of intentional interaction, results from the combination of two systems that are realizations of two clearly distinct semiotic principles.

Notes

1. Semiotic systems can also be components of technical systems, especially of computers. In this case, the exits and the entrances respectively are output and input devices (cf. Figge 1991).
2. For the concept of by-product, cf. Lindauer 1990: 248-252.
3. Sebeok (1967: 88), for instance, defines the field of zoosemiotics as "the ways whereby animals communicate with each other" and then writes about one of its branches: "The basic assumption of zoosemantics is that, in the last analysis, all animals are social beings, each species with a characteristic set of communication problems to solve" (1967: 93).

References

Bonner, John Tyler
1983 "Lockstoffe sozialer Amöben", *Spektrum der Wissenschaft* 6/1983: 110-118.
Eschbach, Achim (ed.)
1986 *Perspektiven des Verstehens*. Bochum: Brockmeyer.
Figge, Udo L.
1986 "Bienentanz und Menschensprache", in: Achim Eschbach (ed.), 42-71.
1991 "Computersemiotik", *Zeitschrift für Semiotik* 13, 321-330.
Frisch, Karl von
1965 *Tanzsprache und Orientierung der Bienen*. Berlin: Springer.
Lindauer, Martin
1961 *Communication among social bees*. Cambridge, Mass.: Havard University Press.
1990 *Botschaft ohne Worte: Wie Tiere sich verständigen*. München: Piper.
Michelsen, Axel – Wolfgang H. Kirchner – Martin Lindauer
1986 "Sound and vibrational signals in the dance language of the honeybee, *Apis mellifica*", *Behavioral Ecology and Sociobiology* 18: 207-212.
Sebeok, Thomas A.
1967 "Animal communication", *Revue internationale des sciences sociales* 19: 88-95.

Shimomura, O. – H. L. B. Suthers – John Tyler Bonner
1982 "Chemical identity of the acrasin of the cellular slime mold *Polysphondylium violaceum*", *Proceedings of the National Academy of Sciences of the United States of America* 79: 7376-7379.

Vogel, Günter – Hartmut Angermann
1984 *dtv-Atlas zur Biologie*. Vol. 1. München: dtv.

Opposition at the roots of semiosis

Winfried Nöth

γίνεσθαί τε πάντα
κατ' ἐναντιότητα

All things come into being
by conflict of opposites.
Heraclitus[1]

The universality of opposition in semiosis has been an essential tenet of the major doctrines of semiotics. Opposition, moreover, seems to be constitutive not only of semiosis, but also of the structure of the presemiotic micro- and macrocosm. The classics of semiotics have mainly argued from a synchronic point of view when investigating opposition as a prerequisite of structure and system in semiosis. The present paper extends this argument from the synchronic domain to the origins and evolution of semiosis.

1. Roots of semiosis: From difference to opposition

Opposition has been a key concept in structuralist semiotics. The term is closely related to the notion of *difference*. Often, both concepts are not clearly differentiated or even used synonymously, but usually, *difference* is the more general concept which comprises *opposition* as a more specific subcategory. The concept of *difference* became established as a key term of semiotics by Saussure, whereas the distinction between difference and opposition has its semiotic foundation in the writings of Jakobson.

1.1 Paradigmatic difference, otherness, and signs

According to Saussure (1916), difference is the source of semiotic structure. The value of any semiotic element can only be discerned with reference to all other structures of the system from which the given element differs. This idea is expressed in Saussure's *principle of differentiation*. It states that "in language, as in any semiological system, whatever distinguishes one sign from the others constitutes it. Difference makes character just as it makes value and the unit" (Saussure 1916: 121). According to this principle, semiotic elements can only be defined negatively. Nothing in semiosis exists positively without reference to otherness. This is the essence of Saussure's (1916: 120) dictum that "in language there are only differences *without positive terms*" and that "everything in language is negative".

The principle of differentiation holds both for the expression and the content planes of a semiotic system. As to the phonetic expression plane of language expression, Saussure (1916: 118) argues: "The important thing in the word is not the sound alone but the phonic differences that make it possible to distinguish this word from all others, for differences carry signification." With respect to the content plane, Saussure (1916: 117) gives this definition of meaning in terms of difference: "Concepts are purely differential and defined not by their positive content but negatively by their relations with the other terms of the system. Their most precise characteristic is in being what the others are not."

Thus, a given phoneme acquires its semiotic value by its difference to all other phonemes of the system, and a concept derives its meaning only from its difference from all other concepts. Difference, in this interpretation, is thus a dyadic relation between a given term and all other elements of the same semiotic plane. In other words, Saussurean difference is a paradigmatic relation between a semiotic element and the other members of the paradigm to which it belongs. Saussure (1916: 118) expresses this idea of difference as paradigmatic otherness by saying "that a segment of language can never in the final analysis be based on anything except its noncoincidence with the rest". The logical nature of this dyadic relation of otherness is the one of a contradictory opposition (*p* vs. not-*p*).

1.2 Syntagmatic difference as a source of cognition

Parallel to Saussure's thesis of difference as the source of semiosis, cognitive science and information theory have developed the theory that difference is the source of cognition and information. A main proponent of this theory, Gregory Bateson (1979: 29, 98-99), more recently followed by Merrell (1992: 207) states:

> Perception operates only upon difference. All receipt of information is necessarily the receipt of news of *difference*, and all perception of difference is limited by threshold. Differences that are too slight or too slowly presented are not perceivable. They are no food for perception ... Precisely because the mind can receive news only of difference, there is a difficulty in discriminating between a *slow change* and a *state*. There is necessarily a threshold of gradient below which gradient cannot be perceived . . . *Information* consists of differences that make a difference.

According to this definition, perceptual differences arise as cognitive figures of otherness against a ground of unperceived samenesses. Differences presuppose a change against the background of static sameness. Mere sameness without difference remains unperceived, but so does mere difference without sameness, which is perceived as chaos. In this definition, difference is also a dyadic relation, but not one of a paradigmatic kind. The terms of this dyad, sameness vs. otherness, constitute the syntagmatic relation between a ground and a figure.

Within the framework of semiotics, Greimas (1966: 19) defines difference as the cognitive root of semiosis when he writes in his *Structural Semantics*: "When we perceive differences, the world 'takes form' in front of us and for us." While, according to Bateson, difference is the root of information, Greimas and Courtés (1979: 79) define it as "the first condition for the appearance of meaning". In this interpretation, difference is also a dyadic structure whose terms are otherness and sameness (Greimas and Courtés 1979: 79): "A difference can only be recognized over against a supporting background of resemblance. Thus, it is by postulating that difference and resemblance are relations . . . which can be gathered together and formulated into a specific category, *alterity/identity*, that one can construct the elementary structure of signification."

1.3 Opposition as a dyadic structure

The concept of opposition is sometimes defined in the Saussurean sense of difference as paradigmatic otherness (cf. 1.1). In the tradition of Prague School phonology, e.g., a given phoneme is in "opposition" to all other phonemes which create a difference of meaning in the same phonological context. For example, the phoneme /k/ is in opposition to /f/, /p/, /r/, /t/, and /w/ in the context of /-u:l/ since these phonemes differentiate the meanings of *cool*, *fool*, *pool*, *rule*, *tool*, and *wool*. Jakobson (1962: 301-303, 421), however, rejects this usage of the term opposition and calls such relations "mere differences". In contrast to the general concept of difference, which relates a given term to an open class of other terms, the more specific concept of opposition in Jakobson's interpretation relates a given term only to a single other term. In this interpretation, opposition is synonymous with *binary opposition* (cf. Holenstein 1976: 121-137; Waugh 1976: 65).

Terms in binary opposition, according to Jakobson (1962: 637), are more closely related than dyads of terms forming a mere "contingent duality". Like the idea of 'whiteness' evoking the one of 'blackness' and the idea of 'beauty' evoking the one of 'ugliness', "the presence of one term in a binary opposition necessarily implies and educes the other, opposite term" (Jakobson 1962: 637). Therefore, "the phoneme by itself is not a term of opposition. E.g., the phoneme /b/ does not call univocally, reversibly, and necessarily for a definite opposite" (Jakobson 1962: 421). Not the phonemes, but only their distinctive features, such as '+/-voiced' or '+/-nasal' are terms of opposition according to this definition.

The plus/minus pattern of oppositional structure, however, does not only represent dyads of mutually exclusive alternatives. Jakobson (1962: 273) admits two types of oppositions, opposition between contradictory and opposition between contrary terms. Dichotomies of the exclusive yes/no or either/or kind represent contradictory oppositions, while polar extremes at the end points of a graded continuum, such as between 'white' and 'black' represent contrary opposition. Further below (cf. 4.0), we will refer to these two types of opposition as digital vs. graded opposition (see also Nöth 1994a).

In the sense of a dyadic structure, the term opposition has meanwhile become generally accepted in the field of structural semantics. Thus, Lyons (1977: 270ff.), while further extending the traditional classification into contraries and contradictories, distinguishes four major types of semantic "opposites", antonymy (*happy/sad*), complementarity (*male/female*), converseness (*parent/child*), and directional opposition (*up/down*). Lyons differentiates between oppositions as binary contrasts of meaning and other semantic relations which he calls nonbinary contrasts.[2]

1.4 Opposites as elements of semiotic systems

The "necessary and reciprocal implication", which Jakobson postulates for the two terms forming an opposition, can only come about by the mediation of a third term, a *tertium*, specifying the kind of relation by which the terms of the dyad are more closely associated than terms of a mere contingent duality. This relation is one of sameness[3] or "equivalence in difference" (Jakobson 1971: 262). In the light of this third term specifying the sameness in difference, the oppositional dyad turns out to be a triadic relation.[4] Let us designate henceforth the three terms involved in an opposition as *primum* and *secundum*, with reference to the dyad, and *tertium*, with reference to the term of equivalence. The question whether the tertium really comes third, as the logical term *tertium comparationis* suggests, or whether it is perhaps a primum, from which it derived by a *principium* (a "first") *divisionis* (cf. Jakobson 1962: 634), will have to be investigated below (2.).

The integration of oppositional dyads into semiotic triads is the process which leads from semiotic elements to semiotic systems. Oppositions are thus the building blocks of semiotic systems (cf. Ivanov 1973, 1983). In the field of semantics, e.g., Greimas (1966) has devised the model of a "semantic universe" whose elementary structures are binary opposites (prima and secunda) integrated via their tertia into the hierarchy of a semantic system. Figure 1 represents a segment of this universe, the semic system of 'spatiality'. In Greimas's interpretation, the dyads of elementary structures of signification are characterized by relations of *disjunction*, specifying the differences between the semes, and of *conjunction*, specifying their tertium of semantic sameness.

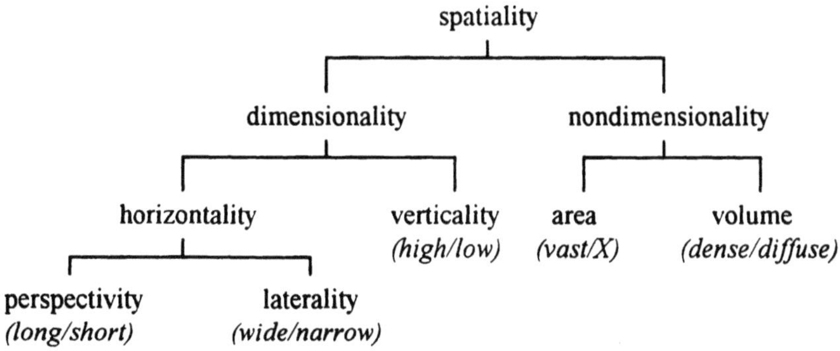

Figure 1. Greimas's (1966: 33) semantic universe of 'spatiality' exemplifying the integration of prima and secunda into tertia.

Whereas semiotics in the tradition of Jakobson considers oppositions and their integration into tertia as the source of semiotic systems, the integration of dyads into triads is a more fundamental root of semiosis in the semiotic philosophy of C.S. Peirce. Semiosis, according to Peirce (CP 6.32), is a process of *mediation*, "whereby a first and a second are brought into relation" by means of a third. Mere difference defined as otherness would not qualify for semiosis. Since "*other* is merely a synonym of . . . *second*" (CP 6.213), difference can only relate a first to a second.[5] "A sign", by contrast, "is a sort of a Third" (Peirce 1977: 31). A dyad without the mediation of a third is only a presemiotic "individual fact, as . . . it has no generality in it" (CP 1.328).[6]

1.5 Antisymmetry in opposition

In spite of their equivalence vis à vis their tertium, the oppositional dyads of semiotic systems are typically not symmetrical, i.e., of equal value within the system. The difference between the dyadic terms is also a difference of value. This kind of dyadic nonequality against a background of sameness corresponds to the geometrical pattern of antisymmetry, i.e., a symmetry of constituents which are equal in one respect, but different in another (cf. Ivanov 1974; Nöth 1994b). In phonology, semantics, and grammar, antisymmetry has been investigated as the opposition between the marked and the unmarked term of oppositions (cf. Waugh 1976: 89-102). Compared to the unmarked pole, the marked member of an opposition has an additional value. It is structurally more complex, provides more specific information, occupies a subsequent position in serial order, is learned later by children and lost earlier by aphasics. This antisymmetry between the dominant and the subordinate, the focused and the unfocused, or the positive and the negative is so deeply rooted in the evolution of culture that Derrida (1977: 236) even ascribes a metaphysical dimension to the idea of the priority of the unmarked, giving the following two indicators to bear witness to this argument:

> 1. The hierarchical axiology, the ethical-ontological distinctions which do not merely set up value-oppositions clustered around an ideal and unfindable limit, but moreover *subordinate* these values to each other (normal/abnormal, standard/parasite, fulfilled/void, serious/non-serious, literal/non-literal, briefly: positive/negative and ideal/non-ideal); and in this . . ., there is metaphysical pathos . . . 2. The enterprise of returning "strategically", ideally, to an origin or to a "priority" held to be simple, intact, normal, pure, standard, self-identical, in order *then* to think in terms of derivation, complication, deterioration, accident, etc. All metaphysicians, from Plato to Rousseau, Descartes to Husserl, have proceeded in this way, conceiving good to be before evil, the positive before the negative, the pure before the impure, the simple before the complex, the essential before the accidental, the imitated before the imitation, etc. And this is not just *one* metaphysical gesture among others, it is *the* metaphysical exigency, that which has been the most constant, most profound and most potent.

2. Monads, dyads, triads, and the emergence of opposites

Are oppositions primary in the evolution of semiosis, as their universal ubiquity suggests (see 3.-5.), or are they secondary, having emerged from structural monads in a process of bifurcation?

2.1 Dyads from monads and triads from dyads

The order of the emergence of dyadic opposites in relation to the monads and triads which they imply is a central topic of Koch's evolutionary cultural semiotics. Logically, oppositions are either primary or they are secondary in the semiogenetic development of primum, secundum, and tertium (cf. 1.4). In the former case, the origin of the triad is a dyad, in the latter it is a monad (cf. Koch 1984: 284; 1993a: 6). In the case of initial monadism, the semiotic tertium is the primordial monad from which the oppositional dyad emerges in a process of differentiation, bifurcation, or bisociation. In the case of initial dyadism, the oppositional dyad comes first, and the common tertium is discovered later in a process of integration, consociation, or generalization (cf. Koch 1983: 375-380). According to Koch (1984: 284, 299; 1993a: 6), initial dyadism is the normal course of biogenesis, whereas initial monadism is the course of epistemogenesis, including mythological accounts of genesis. The biological order of the dyad of a male and a female giving birth to the tertium of their offspring is reversed in mythogenesis, when a monadic creator gives birth to a dyad of man and woman.

The simple twofold distinction between dyads arising from monads and dyads engendering triads requires further specifications in view of a more adequate account of semiogenesis. With respect to the path from dyad to triad, the consideration of the temporal dimension leads to a distinction between dyads originating simultaneously and dyads evolving successively (cf. Koch 1986a: 169; 1987: 4, 23). An example of a simultaneous dyad giving rise to a triad is bisexual biogenesis. The successive development of dyads in the form of a primum preceding the development of a secundum is apparent in Jakobson's law of the ontogenetic development of binary oppositions according to which an unmarked primum precedes the marked secundum (cf. 1.5).

The specification of the evolutionary path from monad to dyad requires the distinction between polarized and nonpolarized dyads. The former are also called *dualities*, the latter *polarities* (cf. Bochner 1973: 350). In dualities, the emphasis is on the aspect of dyadic sameness. Symmetry or parallelism characterizes the relationship between the elements of a duality. In biogenesis, mitosis and DNA replication exemplify the genesis of dualities as far as the reproduction is one of dual sameness (see 4.2). Oppositions, defined as antisymmetries (cf. 1.5), are polarities, but they may originate from dualities. According to Koch (1983; 1986a: 111, 117; 1993b: 171-175), duality and polarity are characteristic of two subsequent phases in the evolution of culture. In the first phase, which he calls *prosphory*, we find dualities, and these dualities then become polarized to oppositions in a subsequent phase, which he calls *diaphory*.

To summarize, we distinguish the following four paths in the semiogenesis of oppositions: (1) from monad (spontaneously) to oppositional dyads, (2) from monads, via (symmetrical) dualities, to polarized dyads, (3) from simultaneous dyads to tertia completing the triad, and (4) from successive dyads of a primum and secundum to a tertium.

2.2 Semiogenesis as the mediation between opposites

The idea of mediation between oppositional dyads (cf. 1.4) has further semiogenetic implications. Peirce, in the framework of his semiotic philosophy of evolution, defined the process of mediating between a first and a second as the very principle of evolution. However, in his account, the order of firstness, secondness, and thirdness does not correspond to the temporal sequence of actual genesis: "The origin of things . . . contains the idea of First, the end of things that of Second, the process of mediating between them that of Third . . . Mind is First, Matter is second, Evolution is Third" (CP 6.32). If the thirdness of mediation comes between the beginning and end in which the origin is negated,[7] the temporal sequence of evolution is firstness, thirdness, secondness. However, this temporal order of events is not the cognitive order in the process of semiosis. From the cognitive point of view, the dyad of firstness (mind, the source of the cognition) and secondness (matter, the object of cognition) precedes thirdness, which is the dynamic (evolutionary) process of cognition and semiosis as the mediation between firstness and secondness (cf. 1.4).[8]

In a different way, the process of mediation between opposites is a principle of evolution according to Koch's evolutionary cultural semiotics (Koch 1984: 136-148, 420ff.; 1986a: 2, 117ff.; 1986b: 49ff.; 1986c: 147). According to this theory, the evolution of culture proceeds by the integration of polarities into new tertia. Koch describes such cultural integration of polarities as the third phase of an evolution beginning with a first phase of simple dualities and followed by a second phase of polarities (prosphory and diaphory, cf. 2.1). The transition from dyadic opposition to triadic mediation brings about "a shift from structure to process" (Koch 1983: 375). The evolutionary law underlying this process is: "In the beginning was the dyad that immediately engendered the triad" (Koch 1984: 283). There are three types of cultural mediation distinguished by Koch (1983: 139-47; 1993a; b). A *tertium mediationis* is the integrator of oppositional dyads in the domain of metaphysics. For example, "Christ" mediates between 'human' or 'mortal' and 'divine' or 'immortal', and Pascal's metaphor of man being a "thinking reed" mediates between the ideas of "natural weakness" and "strength". Mediation in the domain of "concrete metalanguage" takes place by means of *tertia comparationis*. Vehicles of such mediation are the metaphors of everyday language. Koch's third type of mediation occurs in the field of "concrete aesthetics", which comprises such phemomena as metrical recurrence and other repetitive correspondences in art and nature. It is called *tertium attractionis* and describes the special semantic effect associated with a third element continuing, and at the same time breaking, the pattern of a binary recurrence. Examples of this effect are Caesar's *Veni*, *vidi*, *vici* or Richard III's *A horse*, *a horse*, *my kingdom for a horse*.

More recently, the process of the triadic integration of polarities has been investigated by Wilkins (1987) and Spinks (1991: 77-87) in various domains of semiosis. Spinks points out that a triadic integration of polarities seems to be inherent in the structure of the "triune brain" which consists of the dyad of the right and left hemispheres with the limbic system, the "reptilian" brain as a tertium (for this argument see also Koch 1983: 421ff.). In the domain of culture, Spinks exemplifies *psychological integration* from C.G. Jung's mediation between consciousness and unconsciousness in the form of a 'conjunction of opposites', illustrates *divinatory integration* by means of the Chinese *I Ching's* production of signification from semantic polarities, and interprets the ambivalent figure of the trickster as the prototype of *mythic integration* between essential polarities of human life. Spinks further discusses a variety of triadic processes of theoretical semiotic and epistemological integration.

3. Universal ubiquity and cosmogenesis of opposites

Oppositions are omnipresent at all stages of evolution from the origin of cosmos to the development of philosophical systems. Koch (1984: 285; 1993a: 7), distinguishing seven stages of this evolution, gives the following examples of typical oppositional dyads at each evolutionary level:

(1) cosmogenesis	positron	-	electron
	gravitation	-	expansion
(2) biogenesis	life	-	death
	female	-	male
(3) psychogenesis	limbic system	-	cortex
	id	-	ego
(4) sociogenesis	amity	-	enmity
	egoism	-	altruism
(5) semiogenesis	projection	-	introjection
	signifié	-	*signifiant*
(6) glottogenesis	arbitrary	-	motivated
	text	-	sentence
(7) epistemogenesis	concretism	-	abstractism
	genesis	-	metagenesis

The present investigation of the evolution of opposites is restricted to the lower levels of this evolutionary scale. We begin with a few reflections on opposites in presemiotic cosmogenesis before crossing the semiotic threshold to the domain of biogenesis (4.).

As far as our topic is concerned, the most striking feature of our physical universe is its totally dyadic division into particles and antiparticles, matter and antimatter. According to the account of particle physics given by Gardner (1979: 189),

> every elementary particle is now known to have a corresponding antiparticle. The two particles are alike in all respects except that they are opposite in the sign of any quantity (represented by a plus or minus quantum number) that is conserved. If a particle is charged, its antiparticle has an equal but opposite charge. If it has a magnetic moment, its antiparticle has a magnetic moment of opposite sign.

The origin of the universe, as far as it has been traced back by the so-called Grand Unified Theories, began 15 billion years ago. During the earliest phase about which we have any knowledge, from 10^{-43} to 10^{-32} seconds before the Big Bang, there was complete symmetry between matter and antimatter. Particles and antiparticles existed in precisely equal amounts and were completely equal in value and their reciprocal physical effects (cf. Mainzer 1988: 515-516). The origin of cosmos began with the breakdown of this initial symmetry. Antisymmetries and asymmetries disturbed the original universal uniformity and began to create the physical structures of our present universe. In the domain of elementary particles, one of the results of this evolution is an antisymmetrical opposition of matter and antimatter (Bochner 1973: 350-351). For example, the electron and its antiparticle, the positron, evince such antisymmetry. They are structurally symmetrical except for the fact that the positron is much less stable than the electron (Bochner 1973: 351). With an increase in complexity, the structure of physical nature becomes more and more asymmetrical.

As to the role of opposition in cosmogenesis, we can thus conclude that the origin of matter is in symmetrical dyads. The breaking of the symmetry of these dyads meant a creation of *differences*, since nondifferentiated elements of equal value became differentiated in value (cf. Freund 1991: vii). From this differentiation, antisymmetrical *oppositions* emerged, followed by more complex asymmetrical structures. This sequence from symmetrical dualities to antisymmetrical polarities exemplifies the first two of Koch's (1986a: 117) "fundamental phases of structure" (cf. 2.1).

4. Digital oppositions at the biosemiotic threshold

With biosemiotics, we cross the semiotic threshold from the presemiotic physical universe to the domain of semiosis (see Introduction to this volume). Biosemiotics is a field too comprehensive to be adequately reviewed (cf. Sebeok – Umiker-Sebeok, eds. 1992), even from the restricted point of view of the semiogenesis of opposites. Biological opposites that first come to mind as being relevant to semiosis are the binary oppositions of life/death, male/female, parents/progeny, or the antisymmetrical right/left opposition in handedness and cerebral function (cf. Ivanov 1983; Springer – Deutsch, eds. 1985; Spinks 1991: 56-58). However, let us focus on oppositions at some of the lower levels of the biological evolution of semiosis distinguished by Sebeok (1991: 83-96), the genetic code, endosemiosis in the immune system, the metabolic, and neural codes, bacterial semiosis, phytosemiosis, mycosemiosis, and zoosemiosis.

At the lowest threshold of semiosis, opposition already appears in the two main forms distinguished above (1.3), digital and graded opposition. Digital opposites appear at the lowest levels of semiosis, while graded opposites are characteristic of higher forms of biosemiosis (cf. 5.).

4.1 Genetic semiosis

In the biochemistry of life, opposition begins to appear in the form of the digital opposites which exist between the four bases of the two strands of nucleotides in a DNA molecule, adenine (A), thymine (T), guanine (G), and cytosine (C) (cf. Margulis 1982: 10-11; Küppers 1990: 13-15). These four bases form two complementary pairs which are almost identical in shape and chemical properties, but differ such that A vs. T and G vs. C form complementary opposites. The self-reproduction of DNA molecules begins with a separation of the two strands of complementary nucleotides. Then, in a process involving the production of a "negative copy" of hereditary information and its subsequent reversal in a positive form, nucleotides from the environment become attached to their complements on each single strand to form two new double strands identical to the original. This process of linking molecular bases with their complementary counterparts is a process of semiosis since it presupposes a process of base-pair recognition.[9] Prodi (1988a: 194-195) describes this aspect of the origins of semiosis in the recognition of complementary opposites as follows:

> The process of finding each other is therefore a reading or interpretation of reality carried out by A and B according to their constitutional modules. Thus, for A and B, reading reality consists in reviewing and rejecting indifferent objects and choosing significant ones. In its exploration, which terminates in the contact with B, A makes a judgement about reality, since it is moved to change only when it finds its complementary significant reality. This is A's interpretation of the environment: its being moved to change by B. Thus, the material presence of B is detected by A as follows: it elicits a change by means of complementariness. We shall say that A 'reads' reality and 'knows' B. It can only know B through its own structure, which is what makes it reject the indifferent presences.

While molecular semiosis in the recognition of complementarity presupposes the existence of opposites, the process of DNA replication also exemplifies the process of the origins of opposites. In so far as DNA molecules replicate to form new molecules identical to the original, this process is one of creating symmetrical dyads from monads (cf. 2.1). Evolution, however, cannot rely on the mere reproduction of samenesses. As Margulis (1982: 11) points out, "If copying were always faithful, offspring cells would always be exact duplicates of their parent cells. If then environmental conditions altered so that the parent type was no longer suited to the job of making more of itself, the living, replicating system would die out." Therefore, evolution must provide for variety and change. It takes place in the form of mutation or the innovative combination of genetic material from males and females in sexual reproduction. These processes of reproductive innovation result in the creation of opposites in the relation between parent and offspring genes. Thus, mutation, which arises from a mistaken substitution or deletion of one of the DNA bases (cf. Margulis 1982: 13), involves the creation of differences in sameness, resulting in an antisymmetrical opposition between parent and offspring genes.

4.2 Immunological semiosis

The "interpretation of complementary opposites" continues to be the predominant principle of semiosis in the immune system of animals (cf. Prodi 1988b). The senders of immunological messages ("signotopes" according to Bona 1988) in the blood serum of an animal are the antigens, i.e., molecules of foreign substances, such as bacteria or viruses. The receivers of these messages are the leukocytes, which are equipped with a multitude of receptors for this purpose, and the antibody molecules produced by the B-lymphocytes. Because of the great complexity of immunological messages, Jerne (1985: 446) compares the sign repertoire of the immune system to a "lexicon of sentences which is capable of responding to any sentence expressed by the multitude of antigens which the immune system may encounter". As Jerne (1985: 447-448) points out, antigens and antibodies evince the complementary opposition of a picture and its mirror image:

> The immune response to a sentence presented by an invading protein molecule is merely to select, from amongst its enormous preformed antibody repertoire, a suitable mirror image of part of this antigenic sentence . . . [Thus,] the sentences representing antibodies possess partial mirror images of an antigenic sentence. These antibodies are not echoes of the invading antigen, but were already available to the animal in its repertoire of B cells before the antigen arrived.

After recognizing the antigen, the B cell proceeds to interpret it, and this process involves a binary distinction between self and nonself, i.e., autoantigens produced by the own body and foreign antigens. The response of the B cell to the antigen depending on this interpretative evaluation is a yes-no decision (cf. Celada 1988: 74). The response to a foreign antigen is the production of complementary antimolecules (Prodi 1988b: 56), i.e., antibodies, to destroy the invaders. The response to autoantigens is "*no* reaction", which avoids the danger of autoaggression.

4.3 Neurological semiosis

Digital opposition continues to be the basic principle of neurological semiosis. Communication at this level is the transduction of signals in the form of electrochemical impulses between neurons. The binary nature of this process is apparent in two polarities, the opposition between neural activation vs. inactivation, and the opposition between excitory vs. inhibitory signals.

Pribram (1971: 20) refers to the former opposition when he characterizes the neuron as "an 'all-or-nothing' relay". Notice that the step from neural inactivity to activity involves the transformation of a gradual bioelectrical input to a binary pattern of semiosis. When receiving the excitory input, the neuron first remains inactive until a certain discharge level (of -50 minivolts) is reached. Only then does it release its action potential (cf. Trehub 1991: 19-21). Below the discharge level, the bioelectrical input is not yet a signal. It has no "meaning" to the neuron. While the bioelectrical process of stimulation is gradual, the semiotic process is digital, distinguishing only between non-signal below, and signal above the activation threshold.

The second basic digital opposition in neurosemiosis is the one between excitory and inhibitory messages. The transmission of these signals takes place via two kinds of neurotransmitters which have become specialized to perform this task: excitory and inhibitory synapses. The semantic opposition between a signal that excites and one that inhibits a target cell is paralleled by an opposition between a positive and a negative direction of the bioelectric impulse (cf. Trehub 1991: 20-21).

5. Roots of graded opposition in biosemiotic space

In the evolution of semiotic opposition, graded opposites (cf. 1.3) begin to appear at the level of phyto- and zoosemiosis. The emergence of this type of opposition is essentially connected with spatial semiosis, in particular, the semiotic reduction of spatial pluridimensionality to oppositional dyads.

5.1 Semiosis of opposition in space

According to the Peircean definition of semiosis as *goal-directed* (a spatial metaphor!) action (cf. 4.1), orientation in space is sufficient evidence of semiosis. The biological goal which an organism pursues when exerting its capacity for spatial orientation is to enhance its contact with biologically advantageous stimuli and to diminish its contact with sources of danger.

In contrast to endosemiosis, which basically requires decisions between digital opposites, spatial semiosis requires orientational decisions of a multidirectional kind. Nevertheless, semiotic space, unlike geometrical space, is not isotropic in the sense that all directions are equal in orientational value. By contrast, spatial semiosis assigns preferential values to basic directions, such as the vertical and the horizontal, according to biological or cognitive determinants, and this semiotic grid imposed on space is highly oppositional. Furthermore, whenever signs are part of an oppositional paradigm, such as the verbal opposites *up* vs. *down*, these can be graded (as in *further up/down*) and allow for a local tertium (as *neither up, nor down*). These characteristics of spatial orientation seem to be the cognitive root of grading in language (cf. Nöth 1994a).

5.2 Spatial opposition in phytosemiosis

Biologically, the most efficient way of orientation in space is, of course, by free motility, but the capacity for spatial orientation even exists in sessile organisms and plants. Plants move in the process of growth, which is defined as *tropism*. Orientational movements of freely motile organisms, by contrast, are called *taxes* (Kühn 1919: 2-4). The development of taxis is a decisive threshold in the biogenetic evolution from more primitive to higher forms of life. Bonner (1980: 56-75) even describes motility as an "early origin of cultural evolution".

Let us begin with tropism, the phytosemiotics of space. There are two main kinds of tropism, phototropism and geotropism (Kühn 1919: 2-4). Both of these forms of spatial orientation operate according to a principle of directional opposition terminologically expressed in a positive vs. negative dichotomy: positive phototropism is growth towards light, negative phototropism is growth in the direction opposite from the source of light, and positive vs. negative geotropism is the opposition of growth towards vs. away from the gravitational attraction of earth. Both forms of spatial orientation are graded according to the strength of the stimulus and allow for a tertium of stasis which occurs, e.g., when the stimuli of orientation come from opposite directions (cf. Kühn 1919: 3).

5.3 Spatial opposition in zoosemiotic taxis

Due to their capacity for locomotion, animals have an orientational potential which is, of course, more pluridimensional than the one of plants. Nevertheless, spatial orientation of animals is still highly oppositional. Analogous to phototropism in phytosemiosis, e.g., there is in zoosemiosis an oppositional orientation towards vs. away from light, called positive vs. negative phototaxis.

The most general directional zoosemiotic opposition is the one between topic and phobic taxes (cf. Schöne 1983: 64). Topic taxis is positively directed towards a stimulus, while phobotaxis is the negative reaction of stimulus avoidance. As to the evolution of orientational opposition, it is interesting to note that topic taxis is biologically more differentiated and orientationally more direct. Modes of topotaxis include teleotaxis (direct movement towards a stimulus), tropotaxis (orientation towards lateral symmetry in relation to stimuli), and menotaxis (angular orientation in relation to stimulus). In contrast to these positively directed forms of orientation, phobotaxis, the most primitive form of negative orientation in space is unspecific in the direction away from the stimulus. The paramecium, e.g., only succeeds in avoiding a danger zone by a process of a repeated trial and error (cf. Kühn 1919: 5).

5.4 Opposition in semiochemical space

Let us conclude our exploration of opposites in lower zoosemiosis with a brief discussion of some semiochemical patterns of opposition in olfactory semiosis. The most primitive opposition at this level of semiosis is the one between attractant vs. repellent stimuli. In the semiochemical classification of volatile substances, the opposition *kairomone* vs. *allomone* corresponds to this opposition (Nordlund 1981: 22). A system of spatial oppositions also emerges from the more differentiated analysis of response patterns to semiochemicals. Consider the three directional oppositions implied in the typology of semiochemicals set up by Nordlund (1981: 22) in the context of insect semiosis:

(1) *arrestant* vs. *locomotor stimulant*: the former chemical "causes an organism to aggregate in contact with it", the latter "to disperse from a region".
(2) *attractant* vs. *repellent*: the former chemical "causes an organism to make oriented movements towards its source", the latter "to make oriented movements away from its source".
(3) *feeding stimulant* vs. *deterrent*: the former elicits, the latter inhibits feeding in an organism.

Ultimately, these semiochemical oppositions corresponding directly to oppositional motion in space are closely related to the biological opposition between advantageous and disadvantageous stimuli (food vs. nonfood, mating vs. nonmating etc.). At higher levels of chemical semiosis, in the olfactory "language" of mammals, a new spatial aspect of semiochemical communication emerges in scent and territory marking, where odors are used as messages for the recognition of individuals, groups, territories or special sources of threat. In contrast to semiochemical oppositions due to chemically opposed stimuli (kairomones vs. allomones), scent marking produces messages in which one stimulus simultaneously conveys opposed messages to different groups of receivers: the same odor is a signal of safety to the individual within the boundaries of its territory, but indicates danger to the intruder (cf. Ewer 1968: 105). It produces an ambivalent spatial message signifying attraction to members of the own group and repulsion to foreign conspecifics.

6. Prospect: From biosemiosis to psychosemiosis

The biological opposition of territorial defense vs. flight is functionally supported by the psychological opposition between rage and fear. Thus, the "territorial imperative" (Ardrey 1967) which establishes invisible oppositional forces of attraction and repulsion in animal space brings us to the threshold leading from biosemiosis to psychosemiosis. Although, in this paper, the topic of opposition in semiogenesis cannot be pursued much beyond this threshold, we conclude with a few comments on psychosemiotic opposition, restricted to some of the aspects discussed in other contexts earlier in this paper.

The oppositional nature, or "bipolarity", of emotions has frequently been noted (cf. Russell 1979; Nöth 1992). The very concept of emotions has its roots in spatial opposition since its etymology encapsulates the concepts of 'moving' and 'away' (Lat. *emovere*). Dualisms and polarities, such as love vs. hatred, libido vs. destrudo, or pleasure vs. reality principle, are also the foundations of Freud's psychoanalysis (Ciompi 1982: 23, 119-121). Essential features of these psychosemiotic polarities are their gradedness (cf. 1.3, 5.), their antisymmetry (cf. 1.5), and the emergence of a third instance between the opposites (cf. 2.2, and Ciompi 1982: 120). As noted earlier, precursors can be found in biosemiotic processes involving opposed forces of equal strength (cf. 5.2, 5.4), seen particularly in the form of ambivalent and conflict behavior in zoosemiosis.

Gradedness into opposite directions is the feature which emotional semiosis shares with spatial semiosis. The highly antonymic nature of the vocabulary of emotions (cf. Nöth 1992) reflects this characteristic of psychosemiosis. This spatial nature of emotional language is perhaps part of the more general foundation of language in the perception of space currently emphasized in cognitive linguistics (cf. Nöth 1994a). The psychologist Julian Jaynes (1977: 59) ascribes to this tendency of the human mind evolutionary relevance when he calls mental spatialization "the first and most primitive aspect of consciousness". A more specific hypothesis on the phylogeny of evaluative opposition in spatial semiosis has been proposed by Burke (1952: 262), who locates the root of positivity and negativity in the opposition between the animal's "taking what it wants" and its "turning away from what it doesn't want".

Less speculative evidence on the spatial aspect of the genesis of psychic opposites comes from child psychology, where the infant's growing awareness of spatial separation between the "I" and the maternal and environmental "non-I" has been recognized as the ontogenetic root of the self in opposition to the other or even more specifically as the psychological root of the opposition between "yes" and "no" (Spitz 1957). The assumption that the psychogenesis of the I is essentially connected with an experience of spatial opposition is also part of Lacan's psychoanalytic theory of the genesis of the conscious I in relation to the unconscious Other during the mirror phase of child development (cf. Nöth 1990: 303). By the discovery of the opposition between the I in its spatial confrontation with its own external ("mirror") image, the child, according to Lacan, begins to differentiate between the person as an individual subject and an object of social interaction.

Spatial origins of opposition have not only been discovered at the roots of emotional, but also of cognitive child development. For this extension of our topic beyond the threshold of psychosemiosis to the linguistic field of lexical semantics, Lyons (1977: 282) put forward the following hypothesis: "It may well be that our understanding, not only of directional opposition, but of opposition in general, is based upon some kind of analogical extension of distinctions which we first learn to apply with respect to our own orientation and the location or locomotion of other objects in the external world."

Notes

1. according to Diogenes Laertius, Lives IX.8.
2. Thus, Holdcroft (1991: 126) is mistaken when he interprets Lyons as distinguishing between "binary and non-binary opposites", since non-binary semantic relations, according to Lyons, are not opposites, but only contrasts. On the other hand, Holdcroft's subdivision of differences into oppositions and non-oppositions is in accordance with the terminology proposed in this paper. A structural semanticist who rejects Lyons's terminological restriction of opposites to binary relations of meaning is Geckeler (1980: 47). In his view, opposition is "a very general and basic principle of the structural analysis of language". His concept of opposition is presumably synonymous with Saussure's concept of difference.
3. Sameness in difference is not only characteristic of binary semantic opposites, but also of co-hyponyms, terms which are subordinate to the same superordinate term (as cat, dog, or bird to animal). In fact, this characterization is part of Aristotle's definition of differentia specifica: "We call different those things which, though other, are the same in some respect, not only in number but either in species or in genus or by analogy" (Metaphysics 5.9, 1018a). Aristotle considers the difference between such multiple semantic contrasts and binary semantic opposites as a matter of degree: "Since things which differ may differ from one another more or less, there is also a greatest difference, and this I call contrariety" (Met. 10.4, 1055a). Thus, binary semantic opposites (of the subcategory of antonymy, Aristotle's contrariety) differ more in meaning than members of a multiple semantic contrast. Apparently, the reduction of a multiple semantic paradigm to a pair of opposites increases the difference between the terms.

4. Cf. Peirce (CP 1.462): "Qualities cannot resemble one another nor contrast with one another unless in respect to a third quality; so that the resemblance of qualities is triadic."
5. In this respect, Merrell's (1992: 119) interpretation of Peirce, to whom he ascribes the view that "All signs... are a radical alterity", is too Saussurean.
6. Dyadic relations, according to Peirce, are specific of presemiotic reality or existence (cf. Introduction to this volume, 2.). Peirce (CP 1.457) calls such relations oppositions: "Existence is that mode of being which lies in opposition to one another. To say that a table exists is to say that it... dynamically reacts against other things (that is, has inertia)... A thing without oppositions ipso facto does not exist."
7. Cf. Peirce (CP 6.217): "The nothing of negation is the nothing of death, which comes second, or after, everything."
8. Peirce apparently refers to two kinds of evolution. The first, "from the beginning to the end of things", is cosmogenesis, the second, "from mind to matter", is semiogenesis. Notice that the order of the latter reverses the order of cosmic and biological evolution, in which matter is first, mind is second.
9. For the semiotic difference between mere chemical reaction and biosemiotic recognition see also Kull (1992: 224ff.).

References

Ardrey, Robert
1967 *The territorial imperative*. London: Collins.

Bateson, Gregory
1979 *Mind and nature*. New York: E.P. Dutton.

Bochner, Salomon
1973 "Symmetry and asymmetry", in: Philip P. Wiener (ed.), 345-353.

Bogdan, R.J. – Niiniluoto, I. (eds.)
1973 *Logic, language, and probability*. Dordrecht: Reidel.

Bona, Constantin A.
1988 "Structure and direction of message in the immune network and circuits", in: E.E. Sercarz et al. (eds.), 105-117.

Bonner, John Tyler
1980 *The evolution of culture in animals*. Princeton: University Press.

Burke, Kenneth
1952 "A dramatistic view of the origins of language", *Quarterly Journal of Speech* 38: 251-264, 446-460.

Busse, Wilhelm G. (ed.)
1992 *Anglistentag 1991*. Tübingen: Niemeyer.

Celada, F.
1988 "Does the human mind use a logic of signs developed by lymphocytes 10^8 years ago?", in: E.E. Sercarz et al. (eds.), 71-79.

Ciompi, Luc
1982 *Affektlogik*. Stuttgart: Klett-Cotta.
[1988] [Trans. *The psyche and schizophrenia: The bond between affect and logic*. Cambridge, Mass.: Harvard University Press.]

Čmejrková, Světla – František Daneš – Eva Havlová (eds.)
1994 *Writing vs. speaking: Language, text, discourse, communication*. Tübingen: Narr.

Derrida, Jacques
1977 "Limited Inc abc", *Glyph* 2: 162-254.

Ewer, R.F.
1968 *Ethology of mammals*. London: Logos.

Freund, John
1991 *Broken symmetries*. New York: Lang.

Gardner, Martin
1979 *The ambidextrous universe*. 2nd ed. New York: Scribner's.

Geckeler, Horst
1980 "Die Antonymie im Lexikon", in: Dieter Kastovsky (ed.), 42-69.

Greimas, Algirdas Julien
1966 [1983] *Structural semantics*. Lincoln: University of Nebraska Press.

Greimas, Algirdas Julien – Joseph Courtés
1979 [1982] *Semiotics and language*. Bloomington: Indiana University Press.

Hiley, B.J. – F. David Peat (eds.)
1987 *Quantum implications: Essays in honor of David Bohm*. New York: Routledge & Kegan.

Holdcroft, David
1991 *Saussure: Signs, system, and arbitrariness*. Cambridge: University Press.

Holenstein, Elmar
1976 *Roman Jakobson's approach to language*. Bloomington: Indiana University Press.

Ivanov, Vyacheslav
1973 "On binary relations in linguistic and other semiotic and social systems", in: R.J. Bogdan – I. Niiniluoto (eds.), 196-200.

1974 "On antisymmetrical and asymmetrical relations in natural languages and other semiotic systems", *Linguistics* 19: 35-40.

1983 *Gerade und Ungerade*. Stuttgart: Hirzel.

Jakobson, Roman

1962 *Selected writings*. Vol. 1. The Hague: Mouton.

1971 *Selected writings*. Vol. 2. The Hague: Mouton.

Jaynes, Julian

1977 *The origin of consciousness in the breakdown of the bicameral mind*. Boston: Houghton Mifflin.

Jerne, Niels K.

1985 "The generative grammar of the immune system", *Bioscience reports* 5: 439-451.

Kastovsky, Dieter (ed.)

1980 *Perspektiven der lexikalischen Semantik*. Bonn: Bouvier.

Koch, Walter A.

1983 *Poetry and science*. Tübingen: Narr.

1984 "Art: Biogenesis and semiogenesis", *Semiotica* 49: 283-304.

1986a *Evolutionary cultural semiotics*. Bochum: Brockmeyer.

1986b *Genes vs. memes*. Bochum: Brockmeyer.

1986c *Philosophie der Philologie und Semiotik*. Bochum: Brockmeyer.

1987 *Hodos and kosmos*. Bochum: Brockmeyer.

1993a *The biology of literature*. Bochum: Brockmeyer.

1993b *The roots of literature*. Bochum: Brockmeyer.

Kühn, Alfred

1919 *Die Orientierung der Tiere im Raum*. Jena: Fischer.

Kull, Kalevi

1992 "Evolution and semiotics", in: T.A. Sebeok – J. Umiker-Sebeok (eds.), 221-233.

Küppers, Bernd-Olaf

1990 *Information and the origin of life*. Cambridge, Mass.: MIT Press.

Lyons, John

1977 *Semantics*. 2 vols. Cambridge: University Press.

Mainzer, Klaus

1988 *Symmetrien der Natur*. Berlin: de Gruyter.

Margulis, Lynn

1982 *Early life*. New York: Van Nostrand.

Merrell, Floyd

1992 *Sign, textuality, world*. Bloomington: Indiana University Press.

Nordlund, Donald A.

1981 "Semiochemicals: A review of the terminology", in: Donald A. Nordlund et al. (eds.), 13-28.

Nordlund, Donald A. – Richard L. Jones – W. Joe Lewis (eds.)
1981 *Semiochemicals.* New York: Wiley.
1992 "Symmetries and asymmetries between positive and negative emotion words", in: W. Busse (ed.), 72-88.

Nöth, Winfried
1990 *Handbook of semiotics.* Bloomington: Indiana University Press.
1992 "Symmetries and asymmetries between positive and negative emotion words", in: W. Busse (ed.), 72-88.
1994a "The semantic space of opposites: Cognitive and localist foundations", in: Kei I. Yamanaka (ed.).
1994b "Symmetry in oral and written language", in: S. Čmejrková – F. Daneš – E. Havlová (eds.), 97-110.

Peirce, Charles Sanders
1931- *Collected papers of Charles Sanders Peirce*, vols. 1-6 ed.
1958 Charles Hartshorne and Peter Weiss, vols. 7-8 ed. Arthur W. Burks. Cambridge, Mass.: Harvard University Press.
1977 *Semiotics and significs: The correspondence between Charles S. Peirce and Victoria Lady Welby*, ed. C.S. Hardwick. Bloomington: Indiana University Press.

Pribram, Karl H.
1971 *Languages of the brain.* Englewood Cliffs: Prentice Hall.

Prodi, Giorgio
1988a "Material bases of signification", *Semiotica* 69: 191-241.
1988b "Signs and codes in immunology",in: E.E. Sercarz et al. (eds.), 53-64.

Russell, James A.
1979 "Affective space in bipolar", *Journal of Personality and Social Psychology* 37: 345-356.

Saussure, Ferdinand de
1916 [1969] *Course in general linguistics*, trans. Wade Baskin. New York: McGraw Hill.

Schöne, Hermann
1983 *Orientierung im Raum.* 2nd ed. Stuttgart: Wissenschaftliche Verlagsgesellschaft.

Sebeok, Thomas A.
1991 *A sign is just a sign.* Bloomington: Indiana University Press.

Sebeok, Thomas A. – Umiker-Sebeok, Jean (eds.)
1992 *Biosemiotics.* Berlin: Mouton de Gruyter.

Sercarz, E.E., et al. (eds.)
1988 *The semiotics of cellular communication in the immune system.* Berlin: Springer.

Spinks, C.W., Jr.
1991 *Semiosis, marginal signs and trickster.* Houndmills: MacMillan.

Spitz, René A.
1957 *No and yes.* New York: International University Press.

Springer, Sally P. – Deutsch, Georg (eds.)
1985 *Left brain, right brain.* 2nd ed. New York: Freeman.

Trehub, Arnold
1991 *The cognitive brain.* Cambridge, Mass.: MIT Press.

Waugh, Linda R.
1976 *Roman Jakobson's science of language.* Lisse: de Ridder.

Wiener, Philip P. (ed.)
1973 *Dictionary of the history of ideas.* New York: Scribner's.

Wilkins, M.H.F.
1987 "Complementarity and the union of opposites", in: B.J. Hiley – F. David Peat (eds.), 338-360.

Yamanaka, Kei I. (ed.)
1994 *The locus of meaning: A Festschrift for Yoshihiko Ikegami.* Tokyo.

Primate nonverbal communication: Our communicative heritage[1]

Signe Preuschoft and Holger Preuschoft

1. Introduction

Humans frequently use a means of communication not observed in other animals, namely verbal language. At closer inspection, however, the characterization of human beings as *Homo loquens* turns out to be only a very rough approximation. Verbal language is not based on potentials available only to humans and clearly lacking in other animals (H. Preuschoft 1988, 1990). We tend to forget, anthropocentric and obsessed with rationalism as we are, that aside from language there are other means of communication which function, sometimes even better, without language. Our approach will be a biological one, that is, our underlying idea is that communication - by whatever means - is a behavior which helps the individual to survive. In the evolution of humans, verbal language has developed in addition to other communicative systems.

Modern biology is turning away from stating what is, or from sketching ever new evolutionary lines leading from "primitive" forms to "derived" ones, to the question *why* evolutionary changes have taken place. From the same viewpoint, the most fascinating question in this context seems to be: what makes language different from other modes of communication? In other words: what kind of selective pressures have favored the evolution of language in humans? What is language able to achieve that other modes are unable to? This approach is quite different from the one which has stimulated earlier language research in phylogenetic perspective. It is not guided by the idea that what makes humans different from other animals is language. This idea readily leads to the creation of ever new definitions of borderlines and dichotomous differentiations which serve to confirm that we are, indeed, different from, or even "more highly evolved" than, other animals.

To adopt a more biological perspective on the evolution of language, it is necessary to understand the functions fulfilled by earlier communication systems. Intimate knowledge of their functions is the first step to understanding the mechanisms which determine the evolution of communicative displays in general. In this chapter, we intend to present an overview of the means of communication among our closest biological relatives. If the readers find themselves reminded of their own feelings, experiences, or personal situations, this is intended. In the discussion, we will attempt to evaluate present knowledge on primate nonverbal communication with respect to the task demands each mode has evolved to meet in order to see whether we can already catch a glimpse of the function verbal language has evolved to fulfill.

Within the animal kingdom, primates form one order of placental mammals. It is the order to which our own species belongs. Human as well as nonhuman primates are equipped with identical, or at least very similar capacities of perceiving their environment, of responding to it, and of (literally) handling it. All presently available knowledge converges to the idea that all primates have evolved from a common ancestor on a single evolutionary line, independently of the other mammals. As a consequence of this independent evolution, no mammal is as closely related to humans as other primates and vice versa (Huxley 1863).

Primates are divided into two suborders, the prosimians and the simians. The rather conservative prosimians comprise the Lemuriformes, living in Madagascar, the Lorisiformes, and the Tarsiiformes, both of which live in Asia. Simians include the "platyrrhine" monkeys of the New World, and the "catarrhine" primates of the Old World. Within the Platyrrhini, two families, the larger Cebidae and the smaller, squirrel-sized Callitrichidae, are distinguished. Catarrhini are further subdivided into Cercopithecoidea and Hominoidea, the latter containing the lesser apes, or gibbons, the great apes, and our own species, Homo sapiens (simplified after Martin 1990, based on Simpson 1945).

Animal behavior in general is controlled and produced by the brain. Our approach to primate nonverbal communication is based on the sensory "channels" through which sensory input reaches the brain. In doing so, we obtain a link between the study of communication on one hand and morphology and phylogeny on the other. We are aware that other approaches are also possible (e.g., the quality of signals would also yield a basis of classification), but from a biological viewpoint, the former seems more appropriate.

The dominating centers of coordination in the brains of many "primitive" mammals are related to the olfactory system. "Primitive", in this context, means close to the evolutionary roots of all mammals, and living "primitive" forms may be understood as conservative. In contrast to many other mammals, primates have acquired a new, large, and efficient center of coordination, the neocortex, which is completely independent of olfaction, but contains large areas reserved for the processing of visual and acoustic input.

In this chapter, we aim at exploring the design features of each mode of nonverbal communication: On what conditions does successful communication via this channel depend? For transmission of what kind of information is each channel suited? We will investigate the kinds of elements the code comprises in terms of a semiotic analysis and go on to make inferences about the possibilities of deception. Such functional analyses will indicate the range of task demands different channels have evolved to meet.

2. Definitions

We speak of *communication* when information is transmitted between two or more individuals by a *symptom*, a *signal*, or a *symbol*. As a general term referring to either symptom, signal, or symbol we shall use 'display'. Another term for 'display' used in the literature is 'expressive movement'. It is characteristic of expressive movements that they occur before, or even detached from, the consummatory act they are supposed to refer to. This enables us to distinguish between emotionally motivated behavior on the one hand and displays expressing emotions on the other. According to this definition, the act of A hitting B is not communication but interaction because the act of hitting does not *signify* aggression but *is* aggression.

A *symptom* can be defined as something which is causally related to its *designatum*. There is a connexio rerum in the sense of Bühler (1932: 102; cf. Hörmann 1977: 15). As a consequence, a symptom indicates its designatum reliably, in a deterministic fashion, in analogy to a contingent reinforcement schedule. A *signal*, by contrast, bears a looser relation to its designatum. It may be dissociated from its corresponding emotional background, though still occurring in the usual situational context. Consequently, a signal indicates its designatum in a probabilistic fashion, analogous to a schedule with intermittent reinforcement. The distinction between symptom and signal is more a rule of the thumb than a strict borderline, since our understanding of the internal state to which a display refers is always limited, and new evidence might prove that what we took to be a signal is in fact a symptom, i.e., causally produced (in the nomothetical sense) by a certain version of the state. A *symbol* can be even more detached from its designatum, i.e., it can - in principle

- be used independently of any eliciting external condition, and its meaning is not dependent on contextual information (Hockett's criterion of displacement). As Seyfarth (1987: 445) puts it: a symbol "may accurately replace (i.e., elicit the same response as) the object for which it stands". In sum, a display, as we take it, is something that stands for, represents, or refers to, something which is not itself, the designatum.

Detection and recognition of displays as such is largely determined by the observer's own sensory capacities and, along with the latter, his or her ability to perceive gestalten in the sensory input. Understanding the meaning of a display demands thorough scientific examination, if only to avoid unjustified anthropomorphisms. It is the very test case of its being a display. Bloomfield's definition of the meaning of a word proves to be especially helpful in the context of nonverbal signs. According to Bloomfield (1933: 27; cf. Hörmann 1977: 107), meaning is constituted by, or can be derived from, the situation in which the sender utters the sign and the reaction it elicits from the receiver. As a consequence, the intention of a sender to communicate is not crucial to our definition. What is necessary, instead, is predictability of future behavior of sender and receiver contingent upon the performance of a display, or the predictability of a display from a preceding context. In the study of nonverbal communication, this is reflected in the distinction between the points of view of the sender and of the receiver. Accordingly, what a sender expresses by a certain display is termed his *motivation* (Jolly 1985) or *message* (according to Smith 1965). Bühler, in his organon model (1934), labels the sender's message as "expression". The interpretation a receiver attaches to the message is called *meaning* (according to Smith 1965). In Bühler's model it is the "appeal". Under optimum conditions of understanding, message and meaning will be congruent. An analysis of the proximate function of a display will explore its message and its meaning. By contrast, an analysis of the ultimate function of a display aims at a clarification of the role which the display has in the evolution of senders and receivers.

3. Evolution of nonverbal displays

The preparatory considerations of the following section are especially tailored to fit the situation of diurnal primates, but apply to a large extent also to animal communication in general. The term "expressive behavior" implies that an individual expresses an internal state, i.e., its motivation, in overt behavior. As a consequence, early investigations of animal communication focused on the message of a display (e.g., Caryl 1982). In order to detect that message, the antecedents of the display and the senders' subsequent tendencies to act were explored. From these premises, however, it remained a puzzle why threat signals, for example, were followed by attack in only 50% of cases (Stokes 1960, quoted from Hinde 1985a). The reason why an individual should announce its motivation to the public is not at all evident. More recently, Dawkins and Krebs (1978) have emphasized that displays should better be viewed as a cheap device of manipulating receivers. For example, instead of performing a risky and energetically expensive attack to frighten its rival away, a sender could also make use of a display to achieve the same ends. This proposition shifted the focus of research to the exploration of the meaning of a display. Consequently, the receiver's behavior contingent upon the perception of the display was investigated. Of course, experience as much as selection acts on the receivers making them less prone to react to bluff displays if they are not backed up by the designated behavior at least every now and then. We will not delve further into the bulk of literature on the topic of the honesty of agonistic, i.e., aggressive and submissive displays (see, e.g., Caryl 1979, 1981, 1982; Dawkins – Guilford 1991; Guilford – Dawkins 1991). Notice, however, that the possibilities of deception begin to be significantly limited when individuals meet repeatedly and have the capacity of remembering the outcomes of former encounters (van Rhijn – Voedegel 1980). A critical note on this discussion is that it has largely focused on cases where individuals have conflicting interests, which is in line with the general emphasis that sociobiological reasoning has placed on competition. At the same time, models for the evolution of displays referring to a cooperative attitude, e.g., bonding, care-taking, and sexual behavior are still lacking or have been neglected (but see, e.g., Frank 1988).

In his discussion of classical and game theoretical approaches to the evolution of displays, Hinde (1981, 1985a, 1985b) argues that a display expresses a *conditional* tendency to act. The condition upon which the future behavior depends is simply the reaction of the receiver to the display, in the sense of "If you don't stop, I will attack". If the receiver in the example then "stops", what need might there be to "attack"? Hinde (1985b: 110) concludes that there is an "internal state associated with the signal, which persists after its causation, but that the actor's subsequent behavior is also influenced by the response of the reactor." Thus, "although threat postures can be considered as expressive of an internal state, they can also be considered as signals in a process of negotiation between two individuals".

With this in mind, we can attempt to reconstruct the evolution of primate displays as follows: If individuals meet repeatedly, we can expect those individuals to leave the most progeny who can judge or predict the state of a conspecific. If an individual is able to understand, e.g., from fecal deposits that the conspecific occupying a neighbouring territory is in good health, is adult, and of the same sex, the receiver has a clue as to what it should do in case of an encounter (for a similar argumentation see Gosling 1982). Or else, if an individual is unable to predict the behavioral tendencies of its group mates, it will be continuously surprised by their actions and will be vulnerable to all their moods or waste a lot of energy in avoiding where it could just as well have stayed. It is thus plausible that the evolution of displays started with individuals that were able to make themselves receivers of cues symptomatic of an internal state in their conspecifics. Once there were such receivers, bearers of the symptom could become senders by making use of the regularity with which the receiver re-acted to the display, if it conformed with the sender's own interests. Otherwise, involuntary senders might be selected *not* to show a revealing symptom. The consequence of the continuous process of such an "evolutionary arms race" of senders trying to elicit a desirable response with minimal effort and of receivers trying to differentiate properly between bluff and displays worth a reaction is the evolution of relatively trustworthy displays. As in all evolved features, reliability of display is thus never once achieved and then retained, but always the fragile product of a dynamic balance of the selective pressures working on it.

As mentioned earlier, displays may be derived from by-products of functional behavior. Blushing may serve as an example: activation of the sympathicus has the function of preparing the body for ergotropic action. Blushing is produced by dilatation of peripheral blood vessels, and this process is a byproduct of sympathicus activity. Blushing itself is thus a symptom of sympathicus activity.

Displays can also be derived from "intention movements" (e.g., Andrew 1963). For example, when an animal wants to flee from another animal, it has to turn away before starting to run. "To be turned away" or to "show one's back to someone" may be one evolutionary root of "hindquarter presentation" - a display which is found in most Old World monkeys. Displays can also evolve out of "displacement activities" (e.g., Baerends 1975).

A displacement activity occurs when a motivation cannot be executed for some reason or if animals are in a state of conflicting motivations. In such situations, they may perform seemingly "inadequate" behavior, e.g., "nervous" scratching or yawning in primates. All three sources provide the raw material from which signals may arise (e.g., Tinbergen 1952), but they may not be the only ones (see 4.4 C8).

Evolved displays become *ritualized*, i.e., by natural selection, movements or features become conspicuous, stereotyped, and (sometimes) repetitive, thereby obtaining a "typical intensity" (Morris 1957). A ritualized display thus differs morphologically from the motor pattern it is derived from. Ritualization leads to facilitated recognition of the signal by its receiver and to reduced ambiguity (hence to a reduction of information content) in signal production on part of the sender. While facilitated recognition is mainly established by changes in the nervous system of the receiver and is hence difficult to assess, increased precision of performance can be assessed more easily by observation of the display behavior and the role morphological features play therein. To the extent signals are stereotyped and acquired through evolution, they can be subjected to an analysis of homology (see, e.g., Lorenz 1950; S. Preuschoft – van Hooff in press).

4. Channels of nonverbal communication

The major senses of touch, smell, vision, and audition provide a means of classifying nonverbal communication into four types which correspond to the four major channels of entry into the brain. In order to compare these channels more systematically, we will try to assess each of them according to the following criteria C1 to C10:

C1: Does it bridge spatial gaps? Does it travel over large distances, or do senders and receivers need to be in contact?

C2: Does it bridge temporal gaps? Is a dissociation of sender, display, and receiver possible, i.e., can communication be deferred or is it restricted to the given moment?

C3: Is it vulnerable to external influences? Can displays be distorted by poor visibility, noise, wind, water, etc.?

C4: What are the characteristics of display reception?

C5: Are there physical features specialized for display production?

C6: Is a display directed? Is it directed
- indiscriminately (accessible to every possible receiver)?
- selectively (directed towards specific classes of receivers)?
- or specifically (directed towards specific individuals or produced only under certain circumstances)?

C7: What kind of information is encoded in a display?
C8: What is the relationship of the display and its designatum? Is it a symptom, a signal, or a symbol?
C9: Is there room for deception or at least for voluntary production of the display?
C10: What is the functional significance of the channel? What task demands is it suited to meet?

We start with two phylogenetically "old-fashioned" modes of nonverbal communication, tactile and olfactory communication (4.1-4.2) and progress from there to the two more "modern" forms, acoustic and visual communication (4.3-4.4).

4.1 Tactile communication

C1: Tactile information becomes accessible only if sender and receiver are in direct physical contact or at least in immediate proximity, provided that a substrate mediates between them.

C2: Tactile information disappears immediately, i.e., there is rapid decay.

C3: The only external influences able to distort tactile information are superimposed external vibrations and very low temperature.

C4: Organs perceiving tactile information are distributed within the body. Actually, all proprioceptive receptors in the skin, tendons, ligaments, and muscles are involved. Tactile organs specialized to receive tactile information are the widespread vibrissae. Most mammals possess only facial vibrissae. By contrast, lemurs and lorises have additional vibrissae arranged at their wrists and forearms (cf. Hill 1960). However, neither of these organs seems to be specialized for tactile communication. Andres (1967) has argued that dermatoglyphics have been evolutionarily acquired to improve tactile perception. They certainly serve as a means of increasing friction, i.e., to provide a secure hold on a surface (Buck – Bär 1993). No matter whether a contact lasts for a short or for a long time, the information available through the tactile channel is extracted immediately and largely involuntarily.

C5: Tactile information is produced by the muscles and the movements of body segments and by the blood vessels of a sender and is collected from its body surface.

C6: Tactile information may be directed or not: a sender can actively deliver tactile information to a passive receiver, and a receiver can actively collect tactile stimuli from a passive emitter.

C7: Information available through body contact is about temperature, muscle tonus, or kinetic energy contained in a movement. By passive contact, an animal can learn something about whether its interactional partner is relaxed (feels soft) or tense (feels hard or vibrates), whether it is excited (feels warm, due to peripheral vascularization), or "cool". Active touching reveals something about the direction, force, and precision of a movement. Tender movements are usually slow and highly calculable for the receiver. Emergency (fight or flight) movements are brusque, highly energetic, and efficient. Active and passive tactile information is available in nonritualized behavior such as biting, slapping, wrestling, sucking, grooming, or passive body contact, such as in sleeping in groups and huddling. Among gregarious primates, i.e., species living in stable social groups, there seems to be a tendency to ritualize these behaviors. Stumptail macaques, e.g., perform controlled biting (Demaria – Thierry 1990; de Waal 1989), which may qualify as a signal of aggression or dominance. Like other species, Barbary macaques know different forms of embraces, some of which express the relative social status of the interactants; others are symmetrical so that status cannot be recognized from the role each participant plays. Grooming has most probably originally evolved to serve hygienic functions but has later become a primary means of establishing and maintaining social relations (cf. Dunbar 1991). Grooming and certain types of embraces may thus communicate more about the type of a relationship than about the actual emotional state of the sender. In some populations of humans, shoulder patting occurs as a means of communicating a friendly, but also ambivalent attitude in extremely fine shades of meaning. The same holds true for different ways of kissing, where "display rules" determine what kind of kiss is appropriate in view of the situation and the relationship. The latter two behaviors can be assumed to communicate the type of relationship as well as the current attitude held by the sender towards the receiver.

C8: Very often, the display is identical to the behavior it 'designates'. Therefore, it is not easy to draw a line where tactile communication exactly begins. Hence, tactile displays are mainly considered to be symptoms in their communicative function. As interaction, the transmission of tactile information is at least as old as sexual reproduction, and consequently it can be assumed to be used by all animals.

C9: It is difficult to modify tactile information intentionally. Deception is thus largely limited to the extent that an individual can manipulate itself into a certain state. A dissociation of tactile displays from the sender's behavioral or physiological state can hardly be achieved.

Tactile information is therefore highly reliable and free of errors, but provides few degrees of freedom since it consists mainly of symptoms instead of signals.

C10: Tactile communication is very close to the level of facts and serves immediate transmission of reliable information concerning transitory (mood) or long-term (body condition) states of individuals.

4.2 Olfactory communication

The world of smell is highly inaccessible to humans, and, unfortunately, scientific insight is also still limited. For many contexts, it is not clear whether chemical information is processed by olfaction, gustation, or by the vomeronasal organ. Since so little is known, and these sensory channels often work in concert, we have decided to consider them jointly under the heading of "olfactory communication". Olfactory communication is a heritage which goes far back in mammalian evolutionary history. It plays a large role in many vertebrates, including selachians, fishes, and amphibians, whereas in reptiles the olfactory bulb, which processes information entering through the olfactory sense organs, seems to be less developed. This reduction is even more pronounced in birds. In contrast to these animals, many, and among them all "primitive" mammals (extant or extinct), are characterized by the predominance of the sense of smell and a large, well-developed olfactory lobe of the brain. It is assumed that dominance of the olfactory sense was one of the basic characteristics which allowed the ancient mammals to reduce competition with reptiles - the others being homoiothermy and a nocturnal activity rhythm (Starck 1982). Among primates, the often nocturnal "strepsirhine" prosimians have retained olfaction as an important and well-developed sensory system. In the "haplorhine" primates (among them Tarsiiformes), however, the olfactory organ and lobe have undergone a reduction. This is particularly marked among Catarrhines. The olfactory abilities in the apes do not seem to be better than in humans.

Olfactory information may be transmitted actively, or it may be transmitted passively from one individual to another in a more or less hazardous fashion through tactile interaction. Scent marking by leaving odor deposits in the environment is characteristic of the fairly solitary, territorial prosimians, Lemuriformes (e.g., Microcebus spp., Cheirogaleus spp., Lepilemur, Hapalemur, some Lemur spp. [cf. Fornasieri – Roeder 1992], Indri), lorises, galagos, and tarsiers [Schilling 1979], of territorial groups of Platyrrhines (Callitricidae and Cebidae [Epple

1985] including Saimiri [Hopf et al. 1974]). It is also observed in a few gregarious catarrhine primates, especially the drill and the mandrill (Feistner 1991). Allomarking in the sense of active and direct scent transmission from one individual to another occurs among diurnal gregarious prosimians, lemuridae and lorisidae (Schilling 1979), and some New World monkeys (Epple 1985). Self-impregnation, or automarking, can be accomplished by active collection of scent from a conspecific or by usage of own body secretions, as in the case of tail marking (see below). A scent marked individual can be considered a walking odor deposit.

C1: Little is known about the potential of olfactory stimuli to bridge spatial gaps. Prosimians are able to recognize odors at extreme dilutions (Schilling 1979: 523). Schilling (1979: 501) states that prosimians usually "detect odors only at relatively small distances" and sniff "as close as possible to the source". This is a behavior they share with classical macrosomatic mammals like dogs. As allomarking is, by necessity, carried out in a social setting, the bridging of large distances between interactants generally seems less important.

C2: Deposited scent marks function as an "information store", i.e., the information remains accessible over longer periods of time to everybody who happens to sniff. Similarly, the olfactory information carried by an allomarked individual may be understandable to third parties long after the odiferous material has been transferred. Hence, information transmission may take place in the absence of the sender. The length of the period during which olfactory information remains accessible differs between species. Experiments with mouse makis and Lemur fulvus have demonstrated an ability to discriminate between individual conspecifics on the basis of scent marks that are several months old (Schilling 1979: 523; Fornasieri – Roeder 1992: 2075). In contrast, Lemur macaco and tamarins (Callitricidae) stopped to exhibit discriminative responses after much shorter time (L. macaco after 15 days cf. Fornasieri – Roeder 1992, tamarins after three days, cf. Epple 1985: 781). Scent marks are thus a means of dealing with the temporal discontinuity between the emission and the reception of a display. The regular usage of olfactory information to bridge temporal gaps is paralleled only by written language. Or, in Schilling's (1979: 513) words, "scent marking is to olfactory communication what writing is to language".

C3: The transmission of olfactory information can be hindered by external factors, such as high humidity, water washing away the scent, evaporation, or by strong wind. Generally, receiving olfactory information from a conspecific is less prone to the hazards of weather and environment, but only insofar as the marked individual and the receiver are in spatial proximity.

C4: In the strepsirhine lorisiformes and lemuriformes, signal reception is facilitated by a special olfactory adaptation, the *rhinarium*. A rhinarium is "an area of moist, glandular, and essentially hairless skin surrounding the nostrils" (Martin 1990: 327). The tarsiiformes and all simians lack this feature (therefore 'haplorhines'). Another morphological feature involved in the perception of chemical stimuli is the vomeronasal or Jacobson's organ. It is typical for most mammals, and occurs in strepsirhine prosimians, is well developed in the haplorhine tarsiers and in adult New World monkeys, but it is absent in Old World monkeys (Martin 1990). The distribution of Jacobson's organ among primates is roughly paralleled by the distribution of specialized skin glands.

C5: While prosimians and New World monkeys regularly possess skin glands specialized for the production of odiferous substances, there are only few simians from the Old World who possess such organs: drills, mandrills, gibbons and orangutans (Epple 1985: 743, 748, Tab. 18.1). Such glands can be found on different body regions. Common locations are the anogenital region, the upper chest (sternum), neck and mouth region, and the upper extremities. The glands are often surrounded by hair brushes. The antebrachial gland of ringtailed lemurs is equipped with a thorny spur (Schilling 1979). Olfactory information is not only available in specially produced odiferous substances, but is also contained in other products of autonomous functions of the body, e.g., vaginal discharge (see also C8), feces, urine, saliva, or sweat. Likewise, application of odiferous material can be by stereotyped activities or by simple smearing or dragging. 'Urine washing' is an example of a highly ritualized application behavior where a nonspecialized substance (urine) is employed, but its social role remains to be demonstrated (Roeder – Anderson 1991). Schilling (1979, 1990) and Roeder (1985) state that diurnal, gregarious prosimians tend to back up, or even substitute, their olfactory marking behavior by exaggerated visual displays bearing a more or less faint functional relation to marking. This seems to imply a tendency towards ritualized marking behavior in these prosimians.

C6: In olfactory communication we find the whole range from indiscriminate emission to selectively directed sending. Obviously, scent deposits in the environment, on a social partner, and also as automarking differ insofar as the sender has control over who will be his receivers. It is likely that allomarked individuals mediate olfactory messages between several interactants.

C7: Scent may inform about age, sex, bodily condition, such as reproductive and other hormonal states, emotional state, social status, time of depositing, and also about the sender's identity. Even without a personal encounter, a receiver of deposited scent can thus discover who the sender might be and what long- and medium-term attitudes this conspecific holds. Olfactory marks may also aid in group cohesion among gregarious primates. They may be equivalent to notes like "I went this way half an hour ago, Signe." The message may help localizability by group mates in dense forests with conditions of poor visibility. At the same time, scent marks limit the range of undesired receivers to those who are able to "overhear" an olfactory dialogue. Monkeys can thus preclude some predators, e.g., raptors, owls, or humans, from taking notice of them. Similarly, mandrills lead gregarious diurnal lives in dense forests with conditions of poor visibility. They are closely related to other baboons who inhabit open areas and are not known to use olfactory marks at all. Mandrills, however, have specialized sternal glands, and high ranking, sexually mature males can be observed regularly to impregnate objects in their environment (Feistner 1991).

Lemur fulvus and L. macaco do not mark group members at random but prefer high ranking receivers of the same or opposite sex. For each individual, the depositing of marks is highly correlated with the receiving of allo-marks (Fornasieri – Roeder 1993). Little is known about message and meaning of such impregnation, but it seems likely that it communicates attachment. Interestingly, among squirrel monkeys, allomarking can express a completely different relation of sender and receiver. According to Hopf et al. (1974), dominant individuals direct a stereotyped visual "genital display" at subordinate receivers in the course of which a few drops of urine are discharged into the face of the receiver. A similarly offending effect of passing one's own odors to others seem to have the stink-fights of Lemur catta (Jolly 1966). As a general display of aggression, ringtailed lemurs show tail marking, i.e., they impregnate their tails with secretions from different body glands. This functions as a threat, which may be exaggerated by tail waving where the thus perfumed tail is kept in a position above the sender's body pointing at the opponent (Schilling 1974) . In a stink fight, individuals of both bands stand facing each other waving their impregnated tails, presumably to produce air currents transporting their odors to the other group.

C8: Like with tactile communication, it is difficult to draw a line between olfactory interaction and communication. One reason is that odiferous substances are (by-)products of physiological processes.

Another is that most olfactory displays are symptomatic of a certain (often hormonal) state or, in the case of individual odors, are just one aspect, one part of the sender's identity. There is some evidence (e.g., Perret 1992) that odiferous substances may work as true pheromones at least in some prosimians, i.e., as necessary and sufficient conditions, they switch the behavior or physiological condition of a receiver on and off. By contrast, in most Old World monkeys, the influence of olfactory information on the behavior of a receiver seems to be limited though rather complex, as is illustrated by the example of rhesus monkeys. In the fertile period of their menstrual cycles, the vaginal secretions of female rhesus contain lipid acids. Frequent tactile and olfactory inspections by male group members suggest that males derive information about the hormonal state of a female from investigating her anogenital region. Lipid acids are able to attract males as has been shown experimentally: ovarectomized females artificially impregnated with lipid acids became sexually attractive to rhesus males, and the lipid acids per se attracted males to sniff. However, they did not attract every male. Inexperienced young males did not show any reaction to the scent. The smell of lipid acids itself is thus telling only to sexually experienced males and is therefore not a sufficient condition for copulation attempts. In addition, the scent of lipid acids is not a necessary condition: anosmic rhesus males did not hesitate to copulate with females in estrus. The proceptive behavior of the females was more predictive to sexual interactions than was the smell of lipid acids (summarized in Fett 1980; Jolly 1985).

C9: It is evident that such a degree of congruency between odor and designatum sets very strict limits on the possibilities of faking messages or cheating. Nothing is documented on deception by olfactory displays. It seems at least thinkable that in automarking and, to a lesser degree, in allomarking some form of deception may occur: a sender may perhaps allo-mark an "inappropriate" conspecific with its own scent, or an individual may "fake" an odiferous substance by collecting it via self-impregnation from a conspecific, thus transmitting a message it could not produce on its own. The limits to deception are then set by the receiver's ability or inability to disentangle the olfactory "cocktail" it is confronted with. Passive self-impregnation (as in the example of rhesus females) does not seem to leave any possibility for deception. Such odors are merely symptoms of a hormonal state or physical condition.

C10: Olfaction seems to be a slow and inert but rather honest means of communication, which is capable of connecting interactants over temporal or - if conditions are favorable - spatial distances. It provides senders with some protection from being detected.

4.3 Acoustic communication

Together with birds, mammals are considered to be the acoustic experts among the vertebrates. This judgement, however, is justified only by the abilities of sound production and rests largely on the passerine birds. Vocalizations differ from species to species in their acoustic structures and with respect to their contexts. There is an enormous range of different vocalizations used by primates. Assigning names to individual calls is often a clumsy task, and for nonexperts, imagination of acoustic features is aided little by these labels. Words used frequently include grunt, scream, bark, chirp, pant, belch, whistle, whimper, cackle, etc. Other labels are onomatopoeic as, e.g., "whoop-gobble" or "coo". Another way to name vocalizations is according to their acoustic features, such as frequency patterns, e.g., "early high" vs. "late high", or "noisy", "arched", "tonal", "pulsed", "serial" (e.g., Todt 1988), etc. For the purpose of this text, it will suffice to refer to vocalizations by functional terms. In the acoustic repertoire of many primate species we find long-distance calls, alarm calls, contact sounds, and screams. These vocal-izations may be broadly classified in three functional contexts, sounds serving group cohesion and intergroup spacing, antipredator calls, and vocalizations serving immediate communication among group members or within families at short distances (see also Seyfarth 1987).

C1: In all cases, sound carries over large distances and can be received from all directions.

C2: However, acoustic communication is not persistent in time. Decay is immediate, but changing information can be transferred without delay.

C3: Furthermore, sounds are vulnerable to environmental deterrents, such as wind, thick foliage, humidity of the air, and competing noise. Acoustic properties of the long-distance calls of primates are designed to meet these deterrents with an energy in the range of 500 to 2,500 Hz. Sounds of this kind suffer the least attenuation of amplitude with distance, but require expenditure of much energy from the sender. Another way of circumventing deleterious environmental influences is to choose broadcasting times with little sound attenuation. Accordingly, many primates perform choruses at dawn (Seyfarth 1987).

C4: The acoustic receptor organ of primates, like in all mammals, is less highly developed than the ones of some other vertebrates. Like humans, the primates studied so far hear sounds between 16 and 20,000

Hz, many of them even up to 45,000 Hz. Sometimes, their thresholds are lower than those of humans, and the greatest sensitivity may be at different frequencies (Penzlin 1989; Fay 1990). For comparison, insectivorous bats and dolphins are able to perceive up to 150,000 Hz. Some birds seem to perceive 3 Hz as sound, but 8,000 Hz is the upper limit for most. Some species, however, hear up to 29,000 Hz. Still narrower is the range of frequencies accessible to reptiles (20 to 6,000 Hz; Penzlin 1989). Sound perception by birds and reptiles is accomplished by means of ears which differ remarkably from those of mammals. They possess no external ear, no acoustic meatus, only one ossicle, and a short, nonspiraled cochlea (Starck 1982). The fact that mammals are able to perceive a wider range of frequencies than birds is a reflection of the difference in shape of the receptor organ. Reception of acoustic information is quick, as is its production. Since sound is characterized by rapid decay and although an acoustic display calls for attention, important parts of the message may be missed when the receiver's attention is not yet focused on it. This is in line with Waser's (1982) proposal that the first syllable of the mangabeys' "whoop-gobble" may function to draw the receiver's attention to the following "gobble" vocalization, which is more variable and contains more information.

C5: Vocal sound production is accomplished by the larynx, throat, oral, and nasal cavity. The larynx is an organ present in all mammals, but not always used for sound production. Its morphology varies much less than the produced sounds. In this context, the extended discussion about the ability of chimpanzees (or Neanderthal men) to produce speech sounds should be recalled. Lieberman and coworkers stated in several papers (e.g., Lieberman - Crelin 1971; Lieberman - Crelin - Klatt 1972) that the throat, oral and nasal cavity, the "Ansatzrohr" of the German physiology textbooks, is unsuited to produce vowels, which are essential to speech. The latest summary of this discussion of which we are aware was given by Wind (1983, 1990), who concluded that a chimpanzee with a human brain should be able to speak. As a matter of fact, it has been entirely impossible to teach chimpanzees to produce word-like sounds, whereas they are quite competent to handle nonvocal language systems, like American Sign Language or lexigrams (for more details see H. Preuschoft 1990) and even to understand spoken American (Rumbaugh 1992).

In some territorial species, special sound enforcing organs are found. A case in point are the throat sacks of howler monkeys, gibbons, and orangutans that are sometimes developed only in males. The adult male orangutan is equipped with cheek flanges in addition, which may aid in directing emitted sound. In other species, teeth are used for intermale spacing, as in the tooth grinding of langurs. Some species use their hands and arms for sound production, as gorillas do in chest beating, or chimpanzees, when drumming on tree trunks.

Vocalizations can be produced quickly and can be modulated finely even in the course of vocalizing. Vocal information may vary in volume, but there is the additional possibility that different signals merge, and thus create ever finer shades of meaning. In this manner, infinitely subtle blends of messages can be transferred to the receiver. At close range, vocally transmitted information can be tuned "on line" according to the actions of the interaction partner: A male barbary macaque (A), by racing towards a second male (B), elicited intense screaming. Since it was too late for escape, male B performed an appeasement signal, anogenital presentation, while at the same time he was still screaming. Meanwhile, male A had reached him and flung himself on monkey B's back. At the same time, B's vocalization changed gradually from clear screaming through unritualized gargling and then stabilized in a clear snarl and in smacking sounds, which are typically associated with the accompanying facial expression which had gradually changed in parallel to the vocalization (personal observation, S.P.). The audience may interpret that B first communicated his sheer terror. B probably perceived himself a victim of a severe attack. Then, B's attempt at appeasing and inviting an affiliative interaction which turned into an eager confirmation as he became convinced that this was indeed an albeit unconventional, but nevertheless nonaggressive, body contact (namely male-male mounting).

However, graded sounds are not necessarily perceived as graded, as Green (1975) has shown. In an investigation of "coo" contact sounds of Japanese macaques, he found that receivers exhibited categorical perception where sonographs exhibited gradual changes in sound patterns. This is analogous to the perception of stop consonants in language by humans (e.g., /b/ to /g/, cf. Liberman et al. 1957). Vocalizations arranged in series may have meanings to the receiver which differ from those of the individual components of this series (Todt 1988).

C6: Vocalizations may be directed at individuals or may be addressed to the public in general. Duetting in gibbons may serve to strengthen the pair bond, but also proves that the vocalizing couple is still in their territory and also communicates their condition (Preuschoft et al. 1984: chap. 4). An important feature of acoustic displays is that sound can also be perceived by other receivers besides the addressee, e.g., predators. It is intriguing that the vocalizations of mangabeys not used for intergroup communication are less well transmitted in their habitat due to their acoustic characteristics - probably to allow for some more "privacy" in acoustic communication (Waser – Waser 1977; cf. Seyfarth 1987).

C7: Primate sounds are used to refer to a variety of objects in the external and the social world of these animals. Investigations of what they denote or connote to the monkeys have revealed formerly unknown facts about cognitive computation by monkeys:

(a) External referents: Predators: There are exemplary studies on alarm calls in wild vervet monkeys by Seyfarth and Cheney (e.g., 1982, Seyfarth et al. 1980). Similar evidence is available for Lemur catta (Macedonia 1990) or for Saimiri sciureus (Herzog – Hopf 1983, 1984). Vervets use three distinct alarm calls. Each of these vocalizations warns its receiver of a different predator and implies taking appropriate action. There is a leopard alarm eliciting flight to the top of the canopy. An eagle alarm means to flee out of the top of the canopy and to take cover, and the snake alarm warns to stay out of thickets. Appropriate warning behavior has to be learned (Seyfarth – Cheney 1980). It seems as if vervets were born with vague ideas concerning the nature of the designatum of each of these alarms which become more focused in the course of ontogeny. False alarms, given mostly by young individuals are especially telling: a falling leaf elicited an eagle alarm, a mouse rustling on the ground provoked a snake alarm, and a warthog made a juvenile give a leopard alarm. These results are experimentally controlled and still more complex than can be reported here. The findings show clearly that there are primate displays which denote objects in the external world. Such referencing has been termed symbolic or semantic (Marler 1984; Seyfarth 1987). Playback experiments, by artificial separation of a display from its eliciting context, give evidence whether monkeys are able to process the same information regardless of its being elicited, for instance, by a real predator, or by a *symbol* which *stands for* the predator. Alarm calls *refer* to a certain type of danger independent of the context in which they are given.

(b) External referents: Social opponents: It is likely that vocalizations employed in intragroup communication also have external referents. Screams of certain types can elicit support or other interventions from third parties in case of disputes between group members. When juvenile rhesus monkeys are involved in a conflict, they may 'scream'. Such screams can be classified according to their acoustic patterns (Gouzoules et al. 1984). Interestingly, different screams are uttered by juveniles facing different kinds of opponents: higher ranking, lower ranking, or related group members. In conflicts with higher ranking opponents, senders distinguished between aggression with and without body contact. Playback experiments of such screams proved that mothers also reacted differentially: most strongly to contact aggression involving higher ranking individuals and less strongly to aggression by relatives.

(c) Connotative information: Another aspect of vocalizations is that they provide connotative information, too. Even infants are recognized by their voices, not only by their mothers but by every member of the group. This was tested by field experiments on maternal recognition of screams of juveniles from playbacks in vervet monkeys (Cheney – Seyfarth 1982). Such experiments provide valid information only when care is taken that the sender of the vocalization on the tape is out of sight of the receivers. The results indicated that mothers discriminate between their own offspring and other juveniles. However, the monkeys seem to proceed beyond such egocentric classifications: unrelated female bystanders looked significantly more often at mothers when receiving the playback scream, i.e., they behaved as if they were able to match the voices of individual juveniles their respective mothers.

In sum, information encoded in acoustic displays of primates ranges from expressions of emotional state and short- or long-term behavioral dispositions (e.g., estrous call alerting a possible mate or acoustic territory demarcation indicating readiness to defend one's territory) to external referents such as predators, food, or types of social situations.

C8: There is considerable flexibility in the use made of vocalizations: solitary monkeys can withhold alarm calls (Seyfarth 1987). Food calls are given only when finding an abundant food source, but not a scanty one (Dittus 1984). These findings leave no doubt that the displays used are at least signals. Some may even be symbols. In his discussion of toque macaque food calls, Dittus (1984) hints at the possibility that these vocalizations may be signals of a sender's state of elated mood or euphoria (elicited by the sight of a desired object) while at the same time, to the receivers, they may be symbols referring to the availability of a desired object.

C9: The option of withholding acoustic signals and the potential of acoustic signals to be the sufficient cause of a fullfledged response by the receiver allows acoustic signals to be used deceitfully. Whether this is indeed done by primates or not depends further on their capacity to produce the signal voluntarily and to diagnose a situation where this would be advantageous to them. There is room for deception, but to date, positive evidence is based only on anecdotes.

C10: Acoustic displays thus form a very flexible and speedy means of communicating over large distance as well as at close range. They are well suited for reaching a not yet attentive partner, but they are tied to the moment of emission and leave no traces. Their honesty is not beyond all doubt.

4.4 Visual communication

Visual communication is perhaps the most important, certainly the most obvious instrument of communication in all diurnal primates. The most elaborate visual signals are employed by gregarious species in intra-group communication. They differ in several respects from the longer lasting optical markers which distinguish individuals in certain bodily states, members of populations, or demographic classes. Persistent optical markers emphasize certain body regions or extend over the whole body of an individual. Examples are manes, colorful skin or fur patterns (e.g., in many guenons and in the forest baboons), tail shapes (even the basis of name giving: stumptail, longtailed, liontail, pigtail - all of them macaque species!), hair dressing of beards ("Schnurrbart Tamarin") or of heads (e.g., bonnet or toque macaque, or crested mangabey, "Schopfmakak").

There are also features which change their appearance during the lifetime of an individual. In male gorillas, the hair on the back and the dorsal sides of upper arms and thighs turns grey at the age of social maturity (years after sexual maturity), hence the term "silverback" for a mature gorilla male. At the same time, the males also develop a hood at the occipital part of the head and the upper neck. Similarly, in male orangutans, the cheek flanges appear upon reaching social maturity. Most primates are born with a special neonate coat, giving them an appearance which clearly differs from that of their older conspecifics.

Other visual markers undergo cyclic changes. The females in many species of Old World monkeys show prominent anogenital swellings advertising their fertile cycle periods. In gelada baboons, which have bold triangular skin patches on their chests, females in estrous have small areas of cavernous tissue framing these pink skin areas like beads (Alvarez 1973).

Social interactions of primates are usually accompanied by, and based on, gestures and facial expressions which seem particularly understandable to humans. Postural-gestural vocabularies include items of varying degrees of ritualization. Some postures or gestures serve practical purposes. In the "crouch", for example, the sender is the victim of severe aggression and protects himself by pressing his belly to the substrate and concentrating extremities around the body center. "Raised tail", in many species, is indicative of fear sometimes combined with defecation (e.g., for longtailed macaques cf. Angst 1974). In other species (e.g., Japanese macaques) carrying the (short) tail in elevated position indicates top rank in males (Kawai 1960). Presentation of the ano-genital region is very common among primates. While in Old World monkeys and apes it is usually a socio-sexual behavior of inferior or subordinate interactants, in some New World monkeys, the display indicates threat or a superior position. Chimpanzees have a large inventory of manual gestures, one of the most famous being the "begging gesture" with an extended arm (e.g., Goodall 1968). Gestural behavior involving body contact has been dealt with above as tactile communication, although it consists of clearly visual components, too.

Facial displays are more arbitrary than postures and gestures. Their striking similarity to our own facial expressions seems therefore especially telling. Yet, scientific analysis has to demonstrate case by case where true phylogenetic continuities exist and where resemblance is only superficial. For catalogues of primate facial expressions, the reader is referred to van Hooff (1962, 1967), Redican (1975), or Chevalier-Skolnikoff (1973, 1982).

C1: Conspicuous optical markers are visible from a distance. Facial and gestural displays are used primarily for immediate communication at close range and among group members. Hence, they do not bridge large spatial gaps.

C2: Although optical markers do not bridge temporal gaps, their persistence prolongs the time span available to extract the information contained in them. In contrast, facial expressions and gestures decay rapidly. Displays are performed and disappear during fractions of seconds. Greater salience of the signals can be achieved only by prolonged muscle activity or by repetition. As a result, visual displays are not at all suited for bridging temporal gaps. Thus, optical displays require that the receiver is not only close by, but already attends to the body part where the display will appear.

C3: Ubiquity of visual barriers and dim light characterize the evergreen forest habitats of many primate species. Some optical markers seem to convey information which is redundant at close range (neonate markers, sexual markers) but compensate for attenuation of the optical stimuli over distances. Information transmission by transient visual displays is thwarted whenever conditions of poor visibility prevail and is impeded by large distances between sender and receiver.

C4: One of the most characteristic traits of primates are the large, well developed eyes, placed side by side in the head. This allows stereoscopic vision, which is of particular importance for tree living forms: they leap, climb, or brachiate in a three-dimensional environment where correct estimation of distances can be a matter of survival. Nocturnal forms have secondarily increased the size of their eyes and evolved a tapetum lucidum (Martin 1990) to exploit even small amounts of light. All primates investigated so far, from lemurs to apes and humans, perceive colors. There is, however, a widespread weakness in seeing red, which is in fact also the most frequent anomaly in humans.

With respect to vision, humans are again typical representatives of their order. Visual communication of primates is more accessible to us than their signalling in other channels. A limit to this accessibility is set by the speed of movements. For instance, the very small New World callithricids do make use of facial displays (Epple – Smith 1985), but for a human observer, it is extremely difficult to recognize them without the aid of slow motion presentation on film (personal observation, S. P.).

C5: The organic features carrying optically distinctive body marks consist of pelage, skin, bone, connective and fatty tissue, cavernous sinuses and intensely vascularized areas of the skin surface. Such specialized bodily features may be emphasized additionally, e.g., through piloerection or increased vascularization.

Gestures and postures are characterized by movements and the relative positions of the body parts: head, chest, hips, limbs, tail. Facial displays are composed of facial elements, lips, jaws, cheeks, ears, etc., arranged in characteristic patterns by the activity of the facial musculature. Evidently, none of these display constituting body parts exists only for display production. Instead, the existing body parts are often emphasized by color or by hair tufts. For example, the upper eyelids of many cercopithecines are of contrasting color (pale or blue). The body parts decorated in this way may be employed in specific displays, the eyelids, e.g., in the so-called "eyebrow-flash".

C6: Optical markers are conspicuous, but not usually directed at certain receivers. Eye-catching optical patterns may facilitate detection by eliciting an orienting response. Surprisingly little is reported about the usage of optical markers in behavioral displays. By contrast, the majority of facial-gestural displays is specifically directed at one or more group members under precisely defined circumstances. Therefore it can become difficult for a bystander to see exactly what signals are exchanged. In so-called "two frontal behavior" (Kummer 1967), one sender directs two contradictory messages at two different receivers simultaneously. For instance, in the "protected threat", a sender may direct a "presentation" at one interaction partner, while showing a "threat face" to the other.

C7: The information encoded in optical markers indicates the sender's group, population, species, gender, or age. Visual displays have traditionally been classified as having only emotional content. However, the strict dichotomy of emotion (smacking of instinct) and cognition (smacking of reason) is psychologically unwarranted (see, e.g., MacDougall 1928; Lazarus 1984; Zajonc 1984). Its use as a criterion to distinguish between "man" on the one hand and "the animal" on the other is an oversimplification (see Marler 1984) and a hindrance to a more thorough understanding of animals and humans. We do not deny that facial expression may denote emotions or affect states. However, finer analysis of what exactly is communicated is much more informative. In section 3 we reported that displays may express tendencies to act which are dependent upon the reaction to the display by the receiver. Hence, displays may be regarded as indicators of a negotiation process between sender and receiver (Hinde 1985b). More specifically, it might prove fruitful to arrange displays on a continuum ranging from "pure expression" at the one end to "pure negotiation" at the other. This can be elucidated by reports on the silent bared-teeth display (sbt) and the relaxed open-mouth display (rom). These two facial displays are comparably well studied and seem to be fairly universal among the catarrhines:

The sbt is a typical case of a display with a negotiatory function. This display resembles the human smile and is regarded to be its ancestor (van Hooff 1972; S. Preuschoft 1992). It has often been called "fear grin" or "grimace", because in many monkeys it is clearly associated with tense social situations, and its sender is usually the inferior partner. There are, however, remarkable species differences with respect to its exact social function. In some species, sbt is employed as a pacificatory signal, diminishing the aggression of the interaction partner and eventually leading to the reconciliation of the sender and receiver (e.g., in Barbary macaques, S. Preuschoft 1992). The sbt can also have a reassuring function (e.g., Tonkean macaques: Thierry et al. 1989; chimpanzees: van Hooff 1972). It is then performed by a higher ranking to mollify a lower ranking interaction partner. In other species (longtailed macaques: Angst 1974; de Waal et al. 1976; rhesus macaques: de Waal – Luttrell 1985), sbt can be employed by the inferior to communicate his or her acceptance of the subordinate status, irrespective of the situational context. It is then the expression of a sender's evaluation of a long-term dyadic relation rather than an expression of an acute emotional state.

The play face of primates is an example of a typical expressive display. This facial expression with the scientific name "relaxed open-mouth display" is the likely precursor of human laughter (van Hooff 1972; S. Preuschoft 1992). It is characteristic of interactions where both partners exhibit a striking degree of mutual adjustment, both in what they do and when they do it. While these behaviors are often ritualized forms of play-fighting (rough-and-tumble play, gnaw-wrestling), virtually every behavior can be accompanied by a play face and thus be turned into play. Attempts at finding reliable behavioral correlates of rom in chimpanzees have failed so far (Spijkerman, personal communication). This is an interesting result, since it is exactly what one would predict if the play face were a mere expression of 'fun', the playful mood which occurs spontaneously and not contingent upon external conditions. Although absence of bodily needs or presence of play partners stimulate play behavior, they are not sufficient to evoke the playful mood. Even though rom is "only" the expression of an internal state, one should not underestimate the cognitive complexity of its function. Bateson (1955: 179) discovered the metacommunicative function of the play face. He stressed the similarity of the movements in play fighting and real combat, but "it was evident, even to the human observer, that the sequence as a whole was not combat, and evident to the human observer that to the participant monkeys this was 'not combat'". This, Bateson concludes, could only occur if the participants were capable of exchanging some signal which would carry the message "this is not meant seriously". This demonstrates that the seemingly chaotic and moody character of play is only possible by virtue of a special signal functioning as a "key" (or traffic signal) which allows the rules to be broken.

In sum, optical markers correlate with the affiliation of an individual to a taxonomic or demographic class. Visual displays, both facial and postural ones, have been demonstrated to be expressions of emotional states, metacommunicative signals, formal indicators of status, or indicators of negotiation processes.

C8: The occurrence of optical markers is encoded in the genome and depends on hormonal and physiological processes. As a result, optical markers are merely symptoms of the emitting animal's state. Whether such displays can be considered to be of truly communicative value depends on whether they have information content of their own. This can only be examined experimentally, since under normal conditions, too many additional cues are confounded with the display. No such information is available yet.

The correct understanding of species-specific facial expressions is not innate, but their production seems to develop (ontogenetically) fairly reliably, irrespective of the conditions under which a monkey is raised (Miller et al. 1967). It is, at present, unclear whether (or which) primates are able to control their visual signalling voluntarily. Anecdotal evidence suggests that they can do so, at least to some extent (see C9, below). It should be stressed here that systematic investigation of the information content of visual displays lags far behind research carried out on vocalizations. It is extremely difficult to design experiments for visual displays which parallel playback experiments by uncoupling a signal from its situational context or from its sender and which maintain a reasonable degree of ecological validity at the same time. The active participation and seemingly passionate involvement of monkeys in social events obscures the distinction between the external social context of a display and the emotional attitude this context elicits in the sender. In the light of the many parallels between facial and vocal displays, it seems likely that visual displays refer to external objects in the manner described above for screams (cf. Gouzoules et al. 1984), and hence that they qualify as symbols although this still awaits empirical confirmation. In conclusion, we think it probable that among the facial-gestural displays there are many which should be classified as signals at least.

Graded signals are sometimes held to be more honest than those exhibiting a high degree of stereotypy (Maynard Smith – Parker 1976). In visual communication, gradual shifts from one short-lasting display to another can be performed "on line", in the course of the interaction, in exactly the same fashion as described above for vocalizations. The transitory states which occur when prototypical ritualized displays merge can become the raw material for selection to work on. The ritualized facial display of "teeth-chattering" has probably arisen from a blend of silent bared-teeth and "lip-smacking".

C9: Optical markers appear only in animals which have the appropriate genetic makeup. When optical markers have evolved, there is no way of either denying their presence or faking their appearance.

From what has been described so far about transient visual displays it can be inferred that there is room for deception by visual signals. This conclusion is supported by anecdotal evidence only. For instance, the 'play face' is sometimes performed in contexts of teasing, e.g., while putting up the "this-is-not-meant-seriously" expression, the sender approaches and hits the receiver forcefully. It seems highly unlikely that such a sender considers his or her own behavior as play, but maybe he or she thinks "it's fun"?

C10: It seems probable that conspicuous optical markers can lower the signal detection threshold of a receiver and hence facilitate recognition. Moreover, they may serve to catch the attention of an individual and thus prepare the receiver for the reception of further signals. It may also be that relatively persistent optical markers function as a "key", guiding the receiver's interpretation of the activities of the sender. Visual communication by transient signals, on the other hand, resembles acoustic communication insofar as it is highly flexible and can be modified on-line, reflecting the course of the interaction. It is hence well suited for fine-tuned negotiation processes among interactants.

5. Discussion and conclusions

Our investigation of the major channels of communication has shown that each sensory mode is designed to fulfill a specific range of tasks. The overlap in the functions fulfilled by each channel is limited. A rough overview summarizing our report is listed in Table 1. This survey, however, needs a number of additional comments.

Touch is the channel for communicating at close range. As a consequence, tactile communication is least susceptible to interference. Olfactory information transmission can be deferred in time, but scent dilution sets a limit to temporal gaps between scent emission and reception. It also limits the spatial distance between receiver and scent. Sound displays carry farthest in space, but decay rapidly. Their range is limited by distortion and attenuation of sound as well as by background noise. Consequently, acoustic transmission is immediate in time and able to connect interactants over distances. Optical displays work best at moderate distance and when attention of the receiver is guaranteed. Visual communication is immediate both in time and in space.

Specialized organs for the reception of intraspecific displays seem to be very rare in primates. Instead, primates exploit their very efficient mammalian sensory organs to make themselves receivers of intraspecific displays. Adaptation to display reception may exist in the form of lowered signal detection thresholds, increased discrimination capacities,

Table 1. Synopsis of communicative potential of four channels of nonverbal communication according to nine criteria of comparison: "()": may occur or may be absent, "/": or, "?": is possible.

Criterion	tactile	olfactory	auditory	visual
C1 bridging of spatial gaps	no	yes, in principle	yes!	yes, but limited
C2 bridging of temporal gaps	no	yes!	no	no
C3 obstacles	low temperature; vibration	dilution by humidity, devaporation, wind	distance attenuation; distortion by solids, vegetation	solids; vegetation; lack & excess light
C4 reception characteristics	highly developed: proprioception, dermatoglyphics, (vibrissae)	limited: smell, gustation, (rhinarium, vomeronasal organ)	highly developed: (mobile) external ear, inner ear	highly developed eye: stereoscopic & color vision
C5 production characteristics	(ritualized behavior)	(specialized glands, ritualized behavior) (by-)products of body functions, smearing	(throat sacks, cheek flanges) frequency range, advance notice, broadcasting time	(morphological markers; mimic musculature)
C6 receiver specificity	accidental & indiscriminate / deliberate & directed	indiscriminate broadcasting / selectively directed	indiscriminate broadcasting / dialogue-like	(morphological markers, indiscriminate broadcasting: conspicuous, public) behavioral displays: selectively directed
	always private	not private	conspicuous, public	not public
C7 information encoded	body condition; motivation; emotion; social attitude	"signature": taxonomic & demographic, social affiliation; identity; physical, reproductive condition; social, environmental claims; status, motivation, social attitude.	body condition; social, environmental claim; predators; food; social role; spacing; motivation; emotion	physical, reproductive condition; taxonomic, demographic membership; negotiation; metacommunication; social role, attitude; motivation; emotion
C8 nature of reference	symptom	symptom, ? signal, via classical conditioning	symptom, signal, symbol	symptom, signal, ? symbol
C9 deception	not reported	not reported	qualitative evidence	qualitative evidence

or facilitated input categorization, but evidence for such adaptation could not be found in the present survey.

By contrast, it became clear that senders exhibit special adaptations serving display production in three out of four sensory modes of communication. These range from physical features (skin glands, morphological characters, color patterns) to behavioral specializations of display production. Behavioral adaptations are reflected in ritualization of performance and a physical design of displays that shows accommodation to environmental factors and receiver psychology, especially attention. These specializations on the part of the sender suggest that Dawkins and Krebs (1978) rightly emphasized the senders' interests in transmitting information.

To summarize, all primates are equally able to transmit tactile information. Some primates have specialized in producing or applying odor. For acoustic transmission of messages, senders have evolved measures of ensuring that sound arrives. For communication through vision, senders have evolved means to produce more pronounced optical displays.

Information transmission through touch, allomarking, and visual displays is usually highly selective with regard to the receiver, whereas environmental scent deposits, many acoustic, and some optical displays are better characterized as emissions or public announcements. Tactile and visual displays serve private communication, while communication by olfactory, acoustic and some optical displays resembles broadcasting, where sound is often highly conspicuous.

The information content of primate nonverbal displays is very diverse. Aspects of social life are expressed by displays in all four modes. Displays with rapid decay are able to express short-term affective states and motivations. Longer lasting social attitudes and relationships can be communicated through any channel. Identity, as a pattern of the membership of a sender to a taxonomic and demographic class plus individual features is connoted or denoted by displays delivered in any sensory mode. External objects are primarily referred to by acoustic displays or by scent.

Depending on whether they are facts, symptoms, signals, or symbols, displays in the different channels vary with respect to the inadvertence of their emission and reception. The option of avoiding reception, and hence manipulation, is provided by visual displays, because they depend heavily on the receiver's attention. This option does not exist in response to some odiferous materials, especially pheromones. Voluntariness of sending a display is limited in touch and olfaction, where many displays are symptoms. Sending seems to be more voluntary in visual communication and is likely to be under some voluntary control in acoustic communication. As a result, deception is most likely to occur in the visual and especially the acoustic mode.

In the light of these findings, it would be a mistake to view the different communication channels in a progressive succession from old, i.e., primitive, to modern, i.e., sophisticated. Primate lifestyles include extremes. They vary from diurnal to nocturnal and from solitary to gregarious. In spite of these differences, all primates make use of all sensory modalities. Nocturnal forms exploit even small amounts of light. Diurnal primates still process olfactory information. For all primates (including humans) tactile information plays an important role and the acoustic sense is indispensable for all members of the order. Nevertheless, primates do differ with respect to the sensory channels they emphasize in communication. Roughly, small body size, nocturnal activity, or poor vision conditions, territoriality, and solitary life style tend to be correlated with an emphasis on olfactory communication. Among the territorial forms of larger body size, acoustic displays seem to play a greater role. A diurnal and gregarious life style is characteristic of those primates who emphasize visual displays. This trend is stressed by the observation that diurnal gregarious prosimians turn away from mere olfactory marking towards an emphasis on visual aspects of marking behavior (Schilling 1979, 1990; Roeder 1985). In addition, the observation of display transfer between the communication channels opens interesting perspectives for the question of whether animal signals are iconic or arbitrary representations of their designata. What looks like an arbitrary display may in fact be an iconic representation of a display from another communicative channel. When this original display has been lost in the respective species it may be impossible to trace back its relationship to the display still in use.

Many primates perform communicative displays via several sensory channels at once. By doing so, they create redundancy because they do not combine the displays at random, but according to rules. In this way, a certain vocalization, for instance, may indicate an affiliative interaction which is composed of hugging (tactile) and a facial display in addition. Sending the same message in different codes simultaneously does not seem economical, but it increases the likelihood that it will be received properly. Furthermore, when messages in different modalities act in concert, they are able to overcome the restrictions of one code while maintaining its advantages in other domains. For example, an olfactory message will be retained in the substrate, but adding a visual display to marking may reduce response latency in the receiver.

Multi-channel communication and exploitation of the specific virtues of each channel also characterize Homo sapiens: although equipped with a highly flexible and versatile means of communicating in the form of verbal language, humans depend on nonverbal communication to communicate social relations. And even when verbal language is used, the literal message can often be neglected in comparison to the pragmatic function of the "speech act" (Searle 1969). When it comes to the communication of internal states, such as feelings, attitudes, or emotions, nonverbal communication is much more spontaneous, immediate, subtle, and reliable than verbal communication. Furthermore, when verbal messages are ambiguous or when contradictory messages are derived from verbal and paraverbal behavior, all humans rely heavily on the paraverbal code. The typical paraverbal cues come from facial expressions, postural-gestural displays, tactile communication and also from prosody. Of these channels, prosody is the only one which, by definition, does not occur in nonspeaking primates.

We suggest that the use of language is motivated by the evolved psychological need of mutual understanding. Communication is hence assumed to be a primary function of language. But the domain of human language is symbolic reference, which shifts the emphasis on the communication about facts, not about social events (cf. H. Preuschoft 1990). The abstract representational code allows communication out of context, by means of invention, the recounting of events far back in the past, or spelling out future circumstances in advance. In the form of writing, language has the same ability to close the temporal gap between emission and reception that is already found in scent marking. It combines this ability with the flexibility of a symbolic code, but lacks the capacity of immediate adjustment to the partner provided by traditional forms of nonverbal visual communication. Thus, language opens vast possibilities of deception.

Language research in phylogenetic perspective started out with such questions as "Why is it that animals don't speak?" and "(How) do animals speak when we provide them with some developmental aid, i.e., an artificial language?" We want to suggest that more insight into the nature of language (and also of Homo) may be gained when language is considered in its evolutionary context. Language has evolved in an environment of other communicative channels. As mentioned in the introduction, the questions of particular interest to the biologist are therefore: What kind of selective pressures have favored the evolution of language in humans? Why don't other animals speak? What is the acoustic channel of language able to achieve that other channels are unable to do? We now have to ask: what information do humans want to communicate that are disregarded by other primates?

We hope that the information reported in this chapter may help to better focus inquiries into the selective pressures leading to the evolution of communicative displays. The compiled evidence may - as we hope - serve as background from which the biologically important characteristics of language can better be understood.

Notes

1. The present chapter was conceived as a contribution to one of the very stimulating Semiotic Colloquia organized by W.A. Koch in Bochum. These interdisciplinary workshops have drawn attention to the sorts of questions which are raised in neighboring, but also in seemingly remote fields of science. As a consequence, they have stimulated many attempts to answer questions across disciplinary boundaries.
We have to thank the editor of this volume for the invitation to contribute a paper and for his very careful reading and improving of the text. The authors are also indebted to Jean-Jacques Roeder for his stimulating comments on an earlier draft of this manuscript and for his valuable references on olfaction, and to Dietmar Todt who commented on acoustic communication.

References

Altmann, Stuart (ed.)
1967 *Social communication among primates.* Chicago: University Press.

Alvarez, Fernando
1973 "Periodic changes in the bare skin areas of Theropithecus gelada", *Primates* 14 (2-3): 195-199.

Andres, Karl Herman
1973 "Morphological criteria for the differentiation of mechanoreceptors in vertebrates", *Rheinisch-Westfälische Akademie der Wissenschaften* 53: 135-152.

Andrew, Richard J.
1963 "The origin and evolution of the calls and facial expressions of the primates", *Behavior* 20: 1-109.

Angst, Walter
1975 *Das Ausdrucksverhalten des Javaneraffen Macaca fascicularis Raffles 1821.* Berlin: Parey.

Baerends, G.P.
1975 "An evaluation of the conflict hypothesis as an explanatory principle for the evolution of displays", in G.P. Baerends – C. Beer – Aubrey Manning (eds.), 187-227.

Baerends, G.P. – C. Beer – Aubrey Manning (eds.)
1975 *Function and evolution in behavior*. Oxford: Clarendon.

Bateson, Gregory
1955 "A theory of play and fantasy", *Psychiatric Research Reports* A2: 39-51.

Bateson, Paul P.G. – Peter H. Klopfer (eds.)
1981 *Perspectives in Ethology*. Vol. 4. London: Plenum.

Bischof, Norbert – Holger Preuschoft (eds.)
1980 *Geschlechtsunterschiede: Entstehung und Entwicklung*. München: C.H. Beck.

Bourne, Geoffrey H. (ed.)
1977 *Progress in ape research*. New York: Academic Press.

Brown, R.E. – D.W. Macdonald (eds.)
1985 *Social odors in mammals*. Vol. 2. Oxford: Clarendon.

Buck, Christoph – Hans F. Bär
1993 "Investigations on the biomechanical significance of dermatoglyphic ridges", in: Holger Preuschoft – David J. Chivers (eds.), 285-306.

Caryl, Peter G.
1979 "Communication by agonistic displays: What can games theory contribute to ethology?" *Behavior* 68: 137-169.
1981 "Escalated fighting and the war of nerves: Games theory and animal combat", in: Paul P.G. Bateson – P.H. Klopfer (eds.), 199-224.
1982 "Animal signals: A reply to Hinde", *Animal Behavior* 30: 240-244.

Cheney, Dorothy L. – Robert M. Seyfarth
1982 "Recognition of individuals within and between groups of free-ranging vervet monkeys", *American Zoologist* 22: 519-29.

Chevalier-Skolnikoff, Suzanne
1973 "Facial expression of emotion in nonhuman primates", in: Paul Ekman (ed.), 11-89.
1982 "A cognitive analysis of facial behavior in Old World monkeys, apes, and human beings", in: Snowdon – Brown – Peterson (eds.), 303-368.

Dawkins, Marian S. – Tim Guilford
1991 "The corruption of honest signalling", *Animal Behavior* 41: 865-875.

Dawkins, Richard – John R. Krebs
1978 "Animal signals: Information or manipulation?" in: John R. Krebs – N.B. Davies (eds.), 282-309.

Demaria, Christine – Bernhard Thierry
1990 "Formal biting in Stumptailed macaques (Macaca arctoides)", *American Journal of Primatology* 20: 133-140.

Dittus, Wolfgang P.J.
1984 "Torque macaque food calls: Semantic communication concerning food distribution in the environment", *Animal Behavior* 32: 470-477.

Doyle, Gerald A. – Robert D. Martin (eds.)
1979 *The study of prosimian behavior*. New York: Academic Press.

Dunbar, Robert I.M.
1991 "Functional significance of social grooming in primates", *Folia Primatologica* 57: 121-131.

Ekman, Paul (ed.)
1973 *Darwin and facial expression*. New York: Academic Press.

Epple, Gisela
1985 "The primates I: Order Anthropoidea"; Epple, Gisela – A.B. Smith, "The primates II: A case study of the saddle-back tamarin, *Sanguinus fuscicollis*," in: R.E. Brown – D.W. Macdonald (eds.), 739-769.

Fay, Richard R.
1990 *Hearing in vertebrates: A psychophysics databook*. Winnetka, Ill.: Hill-Fay.

Feistner, Anna T.C.
1991 "Scent marking in Mandrills *Mandrillus shinx*", *Folia Primatologica* 57: 42-47.

Fett, Hanns
1980 "Einfluß von Sexualhormonen auf Sexual- und Sozialverhalten von Primaten", in: Norbert Bischof – Holger Preuschoft (eds.), 76-98.

Fornasieri, Isabelle – Jean-Jacques Roeder
1992 "Behavioral responses to own and other species scent marks in Lemus fulvus and Lemur macaco", *Journal of Chemical Ecology* 18: 2069-2082.
1993 "Marking behavior in two lemur species (*L. fulvus* and *L. macaco*): Relation to social status, reproduction, aggression and environmental change", *Folia Primatologica* 220 [in press].

Frank, Robert H.
1988 *Passions within reason*. New York: Norton.

Goodall, Jane
1968 "Behavior of free-living chimpanzees of the Gombe stream area", *Animal Behavior Monographs* 1: 163-311.

Gosling, L. Morris
1982 "A reassessment of the function of scent marking in territories", *Zeitschrift für Tierpsychologie* 60: 89-118.

Gouzoules, Sarah – Harold Gouzoules – Peter Marler
1984 "Rhesus monkey (*Macaca mulatta*) screams: representational signalling in the recruitment of agonistic aid", *Animal Behavior* 32: 182-193.

Green, Steven
1975 "Variation of vocal pattern with social situation in the Japanese monkey (*Macaca fuscata*): A field study", in: Leonard A. Rosenblum (ed.), 1-102.

Grolier, Eric de (ed.)
1983 *Glossogenetics: The origin and evolution of language*. Chur: Harwood.

Grzimek, Bernhard
1988 *Grzimeks Enzyklopädie*. Vol. 2. München: Kindler.

Guilford, Tim – Marian S. Dawkins
1991 "Receiver psychology and the evolution of animal signals", *Animal Behavior* 42: 1-14.

Herzog, M. – Sigrid Hopf
1983 "Effect of species vocalization on the behavior of socially-reared squirrel monkey", *Behavior* 86: 197-214.
1984 "Behavioral responses to species-specific warning calls in infant squirrel monkeys reared in social isolation", *American Journal of Primatology* 7: 99-106.

Hill, W.C. Osman
1960 "*Comparative anatomy and taxonomy - Primates*. Vol. 1: *Strepsirhini*. Edinburgh: University Press.

Hinde, Robert A.
1981 "Animal signals: Ethological and games theory approaches are not incompatible", *Animal Behavior* 29: 535-542.
1985a "Was 'The expression of the emotions' a misleading phrase?" *Animal Behavior* 33: 985-992.
1985b "Expression and negotiation", in: Gail Zivin (ed.), 103-116.

Hinde, Robert A. (ed.)
1972 *Nonverbal communication*. Cambridge: University Press.

Hooff, Jan A.R.A.M. van
1962 "Facial expressions in higher primates", *Symposia of the Zoological Society of London* 8: 67-125.
1967 "The facial displays of Catarrhine monkeys and apes", in: Desmond Morris (ed.), 7-68.

1972 "A comparative approach to the phylogeny of laughter and smile", in: Robert A. Hinde (ed.), 209-241.

Hopf, Sigrid – E. Hartmann-Weisner – B. Kühlmorgen – S. Mayer
1974 "The behavioral repertoire of the squirrel monkey (Saimiri)", *Folia Primatologica* 21: 225-249.

Hörmann, Hans
1977 *Psychologie der Sprache*. Berlin: Springer. 2nd ed.

Huxley, Thomas H.
1863 *Zeugnisse für die Stellung des Menschen in der Natur*. Translated by G. Heberer 1963. Stuttgart: G. Fischer.

Jolly, Alison
1966 *Lemur behavior*. Chicago: University Press.
1985 "Communication", in: Alison Jolly (ed.), 192-217.

Jolly, Alison (ed.)
1985 *The evolution of primate behavior*. 2nd. ed. New York: MacMillan.

Kawai, Masao
1960 "A field experiment on the process of group formation in the Japanese Monkey (*Macaca Fuscata*) and the releasing of the group at Ohirayama", *Journal of Primatologica* 2: 181-253.

Koch, Walter A. (ed.)
1990 *Geneses of language*. Bochum: Brockmeyer.

Krebs, John R. – N.B. Davies (eds.)
1978 *Behavioral ecology: An evolutionary approach*. Oxford: Blackwell.

Kummer, Hans
1967 "Tripartite relations in hamadryas baboons", in: Stuart Altmann (ed.), 63-71.

Lazarus, Richard S.
1984 "On the primacy of cognition", *American Psychologist* 39: 124-129.

Liberman, Alvin M. – Katherine S. Harris – Howard S. Hoffman – Belver C. Griffith
1957 "The discrimination of speech sounds within and across phoneme boundaries", *Journal of Experimental Psychology* 54: 358-368.

Lieberman, Philip – E.S. Crelin
1971 "On the speech of Neanderthal man", *Linguistic Inquiry* 11: 203-222.

Lieberman, Philip – E.S. Crelin – D.H. Klatt
1972 "Phonetic and related antomy of the newborn and adult human, Neanderthal man, and the chimpanzee", *American Anthropologist* 74: 287-307.

Lorenz, Konrad
1950 "The comparative method in studying innate behavior patterns", *Symposia for the Society of Experimental Biology* 4: 455-499.
Macedonia, Joseph M.
1990 "What is communicated in the antipredator calls of lemurs: Evidence form playback experiments with ringtailed and ruffed lemurs", *Ethology* 86: 177-190.
Marler, Peter
1973 "Kommunikation bei Primaten", in: Ilse Schwidetzky (ed.), 39-90.
1977 "Primate vocalization: Affective or symbolic?" in: Geoffrey H. Bourne (ed.), 85-96.
1984 "Animal communication: Affect or cognition", in: Klaus R. Scherer – Paul Ekman (eds.), 345-365.
Martin, Robert D. (ed.)
1990 *Primate origins and evolution*. London: Chapman & Hall.
Martin, Robert D. – Gerald A. Doyle – Alan C. Walker (eds.)
1974 *Prosimian biology*. London: Duckworth.
Maynard-Smith, John – G.A. Parker
1976 "The logic asymmetric contests", *Animal Behavior* 24: 159-175.
McDougall, W.
1928 *Grundlagen einer Sozialpsychologie*. Jena: Fischer.
Menzel, Emil W.
1971 "Group behavior in young chimpanzees responsiveness to cumulative novel changes in a large out-door enclosure", *Journal of Comparative and Physiological Psychology* 74: 46-51.
1972 "Communication about the environment in a group of young Chimpanzees", *Folia Primatologica* 9: 220-232.
Miller, Robert E. – W.F. Caul – J.A. Mirsky
1967 "Communication of affect between feral and socially isolated monkeys", *Journal of Personal and Social Psychology* 7: 231-239.
Morris, Desmond
1957 "'Typical intensity' and its relation to the problem of ritualisation", *Behavior* 11: 1-12.
Morris, Desmond (ed.)
1967 *Primate ethology*. Chicago: Aldine.
Penzlin, Heinz
1989 *Lehrbuch der Tierphysiologie*. (4th edition.) Stuttgart: Fischer.

Perret, Martine
1992 "Environmental and social determinants of sexual function in the male mouse lemur (Microcebus murinus)", *Folia Primatologica* 59: 1-25.
Preuschoft, Holger
1988 "Große Menschenaffen: Einleitung", in: Bernhard Grzimek, 358-399.
1990 "Sprache (oder Schrift?) bei unseren nächsten Verwandten", in: Walter A. Koch (ed.), 77-114.
Preuschoft, Holger – David J. Chivers (eds.)
1993 *Hands of primates*. Wien: Springer.
Preuschoft, Holger – David J. Chivers – Warren Y. Brockelman– Norman Creel (eds.)
1984 *The lesser apes: Evolutionary and behavioral biology*. Edinburgh: University Press.
Preuschoft, Signe
1992 "'Laughter' and 'smile' in Barbary macaques, *Macaca sylvanus*", *Ethology* 91: 220-236.
Preuschoft, Signe – Jan A.R.A.M. van Hooff
in press "Variations in primate affiliative displays: An exercise in behavior phylogeny", in: Ullica Segerstråle – Peter Molnar (eds.).
Redican, William K.
1975 "Facial expressions in non-human primates", in: Leonard A. Rosenblum (ed.), 103-194.
Rhijn, Jan G. van – R. Voedegel
1980 "Being honest about one's intentions: An evolutionary stable strategy for animal conflicts", *Journal of Theoretical Biology* 85: 623-641.
Roeder, Jean-Jacques
1985 *Evolution des systèmes de communication chez les carnivores et les primates*. Paris: Editions du centre national de la recherche scientifique.
Roeder, Jean-Jacques – James R. Anderson
1991 "Urine washing in brown capuchin monkeys (Cebus apella): Testing social and nonsocial hypotheses", *American Journal of Primatology* 24: 55-60.
Roeder, Jean-Jacques – James R. Anderson (eds.)
1990 *Primates*. Paris: Masson.
Rosenblum, Leonard A. (ed.)
1975 *Primate behavior, development, and laboratory research*. Vol. 4. New York: Academic Press.

Rumbaugh, Duane

1992 "Biobehavioral roots of language: Words, apes, and a child", Paper and video presented at the ZIF Conference on Biological and Cultural Aspects of Language Development. Bielefeld.

Scherer, Klaus R. – Paul Ekman (eds.)

1984 *Approaches to emotion*. Hillsdale: Erlbaum.

Schilling, Alain

1974 "A study of marking behavior in lemur catta", in: Robert D. Martin – Gerald A. Doyle – Alan C. Walker (eds.), 347-362.

1979 "Olfactory communication in prosimians", in: Gerald A. Doyle – Robert D. Martin (eds.), 461-542.

1990 "Communications par signaux chimiques chez les prosimiens", in: Jean-Jacques Roeder –James R. Anderson (eds.), 121-143.

Schwidetzky, Ilse (ed.)

1973 *Über die Evolution der Sprache*. Frankfurt: S. Fischer.

Searle, John

1971 *Sprechakte*. Frankfurt: Suhrkamp.

Segerstråle, Ullica – Peter Molnar (eds.)

in press *Nonverbal communication - the missing link between culture and biology?* Princeton: University Press.

Seiler, Rolf

1973 "On the function of facial muscles in different behavioral situations", *American Journal of Physical Anthropology* 38: 567-571.

Seyfarth, Robert M.

1987 "Vocal communication and its relation to language", in: Barbara B. Smuts – Dorothy L. Cheney et al. (eds.), 440-451.

Seyfarth, Robert M. – Dorothy L. Cheney

1980 "The ontogeny of vervet monkey alarm calling behavior: A preliminary report", *Zeitschrift für Tierpsychologie* 54, 37-56.

1982 "How monkeys see the world: A review of recent research in East African Vervet monkeys", in: Charles T. Snowdon – H. Brown – M.R. Peterson (eds.), 239-252.

Seyfarth, Robert M. – Dorothy L. Cheney – Peter Marler

1980 "Vervet monkey alarm calls: Semantic communication in a free-ranging primate", *Animal Behavior* 28: 1070-1094.

Simpson, George G.
1945 "The principles of classification and a classification of mammals", *Bulletin of the American Museum Natural History* 85: 1-350.

Smith, W. John
1965 "Message meaning, and context in ethology", *American Naturalist* 908: 405-409.

Smuts, Barbara B. – Dorothy L. Cheney et al. (eds.)
1987 *Primate societies.* Chicago: University Press.

Snowdon, Charles – H. Brown – M.R. Peterson (eds.)
1982 *Primate communication.* Cambridge: University Press.

Starck, Dietrich
1982 *Vergleichende Anatomie der Wirbeltiere.* 3 Vols. Berlin: Springer.

Thierry, Bernhard – Christine Demaria – Signe Preuschoft – Christine Deportes
1989 "Structural convergence between silent bared-teeth display and relaxed open-mouth display in the Tonkean Macaque (*Macaca tonkeana)*", *Folia Primatologica* 52: 178-184.

Tinbergen, Nico
1952 "Derived activities: their causation, biological significance, origin, and emanicpating during evolution", *Quarterly Review of Biology* 27: 1-32.

Todt, Dietmar
1988 "Serial calling as a mediator of interaction processes: Crying in primates", in: Dietmar Todt – Philipp Goedeking – David Symmes (eds.), 88-107.

Todt, Dietmar – Philipp Goedeking – David Symmes (eds.)
1988 *Primate vocal communication.* Berlin: Springer.

Waal, Frans B.M. de
1989 *Peacemaking among primates.* Cambridge: Harvard University Press.

Waal, Frans B.M. de – Lesleigh M. Luttrell
1985 "The formal hierarchy of rhesus macaques: An investigation of the bared-teeth display", *American Journal of Primatology* 9: 73-85.

Waser, Peter
1982 "The evolution of male loud calls among mangabeys and baboons", in: Charles T. Snowdon – C.H. Brown – M. Peterson (eds.), 115-143.

Wind, Jan
1983 "Primate evolution and the emergence of speech", in: Eric de Grolier (ed.), 15-35.

1990 "The evolutionary history of the human vocal tract", in: Walter A. Koch (ed.), 172-199.

Zajonc, Robert B.
1984 "On the primacy of affect", *American Psychologist* 39: 117-123.

Zivin, Gail (ed.)
1985 *The development of expressive behavior*. Orlando: Academic Press.

PART II

Anthroposemiotic sociogenesis and cultural semiogenesis

The problem of certainty in human communication: An evolutionary view[1]

Peter Meyer

Human communication utilizes various signs, such as words and letters, laughing and crying, body decoration and clothing, as well as others. A common characteristic of these otherwise distinct sign systems seems to be the capability of providing two or more individuals with a message, whose meaning is more or less similar to both sender and receiver. While the everyday practitioner may feel completely satisfied with his communicative success, linguists, epistemologists, communication scientists, and semioticians are eagerly trying to understand the operation of communication, and hence the causes underlying this success. Semioticians agree that a sign is not identical to the thing signified, and furthermore that the messages, as conveyed by one and received by another individual, are never exactly alike. Thus, from a theoretical point of view, the success of everyday communication is rather mysterious, and approaches to an understanding of communication are certainly in need of improvement.

The purpose of this paper is to analyze from an evolutionary perspective the impact which some of the above-mentioned sign systems have in the process of communication and to determine the role of these sign systems in the success of everyday communication. The theory of evolution seems to be particularly suited to such an enterprise because it emphasizes the contribution of any given morphological or behavioral trait to the survival of the living system. Viewed against this background, it is suggested that communication is a fitness maximizing institution. In other words, individuals in any given species may enhance their fitness by putting to use species-specific patterns of communication. Unlike in some recent analyses (Sommer 1993: 439), it is suggested that cheating must not prevail in patterns of communication. Quite to the contrary, only such individuals will be able to reproduce who have the correct signals at their disposal (cf. Helversen – Scherer 1988: 613).

1. Signals in human evolution

A general characteristic of signals is that they must be "definite and discrete" (cf. Scherer 1988: 632) and must provide individuals with means of assessing the future behavior of other animals. Phylogenetically, some characteristics of the phenotype have become optimized and serve as a signal in the behavioral system of the species. An example is the red spot of the redbreasted robin (see 4.). It should be emphasized that such signals add to the time-energy budget of any individual by minimizing the number of necessary trial and error behaviors. As a result, an individual in a bird species will be able to rely upon an alarm call, and it may do so because this signal is part of the evolutionarily tested genetic information of this species.

The point to be stressed here is that high reliability of signals is typical for the lower phyla, whereas communication in the higher phyla, particularly in the human species, superficially at least, is less reliable. In human communication, as pointed out before, there is neither a fixed relation of sign and signified nor does the message convey identical information to the communicating persons. Thus, how can communication enhance the well-being of human individuals and reduce uncertainties or anxieties about the next steps of their fellow beings?

The major purpose of this paper is to provide some clues to this problem. Unfortunately, many interesting questions will have to be left aside, such as the evolution of the gap in reliability and behavioral certainty between the lower and the higher phyla. A short comment must suffice. According to the theory of evolution, the material tissue of the mental and communicative performance of humans is the brain. In spite of its outstanding degree of complexity and its specific human features, its blueprint combines phylogenetically older parts, such as the "reptilian" and the "paleomammalian" brain with more recent parts, the "neomammalian" brain (MacLean 1972). As far as the problem of reliability and certainty of communication is concerned, the structure of the human brain and its functional differentiation brings about the coexistence of older and more recent sign systems (see Schnelle, this vol.). Human infants and adults have various nonverbal signals at their disposal, and the features of these signs show numerous similarities to animal communication, hence to older sign systems with their high degree of reliability. Unlike in some animals, however, the neomammalian parts of the human brain are the material basis of such more specifically human phenomena as language and culture. The elements of language and culture are various types of signs which typically designate objects and facts in an arbitrary and metaphorical manner (cf. Leach 1976 : 15). While this arbitrariness in the relation of sign and signified may well be the major cause of some epistemological problems as well as of the psychological problem of uncertainty, it may also be considered a mainstay of some of the more specifically human achievements, such as language, thinking, and culture.

To cut a long story short: any human individual has a huge empty memory at his or her disposal which may store and process representations of many facts of the world by means of signs. Moreover, the processing seems to employ an inbuilt grammar (Vollmer 1975 : 146) which may not only create growing numbers of sign arrangements, but may also bring novel arrangements into existence unprecedented by the "facts of the world". Viewed from an evolutionary perspective, these achievements of the human mind have enabled humanity to conquer ever new parts of this planet by creating cultural adaptations. From the point of view of social theory, however, this capacity of the human mind poses a major behavioral problem, namely, mutual uncertainty in interaction. Sociologists have provided many insights into the ways human individuals try to cope with this problem, e.g., by establishing social norms (cf. Prewo et al. 1973 : 136). While the impact of such social norms is quite obvious and undeniable, the following considerations emphasize some selective agents inherent in human communication. The major concern of these reflections is the nature of some processes underlying the formation of social norms.

2. Information as an element of evolutionary processes

Typically, evolutionary approaches proceed from assumptions about the phylogeny of any particular trait and from some hypotheses on the selective advantages which this trait may have provided in the past for the survival of the living system. It has been emphasized by some more recent contributions to evolutionary theory that indetermination may well be an important aspect of any behavior as well as of the entire process of phylogeny (cf. Prigogine – Stengers 1981: 22). Unlike in some older interpretations which were rather close to the mechanistic type of explanations of classical physics (cf. Sachsse 1984: 4), these more recent views stress the indeterminate character of natural processes, and hence the probability for the emergence of novelties at any given moment. While a particular biological organ may have developed highly adaptively in an older living system, the blueprint of this old organ may also become integrated into novel and more complex living systems. Consider, for instance, the structure of the human brain. As indicated above (1.), the older parts are structurally the same as those of reptiles. According to some of the experts in this field of research (Swanson 1983: 45; Wuketits 1985: 283), the neomammalian brain, i.e., mainly the neocortex, may be regarded as a sort of superstructure of the paleomammalian and the reptilian brain sections which jointly form the "triune brain" (MacLean 1972: 137). As compared to the reptilian brain, the triune brain throws open a vast array of behavioral options, at the same time liberating the human being to a considerable extent from quasi-automatic reactions to environmental stimuli (cf. Swanson 1983: 49). Last but not least, some features of the triune brain enable human beings to establish mental representations of their environment. For all that, the reptilian brain continues to be a successful organ among the members of the reptile species as well as in its more recent form, among the higher species.

The processes of the emergence of novel features in natural history are a major concern of the theory of evolution. Within their conceptual framework, "variation" is the label applied to this tendency, while the term "selection" is applied to the competition between these variants and finally to their differential reproduction.

According to evolutionary thinking, the number of offspring of any given trait may be considered an indication of evolutionary success. However, it may be assumed that the behavior of the respective organism must also have played an important role in this process. Obviously, the successful variant is primarily linked to behaviors which utilize time and energy in an optimal manner. Since individual lifetime is limited in all higher phyla, it should be expected that selection favors such variants which can utilize available time and energy in such a way that a maximum of copies of the variant will survive. In fact, it is by the inherent laws of heredity that major informational prerequisites of successful behavior will be passed to the next generation. This is basically why not every single organism will have to repeat the total number of trial-and-error processes tested throughout phylogeny. Such information allows for an economical use of time and energy (cf. Callebaut 1987: 80) by providing the organism with a program corresponding with environmental events. In other words, genetically encoded information tends to reduce the degree of indetermination, thereby increasing supplies of time and energy available to a given organism. Supplies of these resources may be further augmented if two or more organisms are capable of exchanging messages. Information, as conveyed by such messages may, however, enhance individual fitness only on the condition that this information is correct, relevant, and "honest". If this is not the case, chances of survival are minimized.

Apparently, selection favors the spread of honest communication in such species where the information conveyed by the message is more or less identical for both sender and receiver. This applies in particular to communication which is based on the exchange of signals. In any given species, such signals are founded on similar affective states and modes of behavioral organization. The combination of these allows for the coordination of individual types of behavior (cf. Swanson 1983: 96). Signals are typical in animal communication (cf. Burkart 1983: 31), but they also play a role in human communication (see 4.). Wilson (1975: 183), defines *signal* as "any behavior that conveys information from one individual to another, regardless of whether it serves other functions as well". As Maturana and Varela (1987: 212) have pointed out, the behavioral coordination of two individuals is not brought about, strictly speaking, by any sort of message received by one individual from the other. According to these authors, the actual coordination should rather be attributed to the internal organization of both of the two individuals concerned. However that may be, similar thresholds in the behavioral program of a species seem to be the basic requirement of communication. According to Bateson (1972: 379), any communicative event presupposes the perception of differences. This difference, therefore, may be regarded as the essential content of any piece of information, which may in turn become the subject matter of a message.

Thus, the information content of any signal is based on the perception of differences which are perceived within an organism. For example, a dog may sense the state of arousal of another dog from his or her odor, and an insect may gather similar information from the movements of a conspecific. In the case of the olfactory and of the visual channel, the major precondition for the perception of differences is the endowment of the respective organism with certain perceptual thresholds. The organism can thereby actively screen out environmental events by utilizing these thresholds. Whenever differences arise between inbuilt threshold expectations and environmental data, information emerges. The organism may then utilize this information for shifting to yet another stage of the behavioral program.

Obviously, such information may enhance the chances of survival of a given species, provided that individual organisms are equipped with some inbuilt cues for adequate reaction to environmental events. For example, an individual bird will sense a state of affective arousal at the perception of a predator. As a consequence, it will give an alarm call. This signal will put its conspecifics into a similar state of affective arousal, which finally will cause common flight. As to the information content of the alarm call, the above considerations suggest that the state of arousal is the essential message which causes similar affective states among the receiving individuals. It can be assumed that selection has favored the combination of certain affective states with signals. This applies, in particular, to stereotyped signals in which "for each signal there is only one response" (Wilson 1975: 181). Regarding individual time-energy budgets, any affective arousal means heightened investments of such resources. Moreover, affective arousal very often prevents the animal from food intake. From an evolutionary point of view, it is to be expected that signals inform in a dependable and "honest" manner so that affective arousal may be restricted to situations where vital interests are at stake. In summary, it must be emphasized once again that species-specific signal systems have been selected to enhance the chances of survival for the individual organism.

As to the definition of information and communication, the concept of survival value may be of use. Consider, e.g., the movement of a rabbit, as perceived by a hawk. Obviously, this instance of perception assures survival value to the receiver, while the sender's chances of survival diminish. According to Helversen and Scherer (1988: 611), it would be ill-advised to apply the term communication to this type of situation. Undoubtedly, there is a flow of information here, but the sender has quite obviously been optimized to suppress such information. Developing this argument somewhat further, we suggest that the term communication should only be applied to processes of information transmission in which the meaning of the message is understood and this information involves an advantage on the part of both the receiver and the sender. As argued above from an evolutionary point of view, communication has been selected in order to enhance the chances of survival of a given species and to reduce behavioral uncertainties. Obviously, signal systems applying stereotyped signals are quite efficient in this respect (cf. Helversen – Scherer 1988: 621). Regarding human communication, where symbols and other types of signs are prominent, the question may be raised as to how the problems of behavioral certainty are being coped with.

3. Symbols and behavior: Some problems of variety

The following considerations on verbal communication return to the topic of indetermination. More specifically, we suggest that human communication entails a sort of internal selection by combining symbols with affective elements. Regarding systems of symbols, it seems that they allow for a virtually unlimited wealth of mental images of future action. The question to be raised here is how agents decide in any given situation whether the system of symbols provides them with criteria for making decisions. From the point of view of the science of behavior, any behavior presupposes decision making. Hence, recipients will have to screen symbolic messages for decision making criteria. It must be emphasized once more that decision making presupposes criteria. The following considerations investigate the selective mechanism inherent in human communication which provides the communicative process with such criteria for decision making. This mechanism employs, on the one hand, certain elements of nonverbal communication. On the other hand, the redundancy of messages and of social behavior is exploited. The following section accentuates certain characteristics of symbols which, in conjunction with grammar, provide the basis for the operation of language.

Human beings seem to be genetically endowed with a faculty of associating facts with names or terms. Consequently, these names designate certain facts, and given the existence of a grammar, numerous discourses on facts may evolve. According to Chomsky (1972), general language competence and the command over a certain set of grammatical rules are products of phylogeny. In spite of this phylogenetic origin, the relation of most symbols to the facts designated by them is arbitrary and metaphorical (cf. Leach 1976: 16; Reimann 1968: 92). During the early epoch of cultural evolution, the vast majority of symbols may well have been icons (cf. Lumsden – Wilson 1983: 109). In the course of cultural evolution, it seems that this close relation between symbols and designated facts has disintegrated and a more arbitrary relation has prevailed.

From a semiotic point of view, symbols become established when given facts become regularly associated with certain names or terms, i.e., when the meaning of a symbol is defined in a conventional manner. Obviously, most mental and communicative processes presuppose such terms. It is noteworthy that only facts which have previously been associated with names and terms may become the object of symbolic communication (cf. Bateson 1979: 61). From an epistemological point of view, it seems to be quite significant that mental systems tend to deal with symbols as if they really coincided with the facts of the physical world (cf. Bateson 1979: 161). Part of this tendency of symbols to develop a life of their own may be accounted for by the increasing distance of signs from their original iconicity. This tendency is further increased by the semiotic fact that the meaning of any given term depends on all other terms of its context (cf. Winner 1989: 380). In other words, any given symbol may acquire different meanings in varying contexts. Moreover, systems of symbols may produce ever new variants of meaning if a grammar is available.

Viewed against this background, human communication certainly may not be regarded as an exchange of homogeneous messages. On the contrary, human communication entails considerable degrees of indetermination and potential for variation. The ultimate sources of variation are to be found in the personality of the communicating individuals (cf. Ivanov 1983: 161). It has been pointed out before that such a situation may be considered advantageous from an evolutionary point of view but raises some major problems for social behavior. With regard to behavior, agents are in need of decision making criteria. Hence, they seek for an interpretation of the "true" meaning of messages.

Unlike in the communication of stereotyped signals, the meaning of symbolic messages is not genetically determined in human communication. On the contrary, it is one of the characteristics of human communication that the message is never completely the same to sender and receiver. Hence, the question is how this structure of human communication may comply with an economical use of time and energy and how some degree of certainty for semiotic agents may be extrapolated from these structural features. First of all, it must be stressed that the high degree of indetermination is quite conducive to the development of selective mechanisms. Much like the natural process in which the formation of ever new variants is a major precondition of natural selection, diverging interpretations of messages create a basis for screening processes. It should be emphasized that the comparison to natural processes does not imply, however, that human communication does not employ novel properties. In fact, some of the emergent properties of symbol systems have already been pointed out. Undoubtedly, symbols evolve from a natural basis. However, it is the interrelation of a number of symbols which gives rise to novel, unprecedented properties (cf. Ivanov 1983: 92-93).

Returning to the ability of the sender and the receiver to agree on the meaning of any given message, the combination of various levels of communication is of particular importance. In addition to language, human communication employs such elements as moaning and laughing as well as various gestures and body postures. According to Bateson (1972: 423), these elements of "paralanguage" facilitate the receiver's task of evaluating a given message by assigning information about the present affective state of the sender. An individual speaking in an emotional tone of voice reveals, consciously or not, some of his or her assessments of the underlying facts as well as of the situation. Obviously, these elements of paralanguage operate as a sort of redundant confirmation or negation of the symbolic language message.

Moreover, some signs may assume the function of contextual boundary markers which in turn will change the meaning of all the signs within these boundaries. This is an example of the operation of metacommunication, which does in fact alter the meaning of any sign, thereby considerably widening the total scope of human sign systems (Bateson 1972: 180). With regard to the operation of human communication as a holistic system, paralanguage and nonverbal communication may help the receiver appraise the meaning of symbolic language by providing redundancy as well as indications of the affective state of the sender. For instance, a smile may introduce a sort of ironical undertone which will prevent those having noticed it from paying too much attention to the jingle of words. Some of these effects of nonverbal communication will be elaborated below.

4. Affectivity as an element of communication

As argued above (1.), paralanguage and various other nonverbal signs refer to the emotional life of the sender. The situation in human semiosis where a given sign releases very similar affective states in both sender and receiver has analogies with certain forms of animal communication. Animal signals may elicit nearly identical affective states in sender and receiver and usually lead to a certain behavior which the expert observer may even foresee. In nonverbal communication among human agents, too, there are numerous indications of close connections to human affectivity. However, it should be stressed that there are no parallels to inborn release mechanisms as they have been brought to light by ethologists. Unlike animal communication, human communication combines two distinct levels. The first is the phylogenetically older one, mainly consisting of nonverbal communication and paralanguage. The second is the more recent level of symbolic language. Furthermore, metacommunication introduces yet another source of indetermination which makes any sign dependent upon its given context. The interaction of these two levels will be further elaborated after a closer look at human affectivity.

In general, the term affectivity is applied to states of arousal of an organism (cf. Izard 1981: 73) which are likely to occur in cases, such as threat by a predator, social bonding, or similar situations. Viewed against the background of evolutionary thinking, it may be particularly interesting to see that affectivity provides some behavioral sequences with higher degrees of affective arousal, whereas in other sequences there are lower degrees. Recalling one of the basic assumptions of evolutionary theory, i.e., the scarcity of time and energy, differential energy expenditure turns out to be a sort of inbuilt selective mechanism. Differential energy expenditure seems to be part of the physiological processing of a given organism which, by utilizing inbuilt thresholds, can shift to various classes of behavior. Some of these classes will put the organism into a state of higher arousal, whereas others are connected to lower levels of energy expenditure. The combination of these thresholds with a sensory system which is suited to screen the environment of the organism will enhance the chances of survival of this organism. The major reason for this assessment is that dangerous situations can thereby be avoided and help can be secured from conspecifics.

The redbreasted robin may serve as an example again. A male bird will attack any red-coloured object, no matter whether it is the red breast of a competitor or a dummy. The red spot operates as a stereotypic signal which reliably elicits affective arousal as well as the corresponding action. The underlying cause for this fixed action pattern is a genetic program. As suggested above, the rigid character of stereotypic signals must be compensated for by a high reliability of eliciting signals in order to prevent cheating. Some of the reasons for this assessment may easily be grasped from an evolutionary perspective.

Stereotypic signals cause major investments of affective arousal which in the case of systematic cheating would result in a physical breakdown of the individuals concerned. Numerous findings by ethologists indicate that inborn release patterns may be activated by dummies in laboratory situations. Some other findings have brought to light that the action specific energies may occasionally be consumed in vacuum activities, i.e., in the absence of the usual releaser (cf. Lorenz 1965: 107). A major disadvantage of this sort of sign-action system is its limited adaptability. Obviously, the system could adapt to major changes in the environment only by natural selection, i.e., the system could only change over several generations.

The dominance of symbols in human communication may be accompanied by the loss of mutual anticipation. As compared to stereotypic signals, this loss is more than compensated for, however, by the increase in variability. Moreover, these symbolic variants are processed internally in the brain, which minimizes the actual behavior of the entire organism. In other words, the ability of testing variants in the mind adds to the time-energy budget of the species by reducing the number of behavioral trials and errors to a minimum. While it cannot be denied that there is a certain number of unavoidable trials and errors in the behavior of any type of organism, the human species may further reduce this number by various cultural institutions. The more successful variants will be stored in individual memories and will occasionally be passed on to the oncoming generation. An example may further illustrate this point. Individual members of the human species are endowed with an affective structure which is homologous to that of other primates (cf. Eibl-Eibesfeldt 1984: 588). Not unlike those animal relatives, human infants sense fear and search for the loving care of their parents. Moreover, according to human ethology, there is a striking similarity of signals employed by primates and human individuals in such situations. Weeping and laughing, e.g., are human universals and also very similar to expressions found in nonhuman primates (cf. van Hooff 1989: 120). Very frequently, facial expressions are linked up with certain states of affective arousal. By ontogenetic learning, human individuals become able to control the expression of affective arousal and to reserve such expressions for certain typical situations. After some time of interaction, mother and child "know", e.g., their mutual affective states, just from habitually doing certain things. As a result of this type of learning by repetition, the child may be able to distinguish some more subtle facets of affective response from others. If, in addition, the child has learned to associate these affective expressions with symbols, he or she has been directed into the right channels.

From a selectionist point of view, it has been argued above that the processing of symbols in the mind, as compared to physical behavior, pays off in terms of time and energy. It has further been suggested that the introduction of novel context markers will increase the time-energy budget and will thereby augment the total number of meanings or variants available for communication. The expression of fear which may appeal in one context to the readiness for help may acquire a comic tinge in another context, for instance, in a stage play. The sense of fear certainly has to be considered a major category of human affectivity. From an evolutionary view, this sense arouses the organism to flight or to other adequate responses to an imminent danger, thereby enhancing individual chances for survival. Unlike in some animal species, the sense of fear in human adults may not only be triggered by signals as is still the case in human infants. In the human adult, symbols replace stereotyped signals. It should be stressed that very different kinds of symbols may have an affective undertone since symbols no longer evince the unity of sign and affective state which prevails in animal communication and, to some extent, throughout early ontogeny. The evolution of symbolic thinking has created new ways of making mental experiments in a rather detached manner. A closer look at the symbols used in such operations brings their amalgamation with affective tinges to light. There is no "pure" information in human signs. On the contrary, all human signs are deeply permeated with emotional evaluation. As Peirce (CP 5.308) puts it: "Everything in which we take the least interest creates in us its own particular emotion, however slight this may be. This emotion is a sign and a predicate of the thing. Now, when a thing resembling this thing is presented to us, a similar emotion arises" (see also Winner 1989: 378). This means that symbolic systems, too, are furnished with standards of evaluation. It may further be concluded that the integration of novel facts into the system of symbolic thinking does not proceed from their objective characteristics. On the contrary, this integration is based on prevailing evaluative patterns (cf. Wurm 1986: 502).

Viewed against this background, signals and symbols seem to differ only gradually. While the relation of signals to their affective-emotional deep-structure and the relation of symbols to their cognitive referents is less apparent, the influence of signals on actual behavior is, nevertheless, beyond doubt. It is the interaction of these two systems which seems to provide behavioral certainty.

As pointed out above, the phylogenetically older system of signals and corresponding affective states must have had a selective effect on the more recent system of symbols, favoring those symbols which are well matched with the affective states of the older systems. The joint operation of these systems provides the system of symbolic information with evaluative standards, and hence with criteria for decision making which originate in the affective structure. While a full agreement on the meaning of any given message may never be fully achieved, the combination of the two systems prevents human communication from total arbitrariness and chaos.

Throughout ontogeny, the human infant learns terms and names and, thereby, comes to associate them with certain emotional connotations. As a result of such ontogenetic learning, schemata are being established which are the basis of a further evaluative integration of signs. In spite of the prevalence of more or less arbitrary sign systems in human communication, signals do not cease to play a role after infancy. Human ethology has brought to light a number of universal expressions, such as laughing, crying, or certain eyebrow movements in greeting rituals (cf. Eibl-Eibesfeldt 1984: 571; van Hooff 1989) which are utilized as a sort of punctuation in symbolic communication. Interestingly, the effects of such expressions are rather similar to those of signals. The receiver tends to regard the emotional content of these expressions as an indication of the "real" intentions of the sender. As a result, the receivers are prepared to shift to the same mood which they feel is being conveyed by these expressions. The impact of nonverbal expressions may again be understood as the use of more clearly visible affective-emotional signs in order to interpret the meaning of symbols in this context.

In the mutual interpretation of the messages of communication, cultural rituals play an important role. Usually, the term ritual designates a certain sequence of symbolic steps in which a certain number of people take part. According to Koch (1986: 36), the term ritual stands for "the most embracing unit of signs ... which is being organized in any given culture". Similar to Koch's view, it has been suggested above (2.) that the concept of "ritual" defines a context, thereby raising the individual sign into a higher order. The understanding of any individual sign hence depends upon this order. Usually, this type of meaning will only last some time and then must be reiterated.

Regarding the consolidation of symbolic meaning, the question may be raised from the point of view of a communication theory whether there are certain types of meaning which become ritually established more often than others. According to Leach (1976: 35), most cultural rituals define certain seasons or boundaries in the course of the year or in the course of the life of an individual. Examples of rituals related to seasonal boundaries are festive days like Thanksgiving Day. Examples of rituals related to boundaries in individual life are rituals accompanying birth, death or the initiation to adulthood. It seems that such boundaries have to be associated with higher orders of signs because this is the proper way of introducing them into cultural discourse.

Goffman (1971) has revealed the impact of rituals on the modern everyday life in modern society. In addition to such rituals of transition as birth, death, and coming of age, there are numerous ritualistic situations structuring the course of everyday life consisting in greetings, clothing, or decorative objects (cf. Wehrspaun 1989: 322). From an evolutionary point of view, the affective-emotional synchronization of individuals engaged in a ritual is quite interesting. By jointly focussing their attention on the respective symbolism, participants seem to synchronize their affective states (Collins 1986: 7).

5. Conclusions

From an evolutionary perspective, it has been argued that the affective-emotional conditions of an individual are determined by the tendency of communication participants to increase their supply of emotional energies. Generally, affective arousal is synonymous with large investments of vital energy. According to the theory of evolution, selection favors the economic use of time and energy. Since social bonding has to be regarded as one of the major categories of human affectivity (cf. Masters 1986: 379ff.), it may be understood why participation in rituals adds to individual time-energy budgets: the primary cause is the synchronizing effect of doing things together, and the secondary cause is the sharing of symbolic meaning. Human social life entails numerous instances of ritual behavior, from a mother rocking her baby to sleep to participation in a banquet. Typically, individuals engaged in any kind of bonding behavior tend to augment their share of affective energies. As to the origin of social bonding and of patterns of sociality, it has been suggested that these phenomena have been selectively advantageous because of their invigorating effects on participants. As a result of natural selection, organisms have become provided with species-specific physiological, affective, and psychological mechanisms which in turn have brought about patterns of sociality. Viewed against this background of evolutionary theory, it may also become evident why all signs that signify membership in a human community are typically highly emotive.

No matter whether this sign is a costume, a flag, or a name which the community may have chosen, individual members will cherish the emotive power of this sign. In fact, affective evaluation is attached to any human sign. Hence, the sign system is provided with a selective mechanism. In addition to that, nonverbal communication and cultural rituals give rise to further selective mechanisms which provide agents with means of decision making: hence the avoidance of behavioral uncertainty.

The process of human communication is not homogeneous. On the contrary, various subsystems may be distinguished, of which the phylogenetically more recent layer of symbols is the most important one. Once existent, these signs create their own regularities and systemic properties which, however, are without bounds. In other words, the system produces numerous variants which present a multitude of behavioral options to the human individual. It has been argued that the laws of phylogeny have in fact prevented the evolution of this sort of dilemma by establishing the system of symbolic processes, i.e., the realm of cognitions, on top of the phylogenetically older system of human affectivity. According to Riedl (1975: 249), it is a general characteristic of phylogeny that new traits and types of order are established on top of phylogenetically older, but successful types of order.

Paralanguage and various aspects of nonverbal communication are closely linked up with the phylogenetically inherited program of human affectivity, offering an additional example for the "order on order" principle. By combining older and more recent faculties, human communication produces variants at a faster pace than the natural process. At the same time human communication tends to select those variants which may provide individual agents with a certain degree of behavioral certainty. This design of human communication has been successful in natural history. It has given a new foundation to the evolution of culture.

Note

1. For an earlier version of this paper published in German see Meyer (1992).

References

Atteslander, Peter
1984 *Methoden der empirischen Sozialforschung.* Berlin: de Gruy ter.

Bateson, Gregory
1972 *Steps to an ecology of mind.* New York: Ballantines.
1979 *Mind and nature: A necessary unity.* New York: Dutton.

Bertalanffy, Ludwig von
1949 *Das biologische Weltbild.* Bern: Francke.

Burkart, Roland
1983 *Kommunikationswissenschaft.* Wien: Böhlau.

Callebaut, Werner
1987 "Why it makes sense to extend the genotype-phenotype distinction to culture". *La Nuova Critica* 1-2: 63-83.
Chomsky, Noam
1972 *Language and mind.* New York: Harcourt.
Collins, Randall
1986 How to incorporate sociophysiology into sociological theory. [Paper presented at Stanford University.]
Eibl-Eibesfeldt, Irenäus
1984 *Die Biologie des menschlichen Verhaltens.* München: Piper.
Goffman, Erving
1971 *Relations in public: Microstudies of the public order.* London: Pelican.
Gruter, Margaret – Roger D. Masters
1986 "Ostracism as a social and biological phenomenon". *Ethology and Sociobiology* 7: 149-158.
Helversen, Otto von – Klaus R. Scherer
1988 "Nonverbale Kommunikation", in: Immelmann et al. (eds.), 609-646.
Hooff, Jan A. R. A. M. van
1989 "Laughter and humour, and the 'duo-in-uno' of nature and culture", in: Koch (ed.), 120-149.
Immelmann, Klaus, et al. (eds.)
1988 *Psychobiologie: Grundlagen des Verhaltens.* Stuttgart: Fischer.
Ivanov, Vjačeslav Vsevolodovič
1983 *Gerade und Ungerade.* Stuttgart: Hirzel.
Izard, Carroll E.
1981 *Die Emotionen des Menschen: Eine Einführung in die Grundlagen der Emotionspsychologie.* Weinheim: Beltz.
Koch, Walter A.
1986 *Evolutionary cultural semiotics.* Bochum: Brockmeyer.
Koch, Walter A. (ed.)
1989 *The nature of culture.* Bochum: Brockmeyer.
Küppers, Bernd-Olaf
1986 *Der Ursprung biologischer Information: Zur Naturphilosophie der Lebensentstehung.* München: Piper.
Leach, Edmund
1976 *Culture and communication: The logic by which symbols are connected.* Cambridge: University Press.
Lorenz, Konrad
1965 *Über tierisches und menschliches Verhalten.* Vol. 1. München: Piper.

Lumsden, Charles J. – Edward O. Wilson
1983 *Promethean fire: Reflections on the origin of mind*. Cambridge, Mass.: Harvard University Press.
MacLean, Paul D.
1972 "Cerebral evolution and emotional processes". *Annals of the New York Academy of Sciences* 193: 137-149.
Masters, Roger D.
1986 "Ostracism, voice, and exit: The biology of social participation". *Ethology and Sociobiology* 7: 379ff.
Maturana Humberto – Francisco J. Varela
1987 *Der Baum der Erkenntnis*. Bern: Scherz.
Meyer, Peter
1987 "Ethnocentrism in human social behaviour: Some biosociological considerations", in: V. Reynolds – V. Falger – I. Vine (eds.), 81-93.
1992 "Evolutionäre Voraussetzungen menschlicher Kommunikation: Verhaltenssicherheit durch selektive Prozesse", in: H. Reimann (ed.), 85-105.
Peirce, Charles S.
1931-1958 *Collected Papers of Charles Sanders Peirce*. 8 vols. Eds. Charles Harsthorne, Paul Weiss, and Arthur W. Burks. Cambridge, Mass.: Harvard University Press.
Prewo, Rainer – Jürgen Ritsert – Elmar Stracke
1973 *Systemtheoretische Ansätze in der Soziologie*. Hamburg: Rowohlt.
Prigogine, Ilya – Isabelle Stengers
1981 *Dialog mit der Natur*. München: Piper.
Reimann, Horst
1968 *Kommunikations-Systeme: Umrisse einer Soziologie der Vermittlungs- und Mitteilungsprozesse*. Tübingen: Mohr.
Reimann, Horst (ed.)
1992 *Transkulturelle Kommunikation und Weltgesellschaft: Zur Theorie und Pragmatik globaler Interaktion*. Opladen: Westdeutscher Verlag.
Reynolds, Vernon – Vincent Falger – Ian Vine (eds.)
1987 *The sociobiology of ethnocentrism*. London: Croom Helm.
Riedl, Rupert
1975 *Die Ordnung des Lebendigen*. Berlin/Hamburg: Parey.
Sachsse, Hans
1984 *Ökologische Philosophie*. Darmstadt: Wissenschaftliche Buchgesellschaft.
Sommer, Volker
1993 "Die evolutionäre Logik der Lüge bei Tier und Mensch". *Ethik und Sozialwissenschaften* 4,3: 439-449.

Swanson, Carl P.
1983 *Ever-expanding horizons: The dual informational sources of human evolution*. Amherst: University of Massachusetts Press.
Vollmer, Gerhard
1975 *Evolutionäre Erkenntnistheorie*. Stuttgart: Hirzel.
Wehrspaun, Michael
1989 "Kommunikation, öffentliche Ordnung und das projektive Selbst". *Zeitschrift für Soziologie* 18: 329-345.
Wilson, Edward O.
1975 *Sociobiology*. Cambridge, Mass.: Harvard University Press.
Winner, Irene P.
1989 "The human sign as an integrative concept in the semiotics of culture", in: W.A. Koch (ed.), 369-387.
Wuketits, Franz M.
1985 *Zustand und Bewußtsein: Leben als biophilosophische Synthese*. Hamburg: Hoffmann und Campe.
Wurm, Wolfgang
1986 "Wertempfinden und Erkenntnis: Zur Verbindung von Evolutionstheorie und Sozialwissenschaften". *Philosophia Naturalis* 23: 477-509.

The culture of nature: The semiotic dimensions of microcosm, mesocosm, and macrocosm[1]

Peter Grzybek

1. The anthropocentric perspective

From the earliest times of epistemogenesis, humans have tried to understand how they are related to their environment, and in their efforts to understand this relationship, they have tried to understand themselves as well. There are two basic possibilities of interpreting this relationship: either one considers oneself as being an integral part of the world by integrating the self into the totality of all natural beings, or one dissociates oneself from the rest of this totality by emphasizing one's uniqueness and by drawing some borderline between one's own self and the environment. As will be shown below, further possibilities may arise with the development of additional or more complex perspectives. The process of self-definition usually also implies an evolutionary perspective in which the difference between the individual and the world is interpreted by means of anthropogenetic or cosmogenetic assumptions.

In attempting to understand the process of self-definition from an evolutionary perspective, i.e., in metagenesis, human thinking seems to be guided by a principle which Walter A. Koch (1986: 26) has called the "metagenetic search for analogy". According to this principle, early attempts at describing and explaining phenomena of our world were derived from a strictly anthropocentric point of view.[2] This anthropocentric bias provides a key to the understanding of many etiological myths of creation, which interpret the origin and evolution of the world as a whole as the creative act of some anthropomorphic primordial being. Let us investigate these mythological concepts in detail. This will lead us to the question of the interaction between humans and nature on the one hand and to the relation between culture and nature, on the other hand.

2. Microcosm and macrocosm in Indo-European mythology

Some of the most intriguing semiotic interpretations of Indo-European mythology have been proposed by Vjačeslav Vs. Ivanov and Vladimir N. Toporov. In a number of studies, Ivanov and Toporov (1970a, b, 1974) have reconstructed an abstract schema of the so-called Indo-European "basic myth". At the center of this myth is the God of Thunder who fights with his opponent, usually a serpent. The scenario comprises the following elements:

(1) The God of Thunder is located on top, e.g., on top of a mountain, in heaven (jointly with sun and moon), on a rock, or on top of the tripartite World Tree;
(2) The opponent is located below, e.g., under a mountain, at the roots of the World Tree, or close to the water;
(3) The serpent then steals the cattle, which represents the most precious good and which serves as a symbol of the other world;
(4) The God of Thunder persecutes the serpent, which changes into the shape of various species (a human being, a horse, a cow, etc.); he shatters the rock under which the cattle is hidden and sets it free; the God of Thunder kills his opponent, tears him to pieces and throws these in all directions;
(5) After the God's victory, it starts raining.

The motif of the appearance of water (5) lets the God of Thunder appear to be a typical culture hero. The motif of dismembering the body of a primeval human being (4), which occurs in many Indo-European myths, is an attempt to explain the origin of the cosmos out of chaos. The various parts of the primeval being's body correspond to the elements of the cosmos. Examples of such primordial beings are Ymir from Scandinavian mythology (*Edda*) and Puruṣa from Ancient Indian mythology (*Rig-veda*). Equivalent figures may also be found in non-Indo-European cultures such as the figure of Pan-ku in Chinese mythology.

The "basic myth" under consideration is associated with ideas which originated much later in ancient Greek philosophy and developed into the well-known concept of microcosm/macrocosm.[3] Philosophers generally agree that this concept is "one of the great ideas by which humans attempt to understand themselves and their relation to the totality of existence" (Allers 1944: 406). Widengren (1954: 20) calls it "one of the most powerful ideas in the history of religion". As Allers (1944: 332), in his thorough analysis of the "microcosm/macrocosm concept", phrases it: "Microcosmism is one of the primary - or, perhaps, even primitive -

forms in which the human mind conceived the nature and position of humans in reality. It satisfies the deeply rooted desire for an all-comprehending conception in which everything finds its proper place within the order of being."

Very roughly speaking, according to this concept, a human being is conceived of as a "small world", just as the world is understood as a "great man". In his ruminations on the history of Iranian religion, Widengren (1954: 48) concludes:

> Cosmos is the body either of some deity or of some cosmic primeval being . . . The elements of the universe are the various parts of his body . . . Man as a microcosm corresponds to this macrocosm; the elements, by which man is composed, are exactly the same as those of which the macrocosm consists. With man's death, his elements return to the universe.

In an attempt to further specify the mythological relationship between macrocosm and the microcosmic parts of the human body, Toporov (1981) has analyzed further relevant, though not always directly related, texts, such as "Wafthrudnir's Speech" from the *Elder Edda*, "Puruṣa's Hymn" from the *Rig-veda* (X, 90), or passages from the *Bundahišn*. The results of this investigation show that the motifs recurring in these texts are not simple or freely variable equivalences. Instead, they constitute a relatively constant schema. In a more recent interpretation of the set of intercultural correspondences discovered by Toporov, Lincoln (1986: 20) states:

> By identifying specific items in the cosmos as alloforms to corresponding parts of the human body, they form a major component of the creation mythology of virtually all peoples of antiquity who spoke Indo-European languages . . . The picture that emerges is one of a fairly consistent homologic system, in which there still existed some room for variation and innovation.

The general schema which emerges involves the following equations:

Flesh	↔	Earth
Blood	↔	Water
Sweat	↔	Dew
Skin, Hair	↔	Plants
Bones	↔	Stones
Eyes	↔	Sun, Fire
Breath	↔	Wind
Thought	↔	Clouds
Head (Skull)	↔	Sky

These mythopoetical equivalences are not to be understood as metaphorical comparisons, which they seem to be from a contemporary point of view.[4] Instead, both body and cosmic elements are derived from a common material, and they represent allomorphic forms of this material. In this respect, too, Lincoln (1986: 5) has confirmed Toporov's earlier assumptions, maintaining that

> these texts do not just call attention, in the manner of poetic imagery, to some perceived similarity between two disparate entities. They state not that 'X is like Y', but rather that 'X was made from Y'. Between the two items linked in such a homology, there is thus posited a fundamental consubstantiality, whereby the one entity may be created out of the material of the other. The two are understood as *alloforms*, alternative shapes, of one another.

However, the relationship between microcosm and macrocosm is not fixed. It can easily change,[5] depending on whether one takes a cosmogenetic or an anthropogenetic perspective (cf. Lincoln 1986: 33):

> For cosmogeny and anthropogeny are seen to be equally creative, each one being but a phase in an oscillating progress whereby whenever the cosmos is created, the body is destroyed, and - conversely - whenever the body is created, the cosmos is destroyed. The material substances common to both microcosm and macrocosm thus pass from one set of alloforms to the other *and back again* as cosmogeny and anthropogeny endlessly alternate.

The mythological prerequisites of the microcosm/macrocosm idea are far from being specifically European or Indo-European: in fact they are nearly universal. Though the motif seems trivial at first sight, it has been extremely powerful and efficient in the process of epistemogenesis. It is not surprising then, that this concept has been taken up and further developed in various philosophies, which have modified it in many respects. Thus, the concrete interpretations of the relationship between microcosm and macrocosm may vary significantly, irrespective of the overall invariable schema. Therefore, as Allers (1944) points out, it would be an inadmissible simplification to speak of "microcosmism" as if it were always of the same kind.

3. Views of the relationship between microcosm and macrocosm

In his survey of various interpretations of the concept of microcosm, Allers (1944) sets up a list of possible relationships between microcosm and macrocosm. The order of this list does not reflect evolutionary chronology, but rather represents an increasing degree of complexity from an analytic point of view. The simplest relationship, which is termed "elementaristic microcosmism", is expressed in the idea that humans contain within their beings all the elements of which the world consists, since they are composed of the same elements which exist elsewhere in the universe. It turns out that it is exactly this kind of microcosmism which characterizes the above-mentioned "basic" mythological concept: whereas this concept is based on the notion of isomorphy (homomorphy), all other microcosm/macrocosm relationships are based on the notion of isology (homology). "Elementaristic microcosmism" furthermore assumes that the components of the human body are arranged in the same manner as the elements of macrocosm. Therefore, this idea can be elaborated into what might be called "structural microcosmism". This version may take two forms, an anthropocentric or a cosmocentric one. In anthropocentric microcosmism, the universe is compared to a human being. In cosmocentric microcosmism, a human being is compared to the universe, and the discovery of the world within the individual becomes possible.

A third relationship between micro- and macrocosm may be termed "holistic microcosmism". This concept, which takes for granted that humans tend to create order around themselves, is based on the idea that the "cosmos", or the order of any ("organized") totality, is always and everywhere essentially of the same kind. In this interpretation, any "organized whole" may be considered as a reproduction of the universal macrocosm. A fourth type of relationship may be termed "symbolistic microcosmism". In this case, microcosm is conceived of as "corresponding to" or "symbolic of" the universe. A fifth type is the so-called "psychological microcosmism". This relationship is based on the idea that humans can internalize the entire universe by knowing it, and by knowing the universe, the human mind in a sense "becomes" the universe. Finally, in "metaphorical microcosmism", the sixth relational type, "cosmos" becomes a general name for every being which represents itself as intrinsically ordered.

As can be seen from these remarks, the more "modern" interpretations of the microcosm/macrocosm concept turn out to be extremely flexible. The more abstract this relationship is conceptualized, the fuzzier the borderlines of microcosm become. The notion of microcosm is no longer restricted to the human body or mind. Instead, the definition may be extended to include various spheres of the human environment. Thus, in a social sense, a family, house, garden, or city can be defined as a microcosm. Theoretically, the borderlines of microcosm turn out to be arbitrary, and it is impossible to define where microcosm ends and macrocosm begins. This definitional vagueness has long been acknowledged by the theoreticians of the microcosm/macrocosm concept: not only the human being but any object considered as a "small world" have been called a "microcosm". Human artifacts, such as buildings, temples, or ritual objects, have been seen as microcosms or have been interpreted as such in symbolic rites (cf. Korvin-Krasinski 1960: 87). As a result, one can conclude that the epistemological efficiency of the microcosm/macrocosm concept can be attributed to this very definitional flexibility, which allows a human being to consider both his or her own integration into the environmental totality and the dissimilarity or separation between the self and the world. This interpretative flexibility results from the fact that both microcosm and macrocosm come into play in these conceptual relationships. Finally, in addition to its epistemogenetic relevance, the microcosm/macrocosm dichotomy turns out to be an instance of what Koch (1974: 200; 1986: 31) has defined as a "metagenetic fallacy", i.e., an explanation of early evolutionary stages in terms of structures which have emerged later in evolution.

4. The biperspective view of nature and culture

In addition to the micro- and macrocosmic dimensions, human epistemology, at least from a particular period of cognitive evolution, has always involved a third, intermediate dimension. This is the sphere of the mesocosm. Scholars have often neglected this intermediate dimension, which is an important sphere in its own right,[6] focusing only on the microcosmic or the macrocosmic spheres and integrating elements of the mesocosm into one of them. Yet, the mesocosm represents a separate sphere, albeit one without any sharply defined borderlines, and it is in this sphere that we see nature as the anticipation of culture or culture as the continuation of nature.

The above chiastic formulation is more than a simple word play. It reflects what Koch (1986: 22-24) has defined as "biperspectivism" in evolutionary cultural semiotics. In explaining nature and/or culture, an individual may adopt either the position of a participant or the position of an observer (P_x vs. P_{an} in Koch's terms). When one analyzes the world from the participant's culturally and individually determined point of view, the perspective is autoanalytic, proceeding by way of introspection or by egocentric and anthropomorphic analogy. The observer, however, adopts a different perspective striving for a point of view from the outside, beyond the domain of description. This heteroanalytic perspective places the analyst in the position of some god-like "ideal observer". As Koch (1986: 145) correctly emphasizes, the heteroanalytic perspective may be subject to significant changes. Every culture and epoch develops its specific ideal about such presumably reliable external perspectives. Thus, in the course of epistemogenesis, two competing options for the description of one and the same phenomenon have emerged. The spheres of nature and culture are hence part of an overlapping biperspectival realm of investigation (cf. Figure 1).

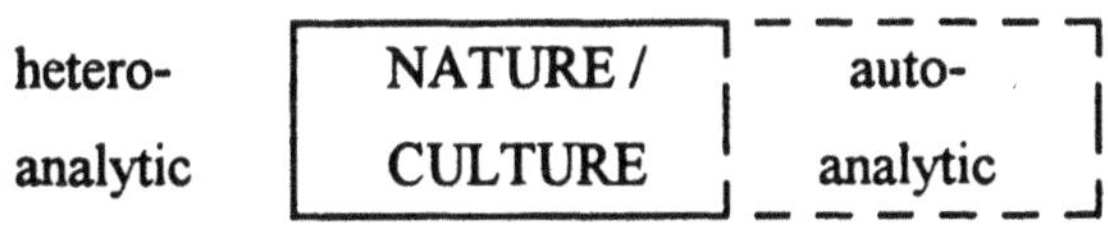

Figure 1. Two perspectives of nature and culture: the distanced heteroanalytic perspective of the external observer and the autoanalytic point of view of the participant.

The apparent symmetry of the two competing perspectives is deceptive, however, because the depicted biperspectivism evinces two important asymmetries. First, the heteroanalytic perspective can easily be unmasked as an autoanalytic projection which remains limited in its approach by the restrictions of the autoanalytic bias. Secondly, as already indicated above, the anthropocentric and thus autoanalytic point of view has always been primary in epistemogenesis. Koch (1986: 58f.) takes this asymmetry into account when he characterizes the first steps of metagenesis, the evolution of human consciousness, as follows:

[1] Man's experience of self;
[2] Projection of [1] onto the environment;
[3] Environment interpreted as "anti-egoistic" and projected back onto [1].

The relationship of these three stages of metagenesis to the hetero- and autogenetic perspectives is shown in Figure 2 (cf. Koch 1986: 58-59):

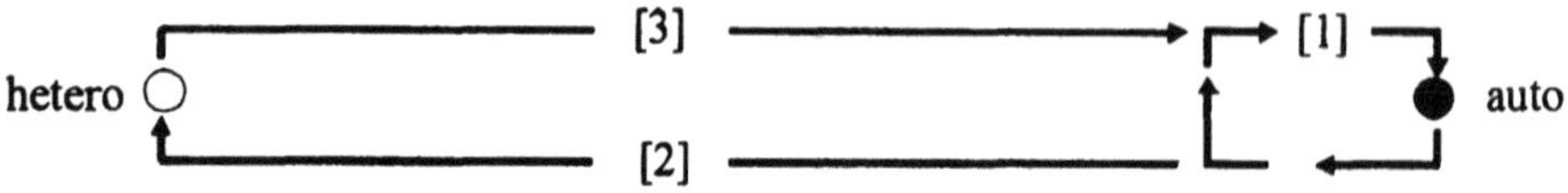

Figure 2. The first three stages of metagenesis from the heteroanalytic and autoanalytic perspectives.

According to Koch (1986: 58-59) two further steps in the evolution of human consciousness are then the exploration of macrocosm [4] and microcosm [5]. Figure 3 shows these stages of metagenesis as a continuation of the stages shown in Figure 2.

As can be seen from Figure 3, autoanalysis and heteroanalysis overlap in an intermediate zone which is accessible from both perspectives. Consequently, the mesocosmic sphere is subject to "biperspectivism". Thus, depending on the perspective taken, Nature explains itself in terms of Culture, or Culture explains itself in terms of Nature (cf. Koch 1986: 4-5).[7] Furthermore, Nature and Culture meet in a "mesocosmic center" (Koch 1986: 54-58), which is located between microanalysis and macroanalysis.

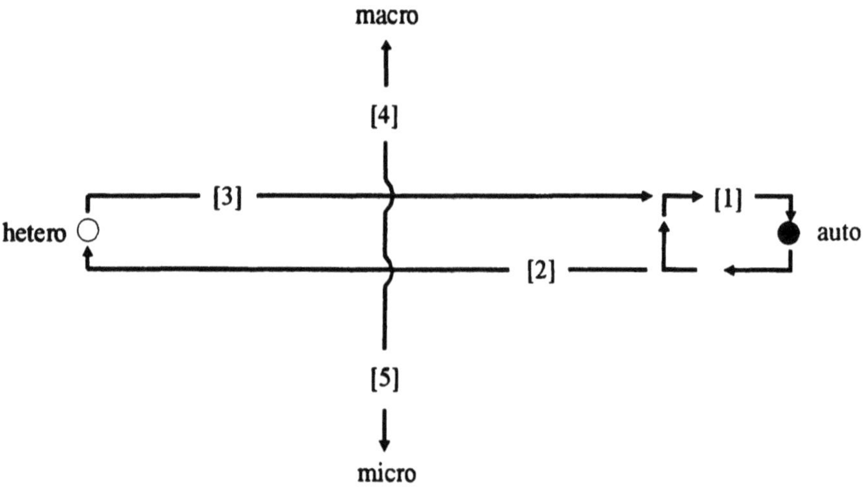

Figure 3. The further evolution of metagenesis by way of microcosmic and macrocosmic exploration.

Thus, the exploration of microcosm and macrocosm leads to such discoveries as nerve cells and genes on the one hand and galaxies on the other. In both directions, discoveries presuppose metagenetic steps [1] through [3] and, consequently, the existence of mesocosmic prototypes. In fact, Koch (1986: 57) concludes that both micro- and macroanalyses are based on mesocosmic prototypes.

What remains unsolved, then, is the question of the metagenetic emergence of mesocosm. An answer to this question is contained in the mythological concepts discussed above. In these concepts, the autoanalytic focus corresponds to the microcosmic perspective, and the heteroanalytic focus corresponds to the macrocosmic perspective. Under these premises, the "elementaristic" microcosm/macrocosm dichotomy actually emerges from the body-oriented inner experience of the individual [1]. After the human body was conceptualized as a microcosm, this concept was transferred to the environment [2], which was then interpreted as a macrocosm and vice versa.[8] Thus, if the assumption of an anthropocentric bias in metagenesis is correct, our mythopoetical views of anthropogenesis and cosmogenesis reflect the following order of cognitive evolution: (1) Human self-consciousness begins with the experience of one's own body. (2) The resulting conceptualization is projected onto the environment. (3) The emergence of a heteroanalytic perspective then turns out to be a self-projection as well, serving as an alternative perspective for the interpretation of the emerging concept of the environment.

This order seems highly probable, although the theoretical possibility of a different or even reverse order cannot be denied. However that may be, in contrast to Koch's assumptions, micro- and macroanalyses should be projected not only onto the vertical, but also onto the horizontal axis of Figure 3, the axis of auto- and heteroanalysis. This reinterpretation results in Figure 4, which turns out to be a modification of Koch's (1986: 54, 1987: 81) model of the interaction between Nature and Culture. This model illustrates the identity of the micro-analytic and autoanalytic perspectives on the one hand and of the macro-analytic and hetero-analytic perspectives on the other. The hatched field symbolizes the intersection of both perspectives in the intermediate mesocosmic sphere as a biperspective area.

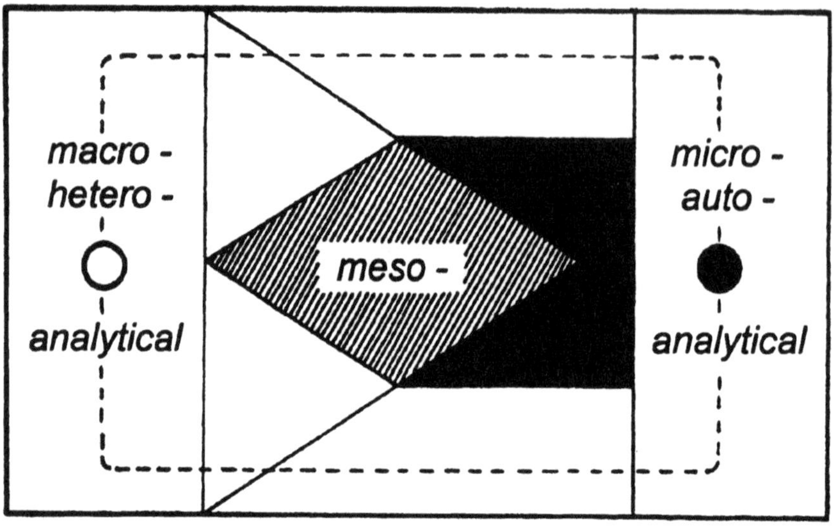

Figure 4. A reinterpretation of the mesocosm in the context of autoanalysis and heteroanalysis

5. The mesocosm and anthropogenesis

It may seem that the additional consideration of a separate mesocosmic sphere is an arbitrary theoretical construct. Yet, there are two convincing reasons which justify the mesocosm as a domain of its own. One is the evidence given by the evolutionary theory of cognition, the other, the relevance of the mesocosm as shown in folkloristic text analysis (see below). The evolutionary theory of cognition is based on the assumption that all structures of the world are closely interrelated, that they all interact in one way or another, and that these interactive relations manifest themselves in evolution (Wuketits 1983: 21f.). Since human beings are not located outside of the world, but are part of it, all our perceptions, cognitions and thoughts are part of the dynamics of this world. In this sense, the real world is not a product of our imagination. Instead, our way of imagining corresponds to the order of nature. Thus, isomorphic principles can be expected to underlie both the real structures outside the human self and the structures of cognition (Vollmer 1983: 22). A further assumption is that the human brain and its cognitive apparatus have been selected in the course of evolution to process only those structures which represent the mesocosmic realm, i.e., the realm of that which has been important for survival (cf. Vollmer 1983: 22). In the context of the evolutionary theory of cognition, Vollmer (1983: 50) defined the human mesocosm as our "cognitive niche". According to this view, the mesocosm is "that part of the real world which we master by way of sensation and action, perceptually and motorically . . . The mesocosm is, roughly speaking, a world of medium dimensions" (Wuketits 1983: 51). In this sense, the concept of mesocosm is anthropocentric by definition since it explicitly refers to the human beings and the spheres of their senses. The mesocosm is subject to concrete experience. Although the mesocosm is anthropometric, its borderlines are vague and not sharply defined. They can be characterized by reference to various concepts, such as

- time, involving the lower and upper limits of seconds (heart beats) and decades (life),
- distances, involving limits from millimeters (dust, hair) to kilometers (horizon, a day's march), or
- temperatures, involving limits from -10 centigrades (below freezing point) up to 100 centigrades (boiling point).

In addition to the assumptions about the mesocosm specified in the evolutionary theory of cognition, we should point out that, from a semiotic perspective, the mesocosmic sphere is not experienced only perceptually or structured only cognitively. It is also a sphere which may be formed semiotically, by practical human influence. Thus, the mesocosmic sphere is characterized by an increasing degree of semioticity. In this respect, Eliade (1957: 42) correctly points out that to the mythical mind, any construct or manufactured object has cosmogeny as its model, because the creation of the world serves as an archetype for any human artifact. Thus, by "inventing" the mesocosm, humans provide themselves with an opportunity to organize their environment. This organization can occur in accordance with a given or assumed sacred principle, when all elements and relations of the mesocosm are considered to be structured with regard to this principle. It can also occur in accordance with an assumption that all entities subject to human influence are intentionally and deliberately organized and semioticized according to anthropocentric principles.

The crucial point is thus the question of the integration or separation of humans into, or from, the natural and/or cultural totality. The separate treatment of a mesocosmic sphere requires a relatively fixed borderline for both microcosm and macrocosm from which the mesocosm is distinct. In this respect, mesocosm turns out to be an intermediate sphere, providing a realm in which biperspective metagenetic analysis becomes possible. In this sphere, nature and culture constantly interfere: it is here, that both "the culture of nature" and "the nature of culture" can be found.

6. The cosmic spheres in folkloristic riddles

Let us conclude our ruminations with a short glance at folklore texts, which display the importance of a separate analysis of mesocosmic structures.

As pointed out elsewhere (cf. Grzybek 1991, 1992, in press), the riddle is a folklore genre of utmost importance in the study of the three cosmic spheres. Based on earlier studies by Frejdenberg (1936), Huizinga (1939), and Kuiper (1960), Toporov (1981) has given convincing evidence that the riddle as a genre goes back to certain archaic rituals which usually took place towards the end of the year. At this time, according to the mythical mind, the universe was in danger of falling apart and dissolving into chaos. Therefore, the world had to be integrated anew by way of a ritual recapitulation of the original act of creation. To achieve this goal, the mythical "*Ur*-Text" was transformed into dialogical texts on topics focusing on cosmogenesis. The dialogues were then exchanged between the participants of the ritual (priests). According to Toporov, these ritual dialogues became desacralized at a later period and developed into precursors of traditional folk riddles, which can thus be considered the profane remnants of the formerly sacred ritual texts. Toporov has also provided evidence that these early ritual dialogues constitute the root of Indo-European poetry and poetic devices.

To return to the topic of the mesocosmic sphere and its representation in folk riddles, let us now investigate examples from various Slavic traditions. As has been argued elsewhere (cf. Grzybek, in press), the emergence of the mesocosm as a separate sphere is an important step in the secularization of initially sacred, ritual texts, and consequently, in the development of poetry in general. In addition to illustrating this secularizing movement, the following riddles exemplify the system of isologies (homologies) which relate the three spheres of microcosm, mesocosm, and macrocosm. Theoretically, isologies are possible between all three spheres and in all directions. One sphere is the expected topic of the riddle question, whereas another sphere appears in the riddle answer. When only interspheric isologies are taken into consideration, six possibilities arise. As can be seen below, almost all of them can be found to occur in traditional folk riddles:

Macrocosm → Microcosm

(1) *Meždu dvukh svetil - posredine odin. - Nos.* (Russian)
'Between two stars I am in the middle. - Nose.'

(2) *Dve zvezdočki malen'kikh vse pole mne svetyat. - Glaza.* (Russian)
'Two little stars illuminate for me the whole field. - Eyes.'

Microcosm → Macrocosm

(3) *Bez ruk, bez nog - čerez tyn polzet. - Mesyats.* (Russian)
'Without hands, without feet, it crawls through the palisade. - Moon.'

(4) *Dva stoyat, dva khodyat i dva minuyutsya. - Nebo i zemlya, solntse i mesyats, den' i noč'.* (Russian)
'Two are standing, two are walking, and two are passing by. - Sky and earth, sun and moon, day and night.'[9]

Microcosm → Mesocosm

(5) *Ja sam oko, neimam vedja ni trevavica, zatvoreno sam pa ipak vidim, a po meni i drugi vide. - Prozor.* (Croatian)
'I am an eye, I have neither eyelids nor brows, I am closed, but still I can see, and others can see through me. - Window.'

(6) *Što bez očej plače? - Vikno.* (Ukrainian)
'What weeps without eyes? - The window.'

Mesocosm → Microcosm

(7) *Polon khlevets belykh ovets. - Zuby.* (Russian)
'A stable full of white sheep. - Teeth.'

(8) *Stoit khata krugom mokhnata, odno okno, da i to mokro. - Rot v borode.* (Russian)
'A hut is standing, mossy all around. It has only one window, and that's wet. - Mouth within a beard.'

Mesocosm → Macrocosm

(9) *Polna povetka vorobyšek, a odin petušok. - Nebo, zvezdy, mesyats.* (Russian)
'The barn is full of sparrows, but there is only one cock. - Sky, stars, and moon.'

(10) *Polno koryto, ogurtsov namyto. - Nebo i zvezdy.* (Russian)
'A full trough, filled with cucumbers floating in it. - Sky and stars.'

Only one theoretically possible relationship is significantly underrepresented: there are hardly any riddles in which a "macrocosmic question" is followed by a "mesocosmic solution". This preliminary finding should certainly not be taken as given: the topic requires a more thorough investigation. However, if a quantitative study of extant traditional folk riddles should reinforce the finding that there exist only a few projections from the macrocosmic to the mesocosmic sphere, we might gather useful information about the evolution of conceptual isologies (homologies) between the microcosmic, mesocosmic, and macrocosmic spheres. More elaborate studies of the semiotic dimension of the mesocosm promise to reveal intriguing results.

Notes

1. A preliminary version of this article was presented at the 3rd conference of the International Society for the Study of European Ideas (Ålborg, 24-29 August 1992), as a part of the workshop "The Construction of Nature: A Discursive Strategy in Modern European Thought". I am happy to express my gratitude for Rachael P. Wilson's and Winfried Nöth's competent editing of this text.
2. This assumption coincides with Barkan's (1975: 8) view of anthropomorphism as the epistemic starting point: "In the life of primitive man, the self, and hence the body, is the only wholeness which can be grasped. Anthropomorphism is, *faute de mieux*, this man's only cosmology."
3. The explicit rise of this concept may be located in the sixth century B.C. Usually, either Anaximenes or Anaximander are credited for having developed it (cf. Allers 1944; Conger 1922).
4. Cf. Barkan (1975: 9): "The primitive belief in a literally anthropomorphic cosmos is partly recapitulated in the literary image of the human body as microcosm, but between these two imaginative conceptions lies a great deal of logic and scientific thought."
5. Cf., among others, Bonfante (1959), Frank-Kamenickij (1938), Hoàng-sy-Quý (1969), Schayer (1935), Wayman (1982), Widengren (1980).
6. At least, this is true as far as an explicit theoretical concept is concerned. Of course, important studies have been done on the comparison of the human being with houses, cities, or social organizations; cf., among others, Douglas (1970), Jager (1985), Lincoln (1986), Civ'jan (1987). Still, elements of this intermediate sphere have usually been attached to the extended sphere of either microcosm or macrocosm, respectively.
7. This concept seems more promising than the assumption of culture as a "second nature" (cf. Glacken 1967: 116ff.).

8. It is not by chance that Koch (1986: 60), referring to Anaximander among others, locates phase [3] around 500 B.C., i.e., the time when the microcosm/macrocosm concept was successively developed into an isological model (see above).
9. There are also riddles in which we find a "microcosmic" riddle question and both a microcosmic and macrocosmic riddle answer, such as: Vidyatsya, a ne skhodyatsya. - Glaza. (Russian) [They see each other, but they don't reach each other. - Eyes.] Khot' i vidyatsya, a ne sojdutsya. - Solntse i mesyats. (Russian) [Although they see each other, they don't come together. - Sun and moon.]

References

Allers, Rudolf
1944 "Microcosmos from Anaximandros to Paracelsus", *Traditio* 2: 319-407.

Barkan, Leonard
1975 *Nature's work of art*. New Haven: Yale University Press.

Bonfante, G.
1959 "Microcosmo e macrocosmo nel mito indoeuropeo", *Die Sprache* 5: 1-8.

Burkhart, Dagmar (ed.)
1991 *Körper, Essen und Trinken im Kulturverständnis der Balkanvölker*. Wiesbaden: Harrassowitz.

Civ'jan, Tat'jana V.
1987 "Das 'Haus' im Weltmodell der Folklore (am Beispiel des Rätsels)", in: Wolfgang Eismann – Peter Grzybek (eds.), 119-134.

Conger, George Perrigo
1922 *Theories of macrocosms and microcosms in the history of philosophy*. New York: Columbia University Press.

Douglas, Mary
1970 *Natural symbols*. New York: Pantheon.

Eismann, Wolfgang – Peter Grzybek (eds.)
1987 *Semiotische Studien zum Rätsel*. Bochum: Brockmeyer.

Eliade, Mircea
1957/ [1990] *Das Heilige und das Profane*. Frankfurt/M.: Suhrkamp.

Evans, J.M.
1966 "Microcosmic Adam", *Medium Aevum* 35: 38-42.

Frank-Kamenickij, I.
1938 "Adam i Puruşa: Makrokosm i mikrokosm v iudejskoj i indijskoj kosmogonii" [Adam and Puruşa: Macrocosm and microcosm in Judaic and Indian cosmogenesis], In: *Pamjati akademika N.Ja. Marra (1864-1934)*. Moskva - Leningrad: Izd. AN SSSR, 458-476.

Frejdenberg, Olga M.
1936 *Poėtika sjužeta i žanra* [The poetics of plot and genre]. Leningrad.

Glacken, Clarence J.
1967/ [1976] *Traces on the Rhodian shore*. Berkeley: University of California Press.

Grzybek, Peter
1991 "Der Körper im Rätsel", in: Dagmar Burkhart (ed.), 195-216.
1992 "Mikrokosmos, mezokosmos, makrokosmos: Model na sveta i poetika văv folklora (Po primeri ot bălgarskite gatanki)" [Microcosm, mesocosm, macrocosm: World model and poetics in folklore (with regard to Bulgerian folk riddles)], *Bălgarski folklor* 18: 5-21.
in press "Poetik und Weltmodell: Mikro-, Meso-, Makrokosmos und V.N. Toporovs Anagrammtheorie der indoeuropäischen Poetik".

Hoàng-sy-Quý, Hoành-son
1969 "Le mythe indien de l'homme cosmique dans son contexte culturel et dans l'évolution", *Revue de l'histoire des religions* 175: 133-154.

Huizinga, Johan
1939 *Homo ludens*. Amsterdam.

Ivanov, Vjačeslav Vs. – Vladimir N. Toporov
1970a "Le mythe indoeuropéen du dieu de l'orage poursuivant le serpent", in: Jean Poullion – Pierre Maranda (eds.), 1180-1206.
1970b "K semiotičeskomu analizu mifa i rituala (na belorusskom materiale)" [On the semiotic analysis of myth and ritual (on the basis of Belorussian material)], in: *Sign - language - culture*. The Hague: Mouton, 321-389.
1974 *Issledovanija v oblasti slavjanskikh drevnostej* [Studies in the field of Slavic antiquities]. Moskva: Nauka.

Jager, Bernd
1985 "Body, house, city or the intertwinings of embodiment, inhabitation and civilization", in: D. Kruger (ed.), 51-58.

Koch, Walter A.
1974 "Tendenzen der Linguistik", in: W.A. Koch (ed.), 190-311.

1986 *Evolutionary cultural semiotics*. Bochum: Brockmeyer.
1987 *Hodos and kosmos: Ways towards a holistic concept of nature and culture*. Bochum: Brockmeyer.

Koch, Walter A. (ed.)
1974 *Perspektiven der Linguistik*, 2 vols. Stuttgart: Kröner.

Korvin-Krasinski, Cyrill v.
1960 *Mikrokosmos und Makrokosmos in religionsgeschichtlicher Sicht*. Düsseldorf: Patmos.

Kruger, Dreyer (ed.)
1985 *The changing reality of modern man: Essays in honour of Jan Hendrik van den Berg*. Pittsburgh: Duquesne University Press.

Kuiper, F.B.J.
1960 "The ancient Aryan verbal contest", *Indo-Iranian Journal* 4: 217-281.

Lincoln, Bruce
1986 *Myth, cosmos, and society: Indo-European themes of creation and destruction*. Cambridge, Mass.: Harvard University Press.

Lorenz, Konrad – Franz M. Wuketits (eds.)
1983 *Die Evolution des Denkens*. München: Piper.

Poullion, Jean – Pierre Maranda (eds.)
1970 *Échanges et communications: Mélanges offerts à Claude Lévi-Strauss à l'occasion de son 60ème anniversaire*. Paris: Mouton.

Schayer, St.
1935 "A note on the Old Russian variant of the Purushasûkta", *Archiv Orientální* 7: 319-323.

Toporov, Vladimir N.
1971 "K strukture nekotorykh arkhaičeskikh tekstov, sootnosimykh s kontseptsiej 'mirovogo dereva'" [On the structure of some archaic texts related to the concept of the 'world tree'], *Trudy po znakovym sistemam* 5: 9-62.
1981 "Die Anfänge der indoeuropäischen Poetik", *Poetica* 13: 189-251.

Vollmer, Gerhard
1983 "Mesokosmos und objektive Erkenntnis", in: Konrad Lorenz – Franz M. Wuketits (eds.), 29-91.

Wayman, Alex
1982 "The human body as microcosm in India, Greek mythology, and sixteenth-century Europe", *History of Religions* 22: 172-190.

Widengren, Geo
1954-1955 "Stand und Aufgaben der iranischen Religionsgeschichte", *Numen* 1: 16-83 and 2: 47-134.
1980 "Macrocosmos-microcosmos speculation in the Rasa'Il Ikhwan al-Safga and some Hurufi texts", *Archivio di filosofia* 50: 297-312.

Wuketits, Franz M.
1983 "Evolutionäre Erkenntnistheorie - Die neue Herausforderung", in: Konrad Lorenz – Franz M. Wuketits (eds.), 11-28.

Evolution of human semiosis and the reading of animal tracks

Gordon W. Hewes

1. Tracking as a semiotic process

The human ability to "read" the tracks made by animals, chiefly in connection with hunting, is based primarily on visual inspection. In most mammal predators tracking depends on the ability to follow a scent trail. Additional trail signs may include dung, urine, disturbances of trailside vegetation, tufts of hair caught on thorns, and other evidence relating to the passage of the animal being sought. The main argument of this paper will be that the development of visual rather than olfactory tracking by human hunters was a species-specific behavior which may have contributed significantly to the natural selective enhancement of human cognitive capacities. Unfortunately evidence for tracking as a component of hunting practice has not yielded archaeological evidence so far, although in principle it might under conditions similar to those which preserved the tracks of early hominids at Laetoli walking over a volcanic ash covered surface. Fossil animal tracks are fairly common, but they rarely reveal whether their makers were being pursued, and if so, the identity of the pursuers.

Previous studies of early human hunting have generally neglected tracking. Mithen (1990) is a welcome exception. The many contributors to Lee and DeVore's *Man the hunter* (1968) ignore it, except for brief remarks, as by Laughlin (chap. 33) where it is termed "scanning". It is worth emphasizing that the semiotic skills involved in tracking need not be limited to hunting behavior. They may be employed in backtracking one's own path out of an unfamiliar area, searching for lost livestock, pursuing human enemies or escapees, and in exploring new territory. Tracking can become a very complicated process, requiring (depending upon the region) a knowledge of the hoof- or footprint patterns of perhaps a hundred species, but also the means of determining their orientation and hence the direction of movement of the quarry being pursued, and from the spacing of the tracks, indications of gait and speed. Further, it is usually very important that the tracker be able to tell how fresh or old the tracks are. Ordinarily it is impractical to expend much time and energy on an old or "cold" trail. Reading the trailside evidence of gnawed or chewed vegetation, of dung, etc., and whether or not this evidence relates to the track making animal, requires further knowledge. Humans (and higher primates) are deficient in olfactory acuity, but the sense of smell may be useful at times. Very recent droppings are likely to be still warm, calling on still another sensory channel. The tracker may have to use hearing, if the animal being pursued is still within earshot, or if it has stirred up other creatures - particularly birds, in its passage. Kinaesthetic cues may be needed, for example to detect (especially in darkness) whether a trail is level, uphill, or downhill. In addition to such direct trail clues, the competent tracker must know something about the typical behavior of the animal he is tracking, from its capacity for evasive action to its ability to inflict injuries if imprudently approached. The tracker may have to avoid letting the quarry pick up the human scent which would drive it to escape at high speed.

2. Tracking capacities in non-human species

That hominids could chase and kill various animals for food even before the emergence of skilled tracking and other hunting techniques seems obvious; chimpanzees and baboons manage to do some hunting without any indication that they use tracking skills. Tracking does not depend simply on the existence of possible signs in the environment, but on cognitive systems which can make semiotic sense of them. A great many animals apparently possess innately-based "tracking abilities", whether or not they use them for predation. Utilization of information in the environment is necessary so that most animals do not wander aimlessly about unable to find their way back to a starting point - which may be a nest, den, lair, or whatever. The complex subject of animal migration presupposes widespread and sometimes amazingly accurate capacities on the part of animals to determine their location and to "navigate" - in some cases on a global scale, between continents, and to isolated oceanic islands. These abilities involve both the familiar senses, and in some species, reactions between environmental phenomena (such as the position of the sun) and internal biological clocks. Tracking an animal of another species in this perspective is an extension of behaviors in general use by mobile animals to orient themselves in relation to the environment, including its food resources. One kind of environmental evidence relating to the location of food consists of discrimination of visual forms, whether of vegetation or of potential prey species, or of the forms of tracks left by them. Many animals can learn to make very fine visual shape discriminations if suitably rewarded, even if these are not normally utilized for tracking purposes. Pigeons can be trained to distinguish among many different shaped leaves, and between photographs with and without fish in them (R. J. Herrnstein in Roitblat et al. 1984: 233-261), even though pigeons do not eat leaves or fish. Presumably, apes could be similarly trained to distinguish hoof- or pawprints, and to identify their makers therefrom. Many primates feeding in the wild use vision as well as taste and smell in determining the edibility of various plants. The point is, with regard to the visual reading of animal tracks for hunting purposes, that it apparently has not emerged in any species other than our own, probably because for "naturally evolved mammalian predators" it would be a superfluous capacity, given their superior olfactory powers. To be sure, humans are "naturally evolved mammalian predators" except that their tracking behaviors appear to depend on cultural factors which we conventionally separate from "nature".

3. Visual tracking

What makes human tracking of animals so different is not that fundamentally different senses are used, but that the visual size and shape of hoofprints or footprints is decoded in two ways: first to determine the kind of animal which made it, and second, from its (normally horizontal) orientation, to determine the direction in which the animal was moving when the print was made. Further, human knowledge concerning these matters is culturally transmitted. Additional information derivable from the prints themselves includes the recency of the print, and from the spacing between successive prints, the speed of the maker, since more widely separated prints of the same animal generally indicate more rapid movement. Nearly all tracks of interest to hunters relate to terrestrial quadrupeds - generally mammals, or to bipedal birds and humans, although in some cases, they may be tracks of reptiles such as iguanas or tortoises, or limbless snakes.

4. Learning and transmission of tracking skills

The ethnographic literature is not very complete regarding the means by which tracking skills are learned, even in studies devoted to specialized hunting peoples. More details about tracking can be found in modern field manuals directly written for field naturalists or ecologists, less interested in hunting than in research. An excellent example is James Halfpenny's *A field guide to mammal tracking in North America* (1986), which provides data on size of tracks, impressions of claws or separate toes, shape differences between fore- and hindlimb tracks, shapes of foot- or pawpads (as distinct from hooves), evidence as to gait, and variations which may occur in snow. Keys and charts for field identification should make the task easy, but in practice it can be difficult, and may require attention to about forty factors and variable dimensions. It is possible that in the earliest stages of language evolution, the number of "keys" needed to decode verbal messages may not have been any greater.

Seen in this way, the total record left by some animal's tracks and other trailside signs within a normal time frame (such as the main part of a day) might be regarded as a kind of unintentional narrative, in which the tracker develops a profile of the kind of creature he is pursuing. The human tracker's cognitive achievement in forming such a "narrative" does not seem inferior to that required for transmittal of a comparably accurate verbal report. Repeatedly successful outcomes would tend to eliminate inefficient tracking behaviors, and sharpen tracking skills.

The process by which a nonhuman predator follows a scent track to its quarry is much simpler, depending mainly on pursuit of increasingly strong stimuli in the same mode. Steen and Wilsson (1990) investigated the tracking abilities of trained hunting dogs to answer the question of how such dogs determine the *direction* of the track makers. They also provide a reference to an event of A.D. 1026, recorded by the Icelandic writer Snorre Sturlesson writing ca. 1230, in which some Norwegians captured by Swedes managed to escape, and to fool their pursuers, tied reindeer hooves pointing backwards to the soles of their boots. Steen and Wilsson point out that this would not have confused the Swedes' tracking dogs, since dogs follow increasingly strong scent trails, and pay no attention to the visual placement of hoofprints. The human tracker does not find the hoof- or footprints of his prey becoming "stronger" as he comes nearer, except perhaps under conditions of rapidly melting snow, where the most recently made tracks may be more distinct, compared to the earlier tracks where the edges have begun to melt. It is also true that some animals emit powerful odors, which even olfactorily limited hominids can easily follow. Certain animals send out scent trails biologically evolved to entice or attract conspecifics, as mating signals, or as territorial markers. Detection of these pheromone trails undoubtedly contributes to the tracking success of predators who can follow them. A few species such as skunks spray strong-smelling secretions to deter attack or pursuit, though this has not prevented them from being hunted by humans for their furs.

5. Tracking in snow

Forrest (1988), in her useful guide to tracking in snow, comments on some adaptations of mammals living in snowy environments. The tracks of many animals look quite different when made in snow of varying depth and density. On the whole, tracks in snow are the most easily "read", especially if the snow has recently fallen, and especially if few animals have disturbed it. Other ground cover conditions are also conducive to tracking, such as wet sand or a fall of volcanic ash (as in the mentioned case of early hominid tracks at Laetoli, Tanzania). It is probable that greater scope for tracking by human hunters emerged when the hominids moved into Eurasia from their probable African homeland, into regions with winter snow but at the same time, with large herds of ungulates.

6. Tracking in connection with trapping and game drives

Hunting drives need not involve tracking; grass and brush fires can flush game into waiting groups of human hunters. However, without using fire, drive-hunters can stampede prey animals into converging fence lines of brush or stones, into pounds or corrals, nets, or over precipices. G. C. Frison (1986: 177-223), a North American archaeologist with wide hunting experience, has examined such communal hunting strategies. Their effective use rests on prior knowledge of habitual game trails, marked by tracks of known species. The setting of traps and snares is not random, but based upon previous consideration of tracks and trails likely to be reused.

7. Use of sounds in tracking

As mentioned, hunters are also guided by sounds, produced not only by vocalizations of prey, and noises made by their movement through vegetation, hoofbeats, noises of eating, but sounds made by birds or insects disturbed by their presence or passage. These are all possible factors in the semiotic setting to which the skilled hunter may have to pay attention. Other signs which may indicate the location of prey (or carrion) are the movements of raptorial birds - eagles, hawks, owls, and of vultures, perhaps flying high above their targets.

8. Tracking for purposes other than hunting

Tracking can also serve for defensive purposes for humans. In some parts of the world humans are still occasionally menaced by large predators, such as lions, leopards, or bears. It is advantageous for the occupants of camps or villages, and especially for their small children, to identify such prowling predators and to track and either drive them away or eliminate them. In regions where raids by human enemies may occur, tracking remains an essential part of community defense. Deliberate marking of human trails is also an ancient practice, to show the way back out of unfamiliar country or to mark the way for others. The applications of tracking beyond hunting., to warfare, border defense, peaceful exploration (including speleology), and law enforcement, are obvious, and can be extended to wheeled and other vehicles. The fact that dogs are still regularly employed to compensate for human tracking deficiencies is well known.

9. The antiquity of animal tracking by humans

How long ago humans came to develop tracking skills, for hunting or scavenging is related to the broader question of the first evidence of significant predation by hominids, on which specialist opinion remains divided. The occasional unsystematic hunting exhibited by chimpanzees and baboons may not be a reliable model for the origin of human hunting behavior. Scavenging is a possible precursor of hunting by hominids, in which stone tools may have been used to butcher carcasses and break marrowbones.

10. Effects of observing prints made by one's own hands and feet

As for the origin of human attention to animal tracks, it could have begun with observations of the effects of impressing human hand- or footprints on suitable surfaces. We do have human handprints (apparently made by blowing pigment through a bone tube) in certain Upper Paleolithic caves, but there is no reason to connect them with the hoofprints of hunted animals. However, depiction of large mammals in the cave paintings of the same period indicates close attention to the realistic depiction of the hooves of many animals. Mithen (1990: 228-230) draws attention to an interest in tracks and hoofprints in the Upper Paleolithic cave art of Europe.

11. Domestic dogs' tracking behavior as a stimulus for human tracking

Without any solid evidence, I suggest that the emergence of really effective visual tracking of prey animals by humans arose only with the domestication of the dog, toward the close of the Pleistocene, when dogs began to accompany human hunters, demonstrating their helpfulness by scent pursuit, and inducing their olfactorily handicapped human companions to examine the tracks visually, which dogs only needed to sniff at. Admittedly this is a highly sentimental scenario. Perhaps, human tracking developed initially from the relatively easy identification of human footprints on a sandy beach, as illustrated in the story of Robinson Crusoe.

12. The importance of feet and footprints in classifying animals

Taxonomic placement of an animal from its tracks is a "pars pro toto" procedure, which happens to agree with scientific practice, especially when it comes to mammals which can be grouped as plantigrade or digitigrade, clawed or hoofed, artiodactyl or perissodactyl. This is not just a modern scientific way of dealing with animal classification. B. K. Smith (1991: 529) noted that ancient Vedic texts concerned with suitability of various animals for sacrifice or human consumption were based on the features of the foot (or footprint): "whole hoofed (horses, asses); five clawed", and so on. The primary distinction made in Leviticus 11 regarding edibility or ritual acceptability of quadrupeds rests on hoof features, most readily seen in the characteristic print or track. The cud-chewing distinction might well have been an afterthought, to rule out swine, or an inexplicable exception as in the case of the cloven-hoofed cud-chewing camel. These hoof- or foot-based food tabus need not go back to the prehistoric concerns of track hunters, but only stress the utility of hoof or foot semiosis applied to the categorizations of land mammals. As a matter of fact, mammalogists pay closer attention to tooth features than to foot anatomy in making finer taxonomic distinctions, but dental formulae and cusp patterns are not very practical criteria for hunters and herders in the field.

13. Similarities between tracks and graphic signs

Although there are some significant analogies between manual gestures used in communication, and animal track semiosis, I shall leave the theoretical implications of such parallels to another discussion. But certain similarities between tracking semiosis and the reading or decoding of scripts should be noted here. Most writing systems are strikingly linear, where the line of symbols runs horizontally, left to right or right to left, or vertically. A few inscriptions are circular, and some used to be boustrophedon - in itself a parallel to the path of an ox plowing a field. In both animal tracking and reading, it is important to know where the "tracks" or symbols begin, their direction, whether they go from left to right, and in same scripts, the significance of size differences (capitals and lower case fonts). Such analogies with animal tracks are particularly crucial in the proper reading of mathematical text. It is even possible that some aspect of dyslexia, which we usually think of as disturbances limited to literate populations, might have parallels in inept performances by trackers in hunting and gathering societies.

14. Tracking and subsistence efficiency

How is tracking related to hunting efficiency or intensification? Usually intensification is related to the use of special meat preservation methods such as drying, smoking, or salting, or to the development of intergroup trade in the products of hunting - from furs to hides to such special commodities as ivory, musk, or quetzal feathers. It may also be considered from the standpoint of hunting weaponry, use of traps, or other special devices, lures, disguises, etc. Intensification can be seen in the emergence of full-time professional hunter specialists.

It should be noted that the land faunal resources are geographically uneven in their distribution, with Africa, Eurasia, and North America most favorably populated with medium to large native ungulates, whereas South America appears relatively impoverished, even before the extinctions of some of its Pleistocene mammal megafauna, and Australia even more so. The Australian aborigines have a probably deserved reputation as excellent trackers, though their skills prior to Colonial times had to be exercised on an unimpressive assortment of marsupials, birds, goannas, etc., with only their fellow humans, dingos, and a few small rodents as representatives of the placental mammals. It seems clear that the full flowering of visual tracking expertise requires hoofed mammals; pawed and clawed mammals are either fairly rare predators, or become important not for food, but for their furs.

15. Pleistocene and modern conditions

Theodore Roosevelt (1910) and other prominent foreign game hunters in Africa in the late 19th and early 20th century encountered what at the time seemed to be "inexhaustible" herds of game, which they were able to collect, mostly without the help of trackers or even dogs, thanks of course in large part to their use of rifles, which enabled them to kill almost any animal which they could get into their gun sights. Roosevelt killed, between April, 1909 and April, 1910, no less than 271 medium to large-sized mammals, and though assisted by scouts and bearers in bringing them back to camp for skinning, etc., mostly on foot. The kill rate of one animal every one and a half days during the American ex-President's monumental safari was not then unusual. Roosevelt (1910: 2) himself commented that the Uganda Railway (completed shortly before his visit to what was then British East Africa) was "pushed through a region in which nature . . . is not and does not differ materially from what it was in Europe during the Pleistocene". Although individual Pleistocene hunters could not have matched what Roosevelt achieved with his rifle, they might not have required much more tracking expertise than his to feed themselves and their families adequately from the herds of hoofed mammals once swarming in Ice Age Europe. The animal remains amassed in the caves at Klasies River Mouth, in the South African Pleistocene might not have required any special tracking skills either (Singer – Wymer 1982).

16. Conclusions

I conclude, pending new and unexpected data, that expertise in tracking first became a major feature of human hunting behavior primarily in the non-African regions of the Old World, and perhaps especially in areas subject to regular winter snow cover, and where the dog was domesticated from the Eurasian wolf.

I think it possible that human substitution of visual tracking for olfactory semiosis emerged in the close collaboration of man and dog, in which the dog provided the stimulus for the human to learn from his canine companion to read trail signs with his eyes which his canine companion could naturally follow with his expert nose. Eventually the human tracker attained a level of semiotic skill surpassing that of the dog or any other predator, in the sense that multiple environmental signs could be worked into a narrative or scenario symbolically communicable to other humans.

Note on sources: Data on tracks and tracking are presented in several manuals or handbooks, usually dealing with the fauna of a particular region. A few are listed below. Additional information data on tracks can also be found in many comprehensive modern zoological works on the fauna or major regions, as in Skinner and Smithers (1990), for Southern Africa, in which there are numerous good illustrations of hoof- and pawprints.

References

Frison, G. C.
1987 "Prehistoric, Plains-mountain, large mammal communal hunting strategies", in: Matthew H. Nitecki – Doris V. Nitecki (eds.), 177-223.

Forrest, Louise Richardson
1988 *Field guide to tracking animals in snow.* Harrisburg: Stackpole.

Halfpenny, James
1986 *A field guide to mammal tracking in North America.* Boulder: Johnson.

Headstrom, Richard
1971 *Whose track is that?* New York: Ives Washburn.

Laughlin, William S.
1968 "Hunting: An integrating biobehavior system and its evolutionary importance", in: Richard B. Lee – Irven deVore (eds.), 304-320.
Lawrence, M.J. – R.W. Brown
1973 *Mammals of Britain: Their tracks, trails and signs*. London: Blandford.
Lee, Richard B. – Irven DeVore (eds.)
1968 *Man the hunter*. Chicago: Aldine.
Mithen, Steven J.
1990 *Thoughtful foragers*. Cambridge: Cambridge University Press.
Murie, Olaus
1974 *Animal tracks*. (Peterson Field Guide Series. 2nd ed.) Boston: Houghton Mifflin.
Nitecki, Matthew H. – Doris V. Nitecki (eds.)
1987 *The evolution of human hunting*. New York and London: Plenum.
Price, T. Douglas – James A. Brown (eds.)
1985 *Prehistoric hunter-gatherers: The emergence of cultural complexity*. Orlando: Academic Press, Inc.
Roitblat, H.L. – T.G. Bever. – H.S. Terrace (eds.)
1982 *Animal cognition*. Hillsdale, NJ: Lawrence Erlbaum.
Roosevelt, Theodore
1910 *African game trails*. New York: Scribners'.
Sebeok, Thomas A. (ed.)
1977 *How animals communicate*. Bloomington: Indiana University Press.
Singer, Ronald – John Wymer (eds.)
1982 *The Middle Stone Age at Klasies River Mouth in South Africa*. Chicago: University of Chicago Press.
Skinner, John D. – Reay H.N. Smithers
1990 *The mammals of the Southern African subregion*. Pretoria: University of Pretoria.
Smith, Brian K.
1991 "Classifying animals and humans in ancient India", *Man* 26(3): 527-548.
Smith, Richard P.
1982 *Animal tracks and signs of North America*. Harrisburg: Stackpole.
Steen, J.B. – E. Wilsson
1990 "How do dogs determine the direction of tracks?", *Acta Physiologica Scandinavica* 139: 531-535.

Analysis of a human releasing mechanism

Paul Leyhausen and Jörg Pfleiderer

Prologue by Paul Leyhausen

The experiments reported in this paper were begun in the years 1944-1945 in a prisoner of war camp in Canada. They were continued in 1947 after my release and the manuscript was finished in 1948. It was then lost for many years. After it had been restored to me I presented it at the International Ethological Conference (Rennes 1969), but somehow never found the time to revise and publish it. Now, after 44 years, I find that still no paper has appeared that covers the same ground using more refined methods. Accordingly, I am now publishing the paper in its original form with only minor changes in style and terminology and a very few necessary additions of more recent publications. For a detailed account of the theoretical background, see Leyhausen (1967, 1973a). My reluctance to publish the paper was also due to a feeling of uneasiness about the rather primitive procedure that made it difficult to prove the validity of the results by the usual statistical operations. Several mathematicians I approached for help declared themselves unable to do so. It required more than mathematical routine and good will; it required ingenuity and the ability to enter into the spirit of the problem and the particularly ethological theoretical approach to find a way of mathematically verifying data collected under the inevitable constraints of a POW camp. When Jörg Pfleiderer, mathematician, physicist, and astronomer, decided to tackle the problem, he could not just apply a few routine calculations. He had to grow into a fully-fledged ethologist and, in a way, to recreate the paper - and my courage to present it in print. Hence, I invited him to become a coauthor without, of course, in any way letting him share the responsibility for my original inadequacies of thought and procedure.

1. Introduction

1.1 The problem

In many cases, animals react to certain situations in a biologically "correct" way without previous relevant experience. However, it is not the whole of the situation or object to which the animal reacts but only a few of its features, which as a rule are sufficient to characterize it reliably but in a rather schematic way. Lorenz (1935) therefore called the perceptive mechanism (cor)responding to these selected features or sign stimuli "Angeborenes auslösendes Schema" (innate releasing schema). Later, when ethology spread to English-speaking countries, this was changed to Innate Releasing Mechanism (IRM) (Tinbergen 1951). It must, however, be noted that with the substitution of "mechanism" for "schema" the emphasis has been shifted: schema relates mainly to the afferent, perceptive function, releasing mechanism to the efferent. It is the former that is relevant in the context of this investigation.

There are IRMs which respond to releasing stimuli presented (a) by inanimate environmental objects, (b) by heterospecific organisms, and (c) by conspecifics.

The IRMs of the third class are by far the most numerous. Class (c) differs from class (a) in that not only are the IRMs selected to correspond ever more accurately to the details of the sign stimuli, but these in turn respond to the selective pressure of the functional peculiarities of the respective IRMs. Morphological structures and movements will thus become increasingly pronounced and unequivocal. We call these specially evolved sign stimuli "releasers". In some special cases, this also happens with (b)-class IRMs, e.g., in the mutual selection of insect IRMs and the pattern of flower design.

Some releaser movements serve only to orient, presenting to the addressee with startling abruptness structural stimuli such as feather crowns or the elongated and/or brightly colored hairs on the neck, back, or cheeks. Originally, all such movements were but expressions of some mood or emotion on the part of the displaying animal. But with the evolution of answering IRMs, they acquire communicatory functions and are thus able to elicit adequate responses from a conspecific. Presumably, this is also true of human facial expressions and expressive gestures and of the impressions they create. It therefore seems plausible to suppose that they had a like evolutionary genesis to those of other animals. Yet, the prejudice that humans do not possess complex innate behavior patterns is rooted firmly in the theories of traditional psychology. Lersch (1932), for instance, opines that ". . . it is far more probable that they (sc., the expressive movements) have to be acquired anew by each individual and that, if they serve any biological purpose at all, this purpose must likewise be experienced anew by each individual." Statements to this effect may be found in many more recent publications.

To decide the issue, a way had to be found to adapt to human subjects the methods usually employed to investigate IRMs in animals. The most direct of these is to rear an animal while withholding from it all experience relevant to the situation or reaction to be tested and then confront it with that situation at an age when the behavior normally becomes functional. If it then behaves in more or less the same way as an undeprived conspecific, it is concluded that the missing experience is not necessary to the elicitation of the normal reaction or, in other words, that the ability to respond adequately to that situation is based on an IRM. Clearly, this type of experiment prohibits itself in the case of humans since there is the risk that such deprivation might do incalculable harm to the mental and social development of the individual. The animal experiments have, however, taught us that IRMs exhibit certain peculiarities which set their perceptive qualities markedly apart from those of individually acquired perceptual experiences. Acquired perceptual experience is structured according to the rules of gestalt (configuration). Its components are not independent, but form a perceptual whole, and any change or omission in part is either not perceived at all (*Gestaltdruck*) or it disfigures or destroys the whole. Therefore, a gestalt is not just the sum of its components (Matthaei 1929). By contrast, the several sign stimuli that arouse an IRM each have their separate releasing power, and their sum adds up according to how many of them are present ("summation of heterogeneous stimuli", cf. Seitz 1940). It follows that they need not all be present; the effect is proportional to their number and relative strength. For each sign stimulus may be varied quantitatively, and its releasing power varies accordingly. This principle - that individual sign stimuli may be quantitatively modified independently of each other and not somehow correlatively - was first discovered by Koehler and Zagarus (1937). It can be applied in order to design dummies which elicit stronger responses than the most splendid natural object could ("supernormal object").

1.2 Procedure

From the above, one might conclude that proof of heterogeneous summation and independent quantifiability suffices to demonstrate that the elicitation of a specific response is due to the existence of an IRM. However, this claim was challenged by Grzimek (1943), who believed he had shown that the perception of individually learned situations also can at least partly be based on the summative effect of component stimuli. Baerends et al. (1965) tried to prove this experimentally. In both cases, the conclusions drawn are not reliably founded as the authors failed to perform the necessary control experiments (for detailed discussion see Leyhausen 1967, 1973a). For the time being, we may safely assume that if a releasing situation is found to satisfy both criteria - heterogeneous summation and independent quantifiability of components - this points to the existence of an IRM. Very early in the emergence of ethology, K. Lorenz was convinced that human understanding of facial expressions was based on IRMs, although he did not publish this explicitly until 1943, and he was, of course, strongly influenced in his thinking by Darwin (1872). In 1941, he suggested to his student W. Schmidt (1957) a simple kind of dummy to test his hypothesis with respect to the expressive and impressive quality of laughter.

A set of laughing faces the size of a postcard were drawn in graphic simplicity and abstractness. They showed the features of a laughing face in varying number and strength. These "dummies" were handed to the experimental persons (EPs) in an unordered pack with the command to arrange them in a series, strongest laughing first, least laughing last, as quickly as possible on first impression, without pausing to reflect. The result seemed to satisfy the above mentioned criteria for an IRM. I remembered this work when, in a Canadian prisoner of war camp, I had plenty of time, was surrounded by any number of potential EPs who had nothing better to do to while away the time, and was thinking how to put all this to good ethological account. Already when I was with Lorenz in Königsberg, mammalian behavior had interested me more than the small birds and fishes which then dominated the ethological scene, and my natural interest in the possibilities of human ethology was, if anything, even keener than that of Lorenz himself. Here then was a rich opportunity to exploit, and I proceeded to exploit it as best I could under the circumstances.

A prisoner of war camp, even when it is managed in such a humane and, I am bound to say, benevolent way as by the Canadian authorities, is a place that subjects its inmates to many kinds of stress. The enforced inactivity is only inadequately alleviated by some sports, learning classes and reading, playing games etc. It all palls with time. The uncertainty about what is happening at home and to the loved ones there, the uncertainty about how long it will last, the crowded conditions, the complete isolation from members of the opposite sex and, last but not least, the fact that most of the inmates are of an age when the temperament is volatile and often intolerant, all contributes to creating an emotional atmosphere in which the slightest or even an imagined provocation may lead to altercations and brawls, although they rarely ended in serious violence. But it gave me frequent opportunity to observe the postures and facial expressions which, in the human male, serve to intimidate or threaten an opponent. And I could hardly fail to see that, despite some variance, these displays were on the whole fairly stereotyped. Hence, as the object of my investigation, I chose the male intimidation display (*Imponiergehabe*, Heinroth 1930).

To find out which postures and facial features were most likely to correspond to an IRM that responds to human intimidating behavior, *if* such an IRM exists, I used three approaches: (a) I asked a number of willing candidates to list those features or behaviors which they thought most imposing or intimidating in another man; (b) I asked people who were good at drawing to draw imposing, intimidating, or threatening men (Figures 1 to 11, overleaf); (c) I used my own observations.

It is, of course, one question whether an IRM exists and another to how many different sign stimuli it responds. However, to test for the two above mentioned criteria, it is not necessary to test them all. I confined myself to those most frequently named, drawn and observed. Using these, I drew sets of "dummies" to test postural and facial stimuli separately. Since it is difficult to create the illusion of movement in still drawings, I did not attempt this in the test sets, with one exception, and that proved a disturbance for reasons to be discussed later. The procedure was the same as that used by Schmidt described above. For better statistical evaluation, it would have been desirable to present the dummies in pairs, each with each, through a tachistoscope, but there was, of course, no tachistoscope to be had in a POW camp. It will be shown that despite this the results are quite reliable.

Figures 1-6. Drawings of imposing intimidating men: 1. Anger (note raised hair and right shoulder); 2. Ready to attack (note elbows and raised hair); 3. Attempted intimidation (note elbow, fist and raised hair); 4. Display of assumed superiority (note left shoulder and elbow); 5. Rising anger, threat (note fist and compressed mouth); 6. Quiet superiority (note elbows, wadded shoulders and straight mouth).

Figures 7-11. Drawings of imposing and intimidating men: 7. Attempted intimidation with underlying defensiveness (note shoulders and elbows, raised "weapon"); 8. "Prelude" display (note angular shoulders and narrow waist); 9. "Authority" (note epaulets and waist slightly taken in); 10. Aggressive threat (note elbows, and how the flap of the jacket is used by the artist to create the impression of a wide, angular shoulder); 11. Suppressed anger (note fist clamped around chair back, compressed mouth and creased forehead with lowered hair line).

2. The experiments

2.1 The A set

A first set of dummies (A set) was drawn and used for preliminary experiments to test the adequacy of the procedure and the details of the features used. The result was an improved set (B1 set).

2.2 The B set experiments

2.2.1. *The B1 set.* The following 6 features were used to draw 11 dummies:

i) piloerection (head) = beret with feather ....(BF)
ii) black beard...(B)
iii) broad shoulders ....(S)
iv) raised elbows...(E)
v) fists ....(F)
vi) chest beating..(CB)

The basic figure of the set is that of a man in medieval attire. To keep within this mode, a beret and feather were substituted for the piloerection on the head. Medieval costume was chosen because the fashion of those times normally exaggerated the features relevant to the investigation. Each dummy was given a name beginning with B. Table 2, block 1 (see Appendix) and Figure 12 show which features were allocated to each dummy. The names were used only for evaluation and were not visible to the EPs during the tests, thus avoiding any emotional or other mental reaction that a name might arouse in an EP. They are printed in the Figures for the benefit of the reader.

The dummies were shuffled and handed to the EPs in a pack with the command: "Lay these cards as quickly as possible in sequence so that the one looking most imposing or threatening comes first and the least intimidating last. Do not stop to think or consider." During the test, I sat back, outside the EP's field of vision, and recorded all his visible reactions and any remarks he made. The result is given in Table 2, block 2. EPs were 60 POWs aged between 21 and 37 (average 31). To get an idea how the dummies were ranked, I added up the number of times each dummy was placed in each column (1 to 11), multiplied by the place number (e.g., 5 times in column 3 = 15), and the sum of all these place numbers was then divided by the number of EPs. In this way, a dummy

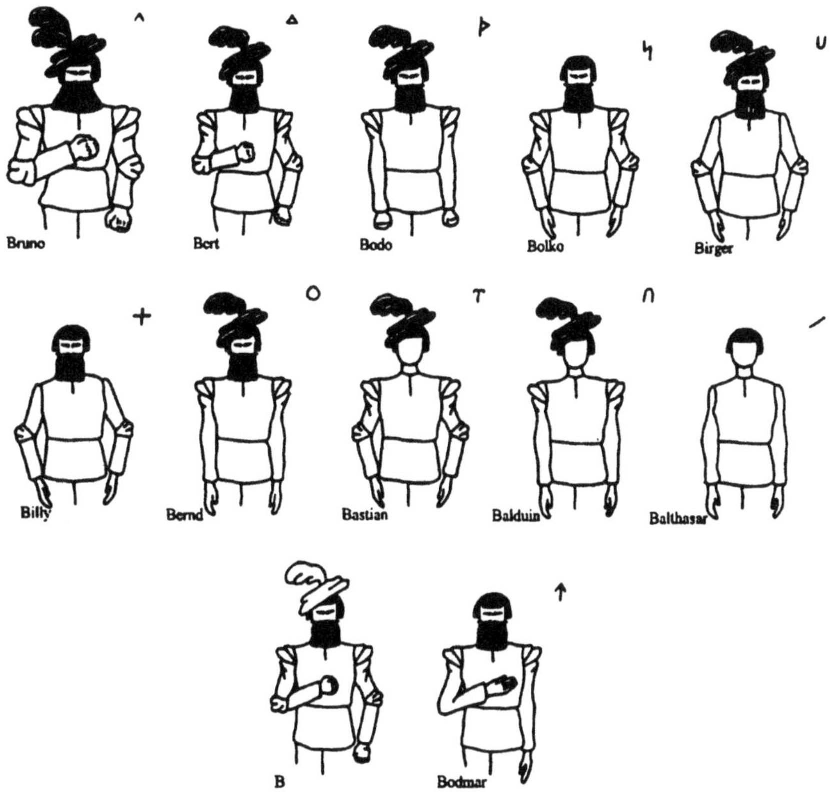

Figure 12. The B1 set of dummies representing men looking intimidating to various degrees. Top and center as in test B1/2. Bottom: B as a sample how BF was presented in test B1/1 and the A set. The ambiguous Bodmar, withdrawn from all later B tests. Each of the drawings on Figures 12-21 was on a separate piece of paper the size of a postcard. The rune-like signs at the top of each dummy were for easy and quick notation.

that all EPs put in place 1 would achieve a ranking number (R) of 1, one that was always placed last a R of 11. The actual ranking is shown in Table 1 (see Appendix).

Table 2 shows a rather broad and uneven distribution for the dummies Birger, Billy, Bodmar, and Bastian (Figure 12). The reason for this was found in the ambivalence of Bodmar. Evidently "chest beating" (CB) was a strong stimulus when perceived by the EP as intended: as a movement. To many, however, it did not appear so, and then it was perceived as a pose and gave Bodmar a look of guileless ingenuity ("He is playacting the innocent", EP no. 4). The ambivalence of Bodmar must naturally influence the placing of other dummies. Another marked ambivalence shows in the placing of Bodo: he appears remarkably often in position 1. Some EPs saw in his posture self-restraint upheld only with difficulty; to them, the arms did not appear limp, as intended, but full of tension. This could not be helped, and throughout the B set tests Bodo retained his ambivalence. Bodmar was eliminated from the B1 test sheets and the rank of the succeeding dummies raised by one. The result of this purge is seen in Table 3 and Table 1/1b. The number of dummies in the subsequent tests was thus reduced to 10.

As a control, the B1 set, minus Bodmar, was presented a second time to the same EPs (3 missing, n=57) after a fortnight, with one alteration: beret and feather, which had been left white in the first test, were now blackened (Figure 12). The reason: in Table 1/1b, the ranking distance between Birger (BF+B+E) and Billy (B+E) is small (0.13), whereas the distance from Bolko (B+S+E) to Birger is four times this (0.59). All three have B+E, Birger adds BF, Bolko S. One would assume S to be a much stronger stimulus than BF. Yet, Bernd (BF+B+S) possesses one stimulus feature more than Billy, including the supposedly strong S, but ranks far below (by 1.20). There are two possible explanations for this inconsistency: (a) BF is not perceived as part of the expressive figure at all and has no stimulus value; (b) the white BF subtracts so much from the threatening effect of the black hair that its own effect is more or less cancelled out (the role of "black" as the color of threat will be discussed later).

If the first is true, a blackening of BF should make no difference. In the second case, however, one would predict that Birger ought at least to draw level with Bolko or even surpass him, and Bernd ought to draw closer to Billy. This is a prediction based on the hypothesis that there is heterogeneous summation of stimulus values. Table 1/1c and Table 4 show that (b) was correct. Birger beats Bolko by 0.63 (!) and the distance between Bernd and Billy has shrunk to 0.56. This change persists in all the subsequent B1 tests (Tables 1/2-8 and 3 to 11). In addition, the first three, Bruno, Bert, and Bodo, who all possess BF, have gained: their Rs have grown smaller, from 1.88, 2.70 and 3.32 to 1.54, 2.67 and 3.02 respectively. As explained above, this means that their positions have become more secure. By comparison, the distributions in Table 4 show less variance than those in Table 3. Only 1 EP (no. 52) noticed the change, and very few missed Bodmar. This goes to show that, on the whole, the injunction to the EPs not to consider the drawings consciously and in detail worked very well. Thus, the ranking was predominantly due to spontaneous reaction, not critical comparison. There were, however, exceptions of the kind discussed below. As with the hand, the fist held in front of the chest also does not appear to be beating the chest; it looks like a pose. That is why Bodo is not as inferior to Bruno and Bert as expected. CB was therefore discarded as a stimulus item in the following B tests.

The tests so far revealed that adults do not always gain a sufficiently strong, instantaneous impression from such small drawings. Especially after looking at the dummies for some time, the EPs tend to analyze the drawings and disregard the direction given at the beginning of each test session to act quickly and not look at details. The record of EP no. 3's first test is typical. He first spontaneously laid the sequence Bruno - Bert - Bodo - Birger - Bolko - Billy - Bastian - Bodmar - Bernd - Balduin - Balthasar. Then, he suddenly rearranged the sequence to Bolko - Bodo - Birger - Bodmar - Billy - Bernd - Bruno - Bert - Bastian - Balduin - Balthasar. His explanation: "On closer examination I feel that this one (he pointed at Bruno) would only frighten small children." He was then asked whether he could restore the first sequence. He did so and remarked: "It's funny that all the beards are in front!" EP no. 42a laid the sequence Bernd - Bodo - Bert - Bruno - Birger - Billy - Bastian - Bolko - Bodmar - Balthasar - Balduin. He explained: "I fancied that the man delivers an address, works himself into a rage and calms down again. The last two I can't place, they don't fit in." Such tests could not, of course, be included in the evaluation since it was obvious that the EPs had not followed directions.

Two years later, the B1 set was tested with school children aged from 10 to 16. The dummies were not handed to the children in a package but haphazardly distributed over the experimental table. Girls and boys were tested separately. The boys' result is shown in Tables 1/3 and 6, the girls' in Tables 1/4 and 7. Distributions and rankings are not appreciably different, but both show some differences when compared with the results of the adult males. Bernd (BF+B+S) is placed well before Billy (B+E) and comes very close to Bolko (B+S+E). The data were then processed separately for the age groups 10 to 12 and 13 to 16 without separating the sexes. Tables 1/5, 8, 1/6, and 9 give the results. With the 10- to 12-year-olds, the difference compared to the adult males is even more marked, and even Bastian (BF+S+E) marginally ranks before Billy. With the 13- to 16-year-olds, Billy has again passed Bernd, and Bastian has sunk back to his old position. When the data for boys and girls of the 13-16 age group are viewed separately, however, we find that the ranking scale of the boys (n = 21) exactly duplicates that of the adult males, whereas that of the girls (n = 29) once more places Bernd before Billy (Tables 1/7, 10, 1/8, and 11). Since n in both groups is small, this is perhaps of doubtful significance, but gains weight when all the results are combined. Incidentally, that young children are more impressionable is shown by Bruno obtaining a ranking of 1.16 from the 10- to 12-year-olds, which comes very close to the ideal 1.0.

The dichotomy between male and female EPs of the 13-16 age group in the ranking of Bernd and Billy rendered an experiment with adult female EPs (n = 33) imperative. In this test (Tables 1/2 and 5), Bernd is once more put before Billy, though not as far as by the 10- to 12-year-olds. Of course, these scant data do not allow quantitative assessment of the significance of this difference between the sexes. It can only be stated that *under the conditions of these experiments* young children and women respond more strongly than adult males to BF and S. When investigated more closely and by more adequate methods, this may prove to be not the only difference between men and women in the way they perceive the stimulus components of the human male's display. Here, I can only indicate the strong probability that, *if* there is a responding innate releasing mechanism, its details are not identical in the male and the female of the species. Table 12 plots the results of all B1 test groups together. Even so, the overall picture is consistent.

2.2.2. *The B2 set.* As outlined above, the individual features eliciting an IRM (sign stimuli) exert their effect through addition (rule of heterogeneous summation), and each may be subject to quantitative variation regardless of the quantitative presentation of the others. This makes it possible to design dummies with a releasing effect stronger than that of any natural object (some or even all sign stimuli quantitatively exaggerated to an impossible degree). The B2 set was to test whether this second principle would work in the case under investigation. For this purpose, the two B1 dummies Bolko and Billy were modified (see Figure 13): Benno was derived from Bolko by drawing a curved line from the shoulder to the groin, creating the optical illusion of broad shoulders and a narrow waist. Benno was further modified to Bertram, in whom the shoulders were broadened excessively. Benno also gave rise to Berthold by removal of the adornment from his shoulders. Barthel was derived from Billy by narrowing the shoulders and extending the waistline.

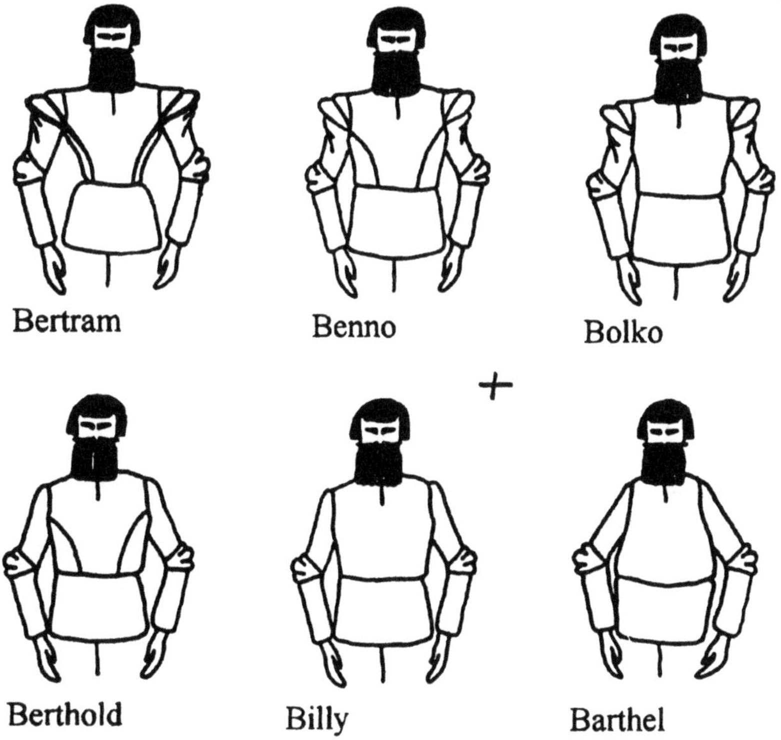

Figure 13. The B2 set

The 6 dummies were given to the EPs with the request: "Arrange these people so that the strongest and best-looking comes first, then the next best and so on." Table 13 proves that the expectation is fulfilled: the broader the shoulders appear, the higher the dummy ranks. Yet the phenomenon again occurs that a number of EPs found Bertram too "blown up" and downgraded him, while at the same time upgrading Bolko. This is discussed below in 2.6.2. However, the broad shoulder is responded to quantitatively. There is also some indication that the feature is really one of proportion – "wide shoulders : narrow waistline". Whether there is an optimal proportion like the one found in young thrushes by Tinbergen and Kuenen (1939) for head/body of the parent must be left unanswered and needs further investigation.

2.2.3. *The B3 set: predictability.* The second essential peculiarity of the stimuli releasing an IRM is interchangeability: the relative eliciting power of each stimulus remains stable whether all or some of the others are present or missing (law of heterogeneous summation). If the results of the B1 tests should indeed rest on this peculiarity, it must be possible to calculate the relative eliciting power of each component stimulus from the ranking of the dummies. This relative stimulus value would, naturally, be confined to the chance stimulus strength given them in the drawings (dummies). This would enable a scale of values of the component stimuli to be obtained and a new test set B3 designed, the dummies representing stimulus combinations not present in any of the B1 set. The outcome of the test(s) could then be predicted quantitatively if heterogeneous summation does indeed exist.

Because the age and sex groups rated some of the dummies differently, the stimulus values were determined separately. The numerical ranking differences between any two dummies may not be interpreted as an exact quantification. The ranking of a dummy depends not only on its immediate neighbors but also on all other members of the set. Even if one presented the dummies in pairs, the result of the comparisons could not be seen as a linear expression of their respective ranks, as even a 100% dominance of one dummy over another could not tell how far superior it really was. As things stood, exact calculation seemed impossible (see, however, sec. 3.1). It was possible to draw up a series of non-equations within each EP group, like the following based on the adult male tests:

Bert	BF+B+S+E+F	>	BF+B+S+--+F	Bodo
Bodo	BF+B+S+--+F	>	BF+B+--+E+--	Birger
Birger	BF+B+--+E+--	>	----+B+S+E+--	Bolko
Bolko	----+B+S+E+--	>	----+B+--+E+--	Billy
Billy	----+B+--+E+--	>	BF+B+S+--+--	Bernd
Bernd	BF+B+S+--+--	>	BF+--+S+E+--	Bastian
Bastian	BF+--+S+E+--	>	BF+--+S+--+--	Balduin
Balduin	BF+--+S+--+--	>	------------------	Balthasar

Bruno is not taken into account because in a "supernormal" dummy the strength of some of the stimuli exceeds what is used as "standard" in this set. Tentatively, the ranking differentials are substituted for the stimulus symbols in the above non-equations. By juggling with these numbers a little, stimulus values were found which, when added up for each dummy, corresponded well with the rankings of Table 1. Table 14 shows the values thus found for each component stimulus for adult males and females, 13-16 and 10-12 age groups respectively. Then the B3 set (Figure 14) was produced, the rank of each dummy determined by adding up the values of Table 14 as pertinent. With heterogeneous summation at work, this should predict the result of the tests with the groups. For obvious reasons there should also be ambivalence in the placing of Beppo and Bosko by the 10- to 12-year-olds, similar to that with regard to Bernd and Bastian in the B1 set, in contrast to the results of the adult EPs. Table 15 presents the predictions and the outcome.

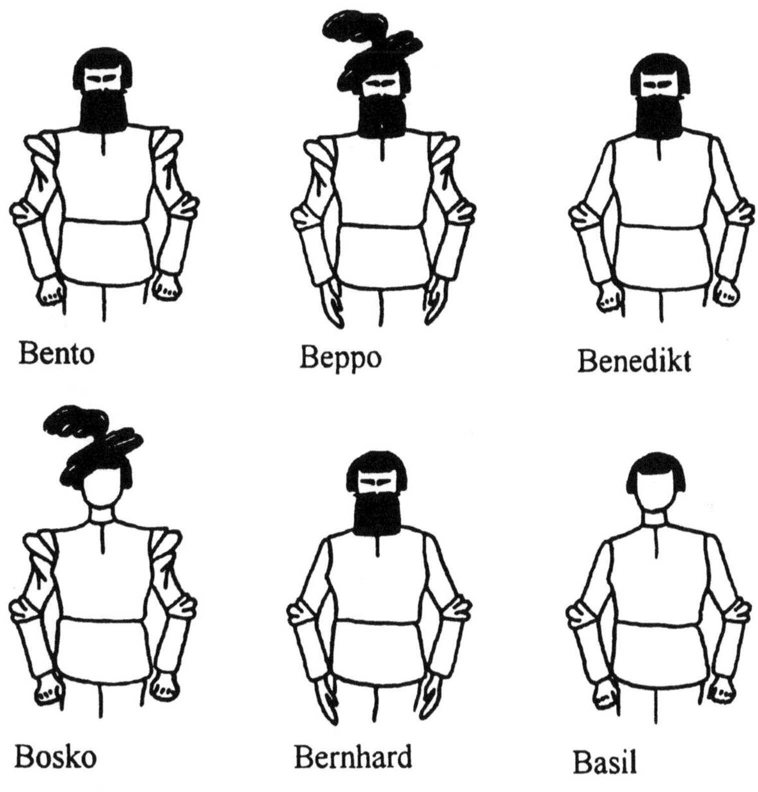

Figure 14. The B3 set

The 10- to 12-year-olds' results are also slightly marred by the younger children's inclination to orient themselves by what might be called a "leading stimulus". In this EP group three dummy sequences were conspicuous:

(a) Beppo-Bosko-Bento-Benedikt-Bernhard-Basil (BF oriented)
(b) Bento-Benedikt-Beppo-Bernhard-Bosko-Basil (B oriented)
(c) Bento-Benedikt-Bosko-Basil-Beppo-Bernhard (F oriented).

This, of course, is a source of error, as the EPs have clearly not followed the instruction to choose quickly and without critical comparison. However, the total group result is remarkably little influenced by it and tallies very well with the prediction.

2.3 The C set

The stimulus feature B(eard) clearly dominated the B1 and B3 sets. To weaken it and so put more emphasis on the other features, I drew a new set of dummies based on medieval armor with closed visor (Figure 15).

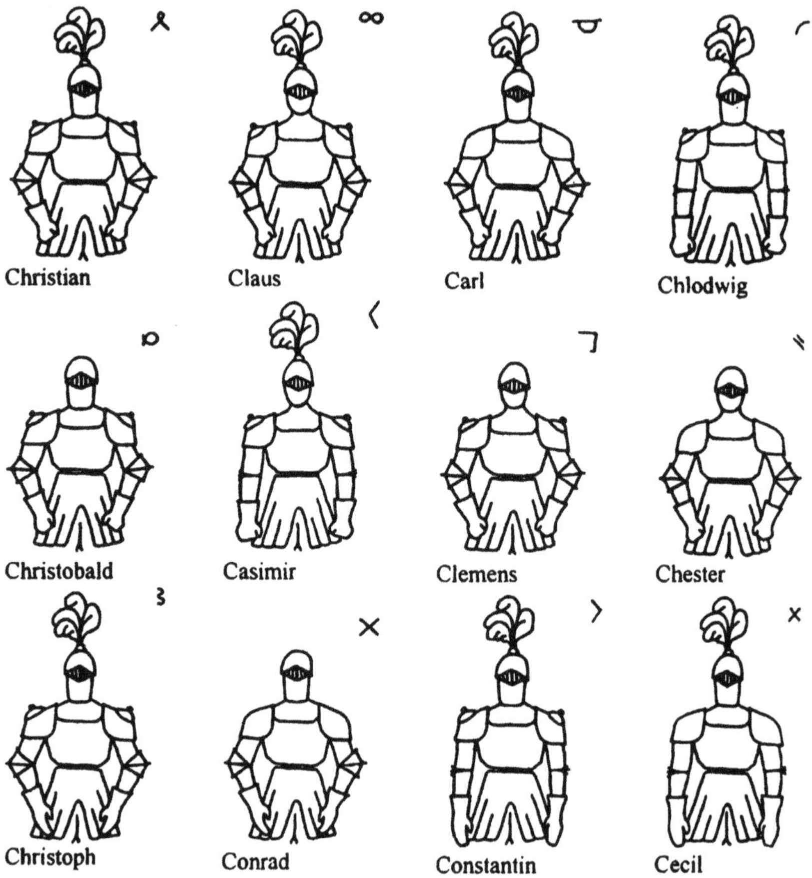

Figure 15. The C set

The stimulus features remained essentially the same as in the B1 set with the exception that no part was blackened. A plume stands for BF, an enlargement of the lower part of the visor for B, a stud on the shoulder for S; E and F are only adapted to the armor but otherwise unchanged. The dummies are:

Christian	BF+B+ S+E+F	Clemens	S+E+F
Claus	BF+ S+E+F	Chester	E+F
Carl	BF+B+ E+F	Christoph	BF+B+ S+E
Christobald	B+ S+E+F	Conrad	B+ E
Chlodwig	BF+B+ S+ F	Constantin	BF+B+ S
Casimir	BF+ S+ F	Cecil	BF+B

The set was tested with adult male EPs only. Some found it difficult to discern at all between what they called "empty shells of armor". Hence, Table 16 presents us with a rather scattered distribution. A ranking, especially for Chlodwig, Christobald and Casimir, would be very uncertain. A considerably larger number of EPs might perhaps rectify this, but that is by no means certain, as dummies whose stimulus totals are close may not achieve significant rank differences. The result, however, is not without interest.

Some of the dummies can be paired off so that the members of a pair differ only in the possession or non-possession of one feature. By plotting the number of times the member of a pair possessing the feature in question is placed higher (+%) or lower (-%) than the non-possessing member we arrive at an assessment of the relative releasing strength of this feature (Table 17). The fact that the proportion of +% : -% remains fairly constant in each group of pairs regardless of the combination of other features the pairs share speaks clearly for heterogeneous summation. Only feature S shows a broader variation, which was to be expected since S provides the weakest stimulus. There may also be other influences causing the phenomenon. This will be discussed together with a similar instance in the results of D set. Apart from this, the constancy of stimulus effect of each feature is all the more remarkable as many EPs observed that the dummies of C set seemed so uniform that they felt unable to arrange them in a definite sequence.

2.4 The D set

It would have seemed logical to test a hypothesis which is based on premises such as evolution, innate behavioral elements and anatomical structures subservient to these behaviors by means of dummies showing the human body without artificial additions and adornments. As mentioned above, this would have made it difficult to emphasize and over-emphasize certain features. But that having been done successfully, we must ask how it is at all possible for artificial additions such as plumes and shoulder padding to stimulate a releasing mechanism that presumably has a much earlier evolutionary history than any clothing or artificial adornment. And while a naked human shoulder, especially in a strong, potentially threatening male, presents a rounded contour to the eye of an opponent, all the drawings of Figures 1 to 11 tend to depict the shoulders as sharp-edged. So a new test set (D) based on the naked male body became necessary.

Again there are 5 stimulus features: (a) BF is replaced by HM: H(air) forming a low, straight, horizontal line on the forehead, knitted eyebrows and a rigidly compressed M(outh). The "neutral" equivalent is a high forehead with wavy hair, arched eyebrows and a mouth curving downwards at the corners. (b) B(eard) remains but in a more natural, "unkempt" form. (c) S(houlder) is drawn in a sharp angle, quite unlike the anatomically correct, rounded contour. However, the width is not greater than in the "negative" dummies with rounded shoulders. (d) E(lbow) as in the previous sets. (e) F(ist) as in the previous sets. For details see Figure 16.

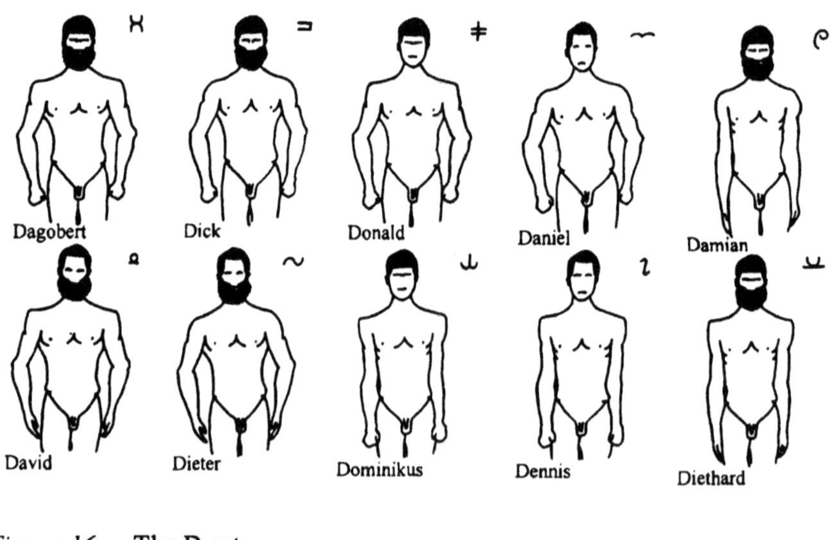

Figure 16. The D set

The features are allocated to the dummies as follows:

Name	*Features*	*R*	*Rank*
Dagobert	HM + B + S + E + F	1.59	1
Dick	HM + B + E + F	2.29	2
Donald	HM + S + E + F	3.69	3
Diethart	HM + B + S	8.45	9
Dominikus	HM + S + F	6.50	7
David	B + S + E	5.71	5
Damian	HM + B	8.76	10
Dennis	S + F	7.43	8
Dieter	B + E	5.83	6
Daniel	E + F	4.71	4

The set was tested with adult male EPs. The result is shown in Table 18. Consistency of stimulus value emerges. Yet, there is also a considerable tendency to arrange the dummies in pairs, particularly evident from the distribution of Dominikus and Dennis (Table 18). This phenomenon was not foreseen when the set was designed and hence not taken care of. With EP n=42, the following pairs were formed:

Pair	*Times*	*Feature difference*
Dagobert-Dick	34	S
David-Dieter	34	S
Diethart-Damian	32	S
Dominikus-Dennis	26	HM
Donald-Daniel	30	HM+S

Of these, the first three are of particular interest because the dummy bearing the anatomically odd feature of the angular shoulder was placed significantly higher in rank than its mate with the "correct" shoulder contour (Table 17). This is highlighted by the remark of EP 52: "Being a doctor, I know, of course, that the angular shoulder is impossible. And yet, looking impartially at the dummies thus disfigured, they are undoubtedly more impressive."

Again we note that the variance between the pairs is relatively large, just as it was in those pairs of the C set which differed only in the possession of S. The fact that S, as shown in the dummies, is the weakest stimulus might appear at least partly responsible for this. Closer inspection, however, reveals that the pairs with the highest total stimulus value show the greatest disparity: Christian/Carl of C set and Dagobert/Dick of D set. The pairs with the lowest overall stimulus value also show the smallest placing difference (C set: Constantin/Cecil; D set:

David/Dieter). One might have expected that the weak stimulus in the company of all the stronger ones would exert less rather than more influence and that it would assert itself better when competing with only few others. But the way it is makes sense: faced with a number of opponents, it is certainly best to concentrate on the one(s) looking most dangerously threatening, and heightened attention renders perception more sensitive to even minor differences. The effect is not confined to the weakest stimulus. Features B and E (C set, Table 17) show it also, although to a lesser degree.

2.5 The F set

A last full-figure test was to decide whether the overall body proportion also plays a part in the spontaneous impression elicited by an opponent. One figure (Franz) displays near ideal proportions of limbs and body. In the other (Fritz), the waistline has been almost imperceptibly lowered so that the torso is just a little too long in proportion to the abdomen and the legs (Figure 17). The dummies were presented to the EPs some 25 cm apart, changing sides irregularly. They were asked, "Which of the

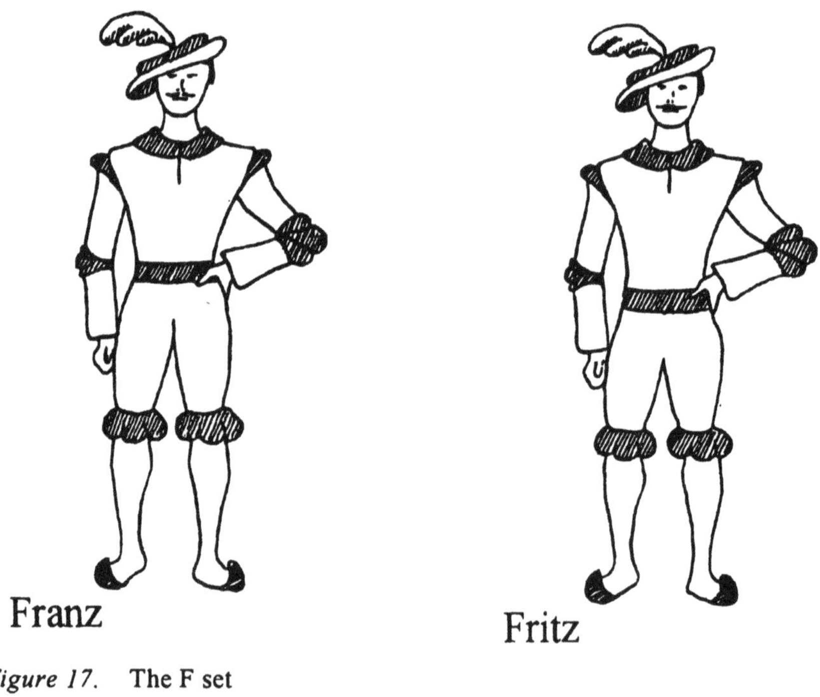

Figure 17. The F set

two looks more imposing?" This question was immediately followed by the order "Quick!" so as to prevent the EPs from recognizing the exact difference between them. This was effective, as the many "no difference" responses prove. In spite of the numerous "undecideds" (20 EPs), Franz (27 EPs) displays a clear advantage over Fritz (17 EPs). It seems certain that the ratio between trunk and underparts influences the impression an opponent creates. But within the scope of this investigation, it was not possible to try and answer the question whether there is an optimum ratio such as the one Tinbergen and Kuenen (1939) found in young thrushes. Their gaping response was elicited most effectively when the parent dummy employed by the experimenters had a head:body diameter ratio of 1:3. Such proportionate ratio features seem to be important in the IRMs of many birds and mammals, as Lorenz (1943) pointed out, but because of the difficulties with which they present the experimenter, they are still virtually unexplored.

2.6 The E sets

So far, the tests were designed almost totally to study the releasing effect of postural and structural features. Facial features received little attention and in some cases were not drawn in. In the B sets beard and eyebrows, in the D set hair, eyebrows and mouth were treated as units. But facial expression indubitably plays an important part in the display of intimidation and threat. As the immediate purpose here is only to test whether the rules of heterogeneous summation and independent quantifiability apply to the perception of facial expressions also, the design of the dummies is once more restricted to a few striking features of intimidation represented in a simplified, schematic way:

H	An angry person wrinkles his (her) forehead; this lowers and more or less straightens the hairline and narrows the expanse between the hair and the eyebrows
EY	Thick, puckered, horizontal eyebrows
FW	Near-vertical furrows on either side of the bridge of the nose
M	Straight, compressed mouth
B	Beard

2.6.1. *The E1 set.* A set of 8 dummies was made (Figure 18):

Ernst		H +	EY +	FW +	M +	B
Edgar			EY +		M +	B
Eugen	*	H +	EY +		M	

Egon	*	H +	EY +		M	
Eduard	**	H +	EY +	FW		
Erwin	**	H +	EY +	FW		
Erich		H +				B
Elmar			EY +		M	

The two pairs Eugen/Egon and Eduard/Erwin share the same basic features but are not identical. In Eugen and Eduard, "neutral" features replaced EY and M: faint, curved eyebrows and a lowered mouth with corners tilted upwards, which was also given to Erich. In Egon and Erwin, however, the features were not replaced, but just omitted (Figure 18).

Figure 18. The E1 set

Gestalt theory claims that our perception should create the whole configuration even when parts of it are missing in the perceived object, or, if the remaining parts are insufficient, the gestalt as such should disintegrate. If this principle were to assert itself in the experiments presented here, gestalt "wholification" ought to occur more readily with Egon and Erwin: their blank spaces would disrupt the configuration of a threatening face less than Eugen and Eduard, whose mouth and eyebrows express a friendly, even cheerful mood diametrically opposed to intimidation, anger and/or threat. But if, in our experiments, the perceptive mechanism works on the principle of heterogeneous summation, the members of a pair should at least achieve identical rank; or, if indeed "black" per se has an intimidating effect, one might even expect Eugen to prove slightly superior to Egon and Eduard to Erwin, as the total area of black in their faces is slightly larger.

The EPs were youths from 10 to 16 years of age. The procedure was the same as with sets C and D. Tables 19 and 20 show the results. Clearly, B(eard) is once more a powerful stimulus item. It is responsible for the ambivalence of Erich, who ranks closest to the two strongest dummies, Ernst and Edgar, and yet occupies the last position 11 times. This is probably due to the more sophisticated approach of the older EPs: the 13-16 years group (Table 19) put Erich last 9 times, the 10-12 years group (Table 20) only twice. A typical remark (EP 15) was: "He looks sinister only because of his beard." Both Eugen and Eduard are superior to their "peers" Egon and Erwin, contrary to the gestalt theory expectation. That this was slightly more pronounced in the younger age group is probably due to a subconscious preference for the "complete" face. The following "eliciting values" were tentatively allocated to the individual features: H = 2, EY = 8, FW = 7, M = 3, B = 16.

2.6.2. *The sets E2 and E3.* To test whether the individual features are indeed valued without regard to the presence or absence of "neutral" features, two more sets of 8 dummies each were made. The somewhat overpowering B(eard) was taken out of the competition by providing all of the E2 set dummies with one, while the E3 set got none. The E2 set was given "neutral" features as described above, the E3 set blank spaces instead (Figure 19 a and b, overleaf). Apart from this, feature distribution is alike in E2 and E3 sets. Thus, the dummies of E3 set could be identified by the first two letters of their E2 opposites:

E2 set	*E3 set*	*features*				*assumed value*		
						E3		*E2*
Erhard	Er		EY +	FW		15	+B(16)	31
Ekbert	Ek	H +	EY +		M	13	+B	29
Enno	En	H +		FW		9	+B	25
Elimar	El		EY			8	+B	24
Emil	Em			FW		7	+B	23
Ewald	Ew	H +			M	5	+B	21
Eberhard	Eb				M	3	+B	19
Esso	Es	H				2	+B	18

Figure 19. (a) The E2 set (Erhard – Eberhard)
(b) the E3 set (Er – Es)

The two sets were tested with the same EPs at an interval of 4 to 5 days. With one exception, the prediction was fulfilled. In the E3 set, El is rated above and not below En (Table 21). However, since the predicted differences between En(no), El(imar) and Em(il), as between Eb(erhard) and Es(so), are extremely small, it is somewhat surprising that there are no more deviations from the prediction. The main result is that the eliciting effect of the individual features has shown itself to be wholly independent both of the presence or absence of "neutral" features and of combination with an overpoweringly strong feature (B in E2 set). When comparing the ranking figures, we note that in the E2 set the ranking differences between the dummies are smaller than in the E3 set. The highest rank is 2.24 and the lowest 6.43, whereas in E3 the figures are 1.42 and 7.42 respectively. This is because the E2 set dummies are all equipped with the powerful feature B, and its effect must, of course, express itself within the available scale of 8 places, leaving proportionately less expressive space to the original four features and making their ranking differences appear relatively small. That same effect, though to a lesser degree, could also be observed in the E1 and D sets. There, the lowest ranking dummies did not, as in the B1 set, lack all stimulus features. Consequently, these dummies ranked "better than nothing" and kept away from the end point of the scale (see Table 1 and column R of Tables 18-20). As has been stated above, the ranking figures are only a relative and not a straightforward, absolute measure of the releasing power of the stimulus combination presented by a dummy. The latter therefore cannot be directly deduced from the former.

2.7 The G set

To test the criterion of independent quantifiability of the individual stimulus features in facial traits also, a further set of 4 dummies was drawn. In these, all the features except one were kept standard (for details see Figure 20 overleaf):

Guido	H + EY +	FW + M,	as in the E1 set
Gernot	H +	FW + M,	EY only half as wide as in Guido
Gert	H +	FW + M,	EY as in Guido but longer
Gustav	H +	FW + M,	EY long as in Gert, but thicker

The set was tested with a group of 39 EPs 12-16 years of age. The result is shown in Table 22. Here it is even more evident than in the B2 set that it is possible to exaggerate a feature well beyond anything that is "natural", yet, far from destroying the effect, proportionately enhance it.

2.8 The GE set

A final test attempted to verify the hypothesis that thickening of the eyebrows enhances the stimulus effect irrespective of their being knitted. The set was made up of Guido and Gernot of G set, Eugen of the E1 set and a new dummy, Gunther, which is equal to Eugen except that its (raised) eyebrows are twice as thick as Eugen's (Figure 21). Guido is predicted to take position number one; whether knitted or thick eyebrows elicit the greater reaction will decide if Gernot or Gunther comes second. If gestalt perception decides the issue, the thickened eyebrows of Gunther ought to disrupt the effect more than the slim ones of Eugen. Table 23 shows the test data with Guido clearly in front and Eugen last. Gunther beats Gernot, but only marginally. Thus it is proven that

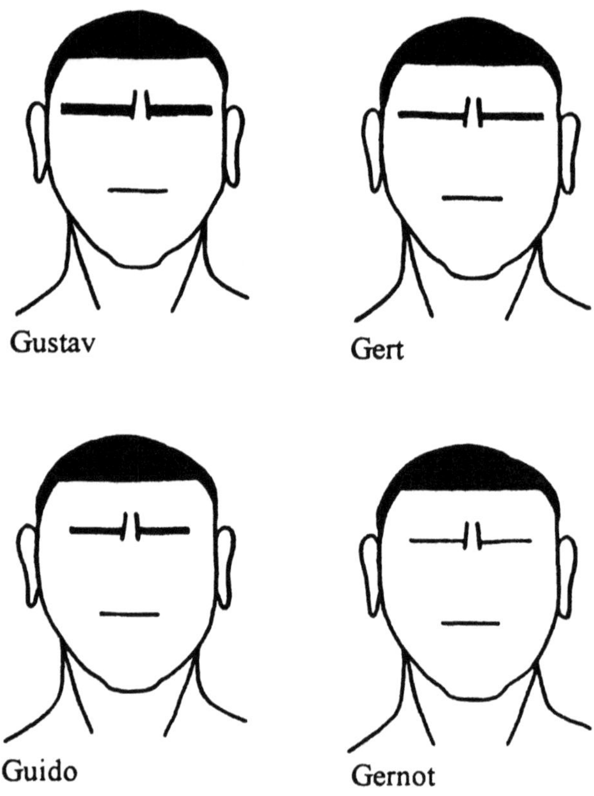

Figure 20. The G set

eyebrow area and eyebrow knitting are independent stimuli. The assumption that in the E1 set Eduard and Eugen were superior to Erwin and Egon because the black areas in their faces are larger is justified.

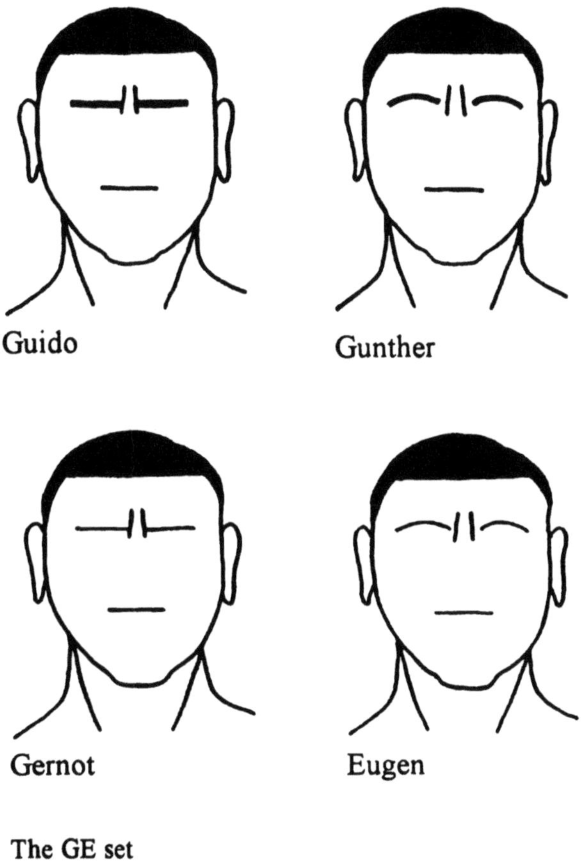

Figure 21. The GE set

3. Discussion

3.1 Reliability

As mentioned before, the method employed was at least partly influenced by the conditions prevailing in a prisoner of war camp, which prevented the use of more advanced technical devices. When the first author (PL) continued this work after his release, the need to ensure the compatibility of the results collected in captivity with those obtained later precluded any change in procedure. Correspondingly, the statistical analysis presented is not a complete one, but restricts itself to a few relatively simple considerations. The phenomena we are looking for should be obvious without refined statistics if an IRM is the dominant factor in the experiments.

The experimental results consist of individual rank orders for dummies. We shall use slightly contracted results, namely the rank order distributions for each dummy as shown, for example, in Table 2. Each rank i occurs m_i times. From these we derive

(a) the median, M, defined by having an equal number of ranks smaller and larger than the median. Since individual rank numbers are integers, the median is *not* uniquely determined.
(b) the rank number R = average rank, defined by $R = \Sigma_i \, i \, m_i / m$ where $m = \Sigma_i \, m_i$ is the number of EPs.
(c) the width W of the distribution (estimated variance), defined by $W^2 = \Sigma_i \, m_i \, (i-R)^2 / (m-1)$;
(d) the estimated uncertainty E of R (mean error of mean), defined by $E^2 = W^2 / m$.

These numbers are given in those Tables that contain rank order distributions (e.g., Table 2).

The implications of our hypothesis that the rule of heterogeneous summation applies are as follows: For each experiment, each stimulus has a certain (average) strength that is independent of all other stimuli. The average response of the EPs is proportional to the sum of these strengths. If our hypothesis makes it possible to predict the average response to each dummy within a reasonably small margin of error, the existence of an IRM is proven in the sense that it does not contradict the experimental results. Some errors in the predictions (i.e., uncertainties in the average responses to individual stimuli) are to be expected. The main reason is that - as is well known of IRMs - every EP responds differently to the stimuli. Some discrepancies between a prediction from average responses and the average outcome are only natural.

The possibility that other hypotheses might reproduce the data as well cannot be ruled out. However, an IRM is complicated enough to make the probability of finding another explanation with comparably few parameters small. Thus, the problem is to find strengths of stimuli which, when summed up for each dummy, reproduce the rank numbers. In fact, the rank numbers represent only a rough approximation of the actual total strength of the combined stimuli. The experiment allows only consecutive integers for the individual rank positions. Sometimes,

EPs might have preferred to give two adjacent dummies almost the same rank, or they would have made the rank difference larger than 1, but were unable to do so. More generally, the difference between two dummies might be exaggerated because the experimental minimum is 1, or larger if other dummies rank in between; or the difference might be diminished because there is a maximum that is equal to the number of stimuli minus 1, or smaller if there are not enough dummies that rank in between. A certain function of the rank numbers might, therefore, be a better approximation to the response than the numbers themselves.

For each dummy we need an equation of the form

$$\text{sum of strengths of stimuli} = \text{function of rank number}$$

to be obeyed with sufficient accuracy. In reality it is, for practical reasons, more convenient to use differences between two such equations, i.e., to compare all pairs of dummies. In all cases in this paper, we have more dummies than features, i.e., we have more equations than parameters to calculate. Thus, it is not generally possible for each equation to be fulfilled exactly: an optimum result is found by a fit (regression). In the case of linear equations, such as we have here, a good choice is a least-squares-fit (LSF), which gives a formally *unique* result for the relative strengths of the stimuli once the above-mentioned function of the rank numbers is chosen. We give examples in Tables 12, 14, 15, and 16.

Even if the hypothesis of the IRM were completely wrong, the LSF would still give a *unique* result. It would, however, be an unreasonable result in at least one of three respects: (1) The strengths would prove to have unreasonable values, for instance a large negative value for a stimulus which is known (e.g., from another similar experiment) to act positively on EPs. (2) The original equations as described above would not be well reproduced. The deviations would have large absolute values, and only their average would be zero. Since average rank number differences are of the order of a small integer, it will suffice, for a "good" result, to reproduce them within an error definitely smaller than 1. (3) The result of the LSF would depend strongly on the experimental input, so that a repetition of the same experiment would produce completely different results. One method to check this is to divide the experimental results into two groups. We did not do this but instead compared different but similar experiments, for instance the B1 series results of males and females. Also, changing the function of the rank numbers which is chosen as input would appreciably change the outcome.

It is therefore reasonable to try different functions. As rank number functions we have used (a) the difference in the rank numbers, (b) a reduced difference, depending on the overlap between the two rank distributions considered, (c) unity with the sign of the difference (i.e., we only ask whether the rank number of one dummy is larger or smaller than that of the compared dummy), and (d) the value of (c) reduced in the same way as the difference was reduced in (b).

The outcome of all experiments proves to be reasonable. Negative strengths do not occur except with weak stimuli, where a small negative result is within the expected uncertainty. The original equations are reproduced well within unity, as verified by the average squared residuals (Tables 12, 16). Also, the different functions (a-d) give similar results, as do comparable experiments (B1, different groups, Table 14). In other words, the hypothesis of an IRM is well corroborated. Nevertheless, even good results are not necessarily exact. They should be taken for granted only within a certain margin of error. The diversity of the results found for the different rank number functions (a) to (d) gives a rough idea of the uncertainty. A thorough calculation of expected errors was not made as it would not have been worthwhile: The aim of our calculations is not to find the exact strengths of certain stimuli but rather to show that the underlying hypothesis of an IRM is a sound one.

Further proof of the reasonability of the hypothesis comes from a comparison between pairs of dummies that differ only in one stimulus. These order experiments are not statistical experiments in the usual sense, where the outcome depends only on a given probability, but they are effectively similar to such experiments. The two order possibilities are represented by two outcome numbers. In a true statistical experiment (such as the tossing of coins), Poisson statistics applies, and these numbers have uncertainties roughly equaling their square roots. The corresponding uncertainty e_1 of the ratio (in %) is expressed by

$$e_1^2 = (+\%) * (-\%) / m$$

where m is the number of casts (number of EPs * number of pairs). If the IRM hypothesis is correct, the outcome should depend little on additional stimuli, or on which pair is used, i.e., all pairs should produce roughly the same result. This can be checked by deriving an expected error e_2 from a comparison of different pairs, expressed by

$$e_2^2 = \Sigma\{(+\%)\text{-average}(+\%)\}^2 / n(n-1)$$

where n is the number of pairs. The same result is obtained if -% are used. The formula is only a rough approximation if n is small. The IRM prediction is that $e_1 \approx e_2$. We have checked this prediction with the data of Table 17 and find

(*series/stimulus*)	e_1	e_2
C/BF:	4.0	0
C/B :	3.8	3.6
C/S :	4.2	6.5
C/E :	3.0	1.4
C/F :	1.3	1.3
D/S :	4.3	4.2

There is a statistical possibility - which is, however, nonsignificant - that e_2 may prove much smaller than e_1 (as indeed occurs in C/BF). If e_2 were much larger than e_1, this would contradict the IRM hypothesis. In the only case occurring (C/S), the result is, however, well within the statistical expectation.

Another way of checking the data uses the assumption that the response to a stimulus follows (approximately) a Gaussian distribution. Let a sensory example illustrate the underlying philosophy. It is well known that the human ear has a limited ability to distinguish different tones. Two tones differing sufficiently little are heard as just one, perhaps giving some impression of roughness. Two tones are heard only if the difference is sufficiently large. Even then, it is often difficult to decide which tone is higher. If two instruments are playing together – particularly effective with a string instrument and a wood instrument – some EPs might find one instrument tuned higher, some find no difference, while others might even have the impression that that same instrument is tuned lower. In other words, even the direction or sign of the response to a stimulus may be uncertain.

In our case, pairs of dummies differing in one stimulus are experimentally placed in either order. The stimulus evokes a certain average response. It is reasonable to assume that most EPs will show a response not too far from the average, about half a weaker one, the other half a stronger one. Greater deviations will probably occur less often, and very large deviations may be expected to be very rare. Such distribution will look rather similar to a Gauss curve. It will certainly not coincide exactly, but let us assume that a Gauss curve is a good approximation, which means that all the usual formulae of Gaussian statistics can be used. There will be EPs with no or practically no response, and even some with a negative response. Cases in which the order of two dummies has been reversed as compared to the average can be interpreted as showing a negative response or, more strictly speaking perhaps, an effective negative response. It would be wrong to say that a stimulus is effective if a dummy possessing it is placed before another not possessing it, but ineffective if the order is reversed. One should rather say that, in the first case, the stimulus is (effectively) positive and in the second (effectively) negative.

As an example, let us take C/BF Christian-Christobald (40 EPs) from Table 17. The ratio is 85%:15%, that is, 34:6 EPs. The response to BF is positive for 34 EPs and negative for 6 EPs. If the numbers are considered to be proportional to the positive and negative wing of a Gauss curve with a maximum at the average response r, then one can find r in relation to the variance s^2 of the Gauss curve. Let us assume that s is roughly constant irrespective of the strength of the stimulus. This would have the plausible consequence that with strong stimuli a negative response is rare but with weak stimuli a negative response occurs much more often. Applying this to the five stimuli of set C (Table 17) and comparing it to the computed average responses r^* to the features (Table 16), we find

stimulus	*ratio*	*r/s*	*r*	r^*
C/BF	85:15 ±4.0	1.04±.16	9.2±1.4	9.7±0.5
C/B	78:22 ±3.8	0.77±.14	6.8±1.2	6.5±0.6
C/S	69:31 ±4.2	0.47±.11	4.2±1.0	2.3±0.3
C/E	87:13 ±3.0	1.13±.15	10.0±1.3	11.9±1.2
C/F	99: 1 ±1.3	2.25±.35	19.9±3	19.5±1.3

where the values in column 4 are those of column 3 multiplied by the best guess of s, namely, the average ratio (=8.8) of columns 5 and 3. There is near-perfect agreement between the strengths of the stimuli in columns 4 and 5. This may be taken as supporting our assumptions of near-Gaussian response distribution, of a near-constant variance of it, and of the IRM hypothesis. Interestingly, the total width of the response distribution (about $2s \approx 17.6$) is about as large as the strongest average response, r=19.9, to a stimulus in the C series (F). We take this as proof that the widths of the rank order distributions can be largely explained within the IRM hypothesis by the spreads in the individual responses. That the errors derived in the fourth column are larger than those in the fifth column can be taken as an indication that the errors estimated from different rank number functions (a) to (d) (Table 16) may be consistently too low. If one required better error estimates, which we do not, a more complete statistical analysis would be necessary.

If the response to a stimulus may be positive or negative, one can expect that, if experimental conditions allow, the response may also be zero or so near to zero that no response is recognizable. In terms of ranking orders, this would mean that EPs would sometimes consider the dummies alike and not take a decision as to the order. There is only one such experiment, the F series. Its result is: (a) 27 EPs prefer Franz, (b) 20 have no preference, and (c) 17 prefer Fritz.

Let us begin with a qualitative explanation to illustrate the numbers given below. The stimulus Franz possesses as compared to Fritz is weak since (a) and (c) differ very little. The range of neutrality is rather large - nearly one-third. Since Franz is preferred in less than half the cases, the median of the distribution (coinciding with the average and with the maximum in a Gauss distribution) lies inside the neutrality range, i.e., the strength of the stimulus is smaller than the half-range of neutrality.

Quantitatively, we assume that the response to the stimulus is again distributed like a Gauss curve with average r and variance s^2. The height of the curve is chosen so that the total area is 64, the number of EPs. The curve can be divided into 3 parts so that the left area is 17, the middle one 20, and the right one 27. The dividing lines lie (slightly approximated) at positions -0.6s and +0.2s as counted from the maximum. Thus it follows that the neutrality range is 0.8s. If we arrange this symmetrically around zero, the dividing lines are positioned at -0.4s and +0.4s. The maximum, corresponding to the average strength of the stimulus, thus lies at +0.2s, i.e., within the neutrality range: Franz is preferred only by EPs who respond more strongly than average to the stimulus.

The result cannot be used to support the IRM hypothesis directly because there is only one stimulus, but it again indicates that the observed width of the rank order distributions is due to the natural spread of the IRM responses and not so much to EPs whose responses are elicited differently.

3.2 Are the criteria used sufficient to prove the existence of an IRM?

As mentioned in the introduction, several authors felt they had proved that the composite stimuli of a learned situation or object were also capable of exerting a summative effect. While not discussing this here in detail, we would only mention that none of these authors performed precise control experiments to prove heterogeneous summation and independent quantifiability and predict the results quantitatively. It must, however, be conceded that relatively few animal IRMs have been soundly analyzed. But so far, no IRM has been found that, when properly investigated, did not prove to function according to the principles of heterogeneous summation and independent quantifiability. This applies to a number of invertebrates as well as to representatives of all classes of vertebrates. Thus, although the basis of inference is still far from universal, it appears fairly trustworthy.

The two "supernormal" dummies, Bruno and Bertram, though top ranking in their respective sets, did not achieve the overwhelming superiority they were expected to show. One might argue that this failure renders the IRM hypothesis doubtful. However, remarks such as "He looks overdone" or "He could perhaps intimidate children" indicate that some EPs quite consciously put these dummies farther down the scale *after* first having spontaneously put them uppermost. It is characteristic of these cases that the subsequent downgrading is by not just one but several positions. The phenomenon does not invalidate the IRM hypothesis, but throws some light on the way experience and IRM collaborate. The former does not alter the latter or become incorporated in it. It establishes an additional filter which the stimuli have to pass before they can reach the IRM. What our EPs experienced was in fact a mild case of "spectre reaction" (*Gespenstreaktion*, Seitz 1940; see Leyhausen 1967, 1973a). This interpretation is supported by the fact that the effect is strongest in the adult EPs and disappears completely in the 10- to 12-year-olds: from them, Bruno obtains nearly the ideal ranking number of 1 (cf. Table 1/1-6).

3.3 The significance of artifacts

We have already remarked on how strange it is that artificial additions to the human body such as items of clothing or head-gear should arouse some innate perceptive mechanism that functions prior to and independently of any individual experience. And is it not astonishing and indeed paradoxical that features such as a plumed beret (BF) or a beard (B) can have greater eliciting power than clenched fists (F)? For a man who gets into a rage and is about to attack another human normally clenches his fists, but does not stop to don a beret, let alone grow a beard.

It is common in all but very few vertebrates to enhance the contour of the body when displaying to intimidate, threaten, or court a conspecific. The physiological and structural mechanisms employed to achieve the illusion of (sudden and) extraordinary increase in body size are as diverse as the animal species concerned. Humans bind up the hair of their head or cement it into a towerlike structure with loam, or even cow dung and urine. They wear feather crowns, plumed helmets, high hats, or whatever, all to create the impression of a tall figure. But it is not that alone, for even on stilts they still do not leave the head without some heightening device. Head-gear of this kind is common to all populations and periods. Equally widespread is the fashion of making the shoulders appear broad and angular by means of padded coats, epaulets, and other adornments and draperies, the latter even in tribes who mostly go naked. Other kinds of draperies or adornments, such as crossbelts, serve to emphasize broad shoulders and a narrow waist just as we did in the B2 set. All this can be easily observed in the drawings made by the EPs (Figures 1 to 11). Even the "supernormal" exaggerations in some of the drawings, although intended to appear ridiculous, retain their impressiveness at first glance. Figures 1, 2, 3 also show the proverbial "hair standing on end" on the head. There are indeed some rare people whose hair will visibly stand on end when they are excited, but the rest of us have lost the ability although we still experience the feeling. There must, however, have been a period in our evolutionary history when the scalp of an excited human quite normally rose, as it still does in the anthropoids. They can also erect the hair on their neck, back, shoulders and arms.

There can be little doubt that, even as late as the Aurignacian, the human body was still covered with at least terminal hair (Kuehn 1936). Speculations that the prehominid ancestors of humans had already become hairless, as in the Aquatic Ape Theory (Hardy 1960), are ill-founded (Leyhausen 1991). However, in monkeys and apes the lie of the hair runs evenly from the head down over the neck and shoulders. When erected, it certainly enhances the contour, but does not make it appear sharp-edged or angular. The exception is the gorilla. On the outer side of its shoulder, the hair forms a whorl, barely noticeable in most females, but prominent in adult males (Figure 22, overleaf). In typical cases, the feature makes the hair stand out from the shoulder like an epaulet (Figure 22 b, c). The older anthropological literature contains some data on excessive hair covering the face and body in humans (Hypertrichosis lanuginosa and H. terminalis), comparing it with the structure, distribution, and lie of the hair in monkeys and apes, but as far as we could establish it contributes nothing to the point being discussed here.

Therefore, I (PL) inspected my fellow prisoners of war at a health parade and found 12 specimens with well developed, sleek terminal hair on the neck, shoulders, back, and breast. Men with curly hair were excluded because it is difficult to trace the true lie of curly hair with accuracy. I asked the twelve selected for permission to chart their torso hair. By copying the 12 charts into one (one over the other), a complete picture of the hair covering the human torso was obtained (Figure 23, overleaf).

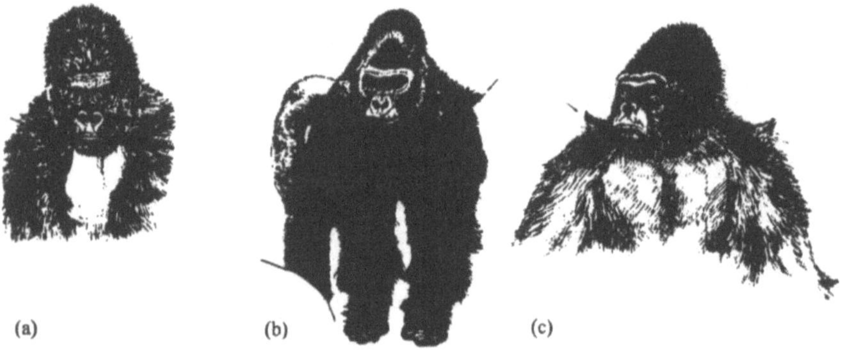

Figure 22. Lie of the hair in gorillas: (a) young adult male (National Zoo, Washington D.C.), the whorl at the shoulder is not noticeable; (b) adult male with moderate "epaulet" (Zoo Munich); (c) adult male with very well developed "epaulets" (Inuyama Monkey Centre, Japan). Drawings from color slides by M. Pfleiderer.

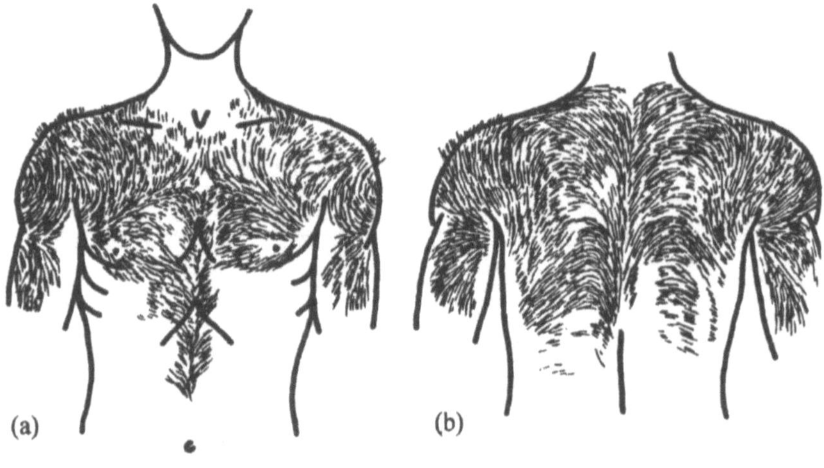

Figure 23. The hair on the torso of 12 adult human males superimposed on each other: (a) chest, (b) back.

While there was considerable variation and asymmetry on the chest (which cannot be adequately shown in a single drawing), there was practically none on the neck, back and shoulders. Checking the vestigial terminal hair and down (lanugo) of the other camp inmates (n > 500) very closely, I could not find a single even slight deviation from the pattern shown in Figure 23. Later, I was able to inspect many more people, both men and women, and have still found no deviation. Considering the extreme variability of *Homo sapiens* in general, such invariance must certainly appear remarkable. It required a very powerful selective pressure to fixate the pattern so firmly and narrowly. Therefore, we feel justified in assuming that the drawing represents a true picture of the way the hair lay over the torso of our not yet hairless ancestors.

Let us now look at it in detail. Viewed from the rear, the difference from the anthropoid pattern, even that of the gorilla, is immediately apparent. The stream of hair springs from the armpit and then parts, one part descending to the elbow, the other ascending the *Musculus deltoideus* to the top of the shoulder where, in a wide U-turn, it joins the stream from the opposite side to flow down the middle of the back. In front, the stream circling the arm also climbs the shoulder, there to meet the hair coming up the back and form a kind of crest caused by the streams from front and back straining against each other (Figure 24).

Figure 24. Maldhari (herder) from the Gir Sanctuary in Gujarat (India); note the little hair crest on the left shoulder. Drawings from color slides by M. Pfleiderer.

Reconstructions of early hominids are usually completely or almost hairless. There are mainly two reasons for this: (1) The anatomist inspiring the reconstruction is interested in demonstrating the details of bone structure and musculature, which would mostly be hidden by covering hair. (2) There is a conscious or subconscious desire that our early ancestors should not present too apelike an appearance. If a reconstruction of this kind is decked with hair in the manner depicted in Figure 23, it looks as in Figure 25a. Supposing the piloerectors were still working in this specimen of early *Homo*, he would probably look like Figure 25b when in a temper. The contour of the shoulder would be exactly like the angular shape we found effective in our experiments.

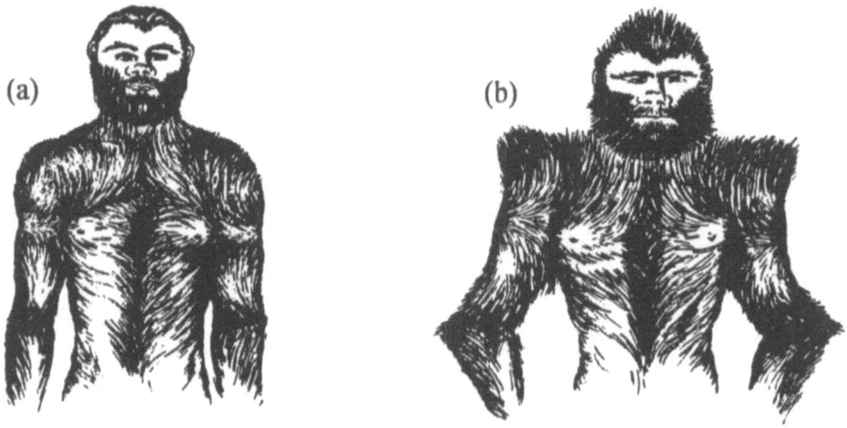

Figure 25. Reconstruction of the hair covering the body of our early hominid ancestors

The age-old and ubiquitous predilection of fashion designers for broad and unnaturally angular shoulder contours cannot be dismissed as insignificant. Even the fact that it waxes and wanes periodically does not invalidate the conclusion that it must have its origin in some kind of releaser man has lost along with his pelt. But the corresponding IRM survived unimpaired, remained functional and active. Thus, while *Homo* became the Naked Ape, in the same process he became the Clothed Ape. The IRM, once established, is not just a passive registering instrument: it "requires" the social partner to present certain specific stimulus conditions and so instigated, indeed forced the development or design of substitutes when the releasing stimuli originally provided by nature got lost (Leyhausen 1983). As is the case with many other products of culture (Leyhausen 1973b), head-gear and shoulder wear of the kind described are by no means "free inventions", but grow out of a motivational system and a behavioral equipment far older than any culture (Leyhausen 1989).

3.4 Broad shoulders and narrow waist - why?

The hair stream pattern on the human neck and shoulders is undoubtedly a specialization that owes its evolution, if not solely at least largely, to the selective pressure of an innate releasing mechanism. But what could be the selective advantage? As the relevant part of human evolutionary history is still in the dark, we can only speculate, but we believe that there is a very good case for the assumption that the interaction of the releaser and the releasing mechanism discussed here played a role in the evolution of walking erect.

The trunk of an anthropoid ape does not go in at the waist and its shoulders are hardly broader than the hips (pelvis). It is a very compact, stiff-rumped body. In a brachiator, the center of gravity is far below the point of support, and its equilibrium is stable. An anthropoid on the ground usually walks on his hindfeet and the knuckles of his hands. On four points of support he is still fairly stable. When, however, he walks erect on his hindfeet, the point of gravity is high off the ground, the support area small, and his equilibrium becomes unstable. To maintain it, he uses his arms much in the way a tight-rope artist uses a pole. Human babies in their first attempts at erect walking do the same, and for the same reason: their waist is still rather stiff. A supple waist allows the upper part of the trunk to move fairly freely in relation to the pelvic region. This renders it possible to shift the center of gravity quickly and keep it perpendicular to the small supporting foot area. At the same time, the arms need no longer serve as a balancing device and are thus freed to perform other tasks, such as the transport of goods or the use of tools or weapons. Hence, a releasing mechanism that helps to breed broad shoulders and narrow waists would be of considerable selective advantage to an erect walking ape or prehominid. A releaser that emphasized these features would presumably influence sexual selection and help to evolve them further. We are convinced that the structures and behaviors investigated in this paper owe their origin to these forms of natural selection.

4. Conclusions

The display of vertebrates normally has three phases: a "neutral" prelude develops into either courtship or intimidation and fighting, depending on the current mood of the displaying animal and the reaction of the conspecific encountered. The prelude display shows some traits that are similar throughout the tribe of vertebrates. Invariably, they present the form of the displaying animal as big as possible, often literally "blown up", to the conspecific opponent. The structures and postures are different according to whether the opponent is met frontally or laterally. Fishes spread their fins maximally or open wide their opercula and/or their interstegal membranes. Some, like many amphibia, inflate themselves. Reptiles extend skin folds or other structures on head, neck, back and tail. Birds erect feathers on head, neck or tail, and mammals special fur areas. These are but a few examples of the almost innumerable devices that all serve the same end: to appear big and impressive to the other animal.

Humans are no exception. They erect themselves as best they can. The diversity of the means employed strongly suggests that they cannot be homologous. But their unmistakable analogy suggests equally strongly that the perceptive function they stimulate must indeed be very ancient and certainly homologous from fish to man. As this basic IRM was evidently able to force the selective evolution of the most diverse and yet convergent structures and displays, one cannot wonder at its also instigating the development of analogous artifacts.

The skeptic who is still unconvinced that our experiments are proof of the existence of a veritable IRM in humans will perhaps present us with some other explanation for the fact that the results of our experiments forced us to postulate a hair pattern in humans very different from that of all other primates and so far not adequately described by physical anthropologists, and that these results led to the discovery of exactly such a pattern.

It has often been said that it is futile to discuss the roles innate and acquired traits may play in adult human behavior because they are so intricately interwoven that any attempt at analyzing them separately must fail. Here, we claim, is at least one way that allows us to do just this, though in a limited field of perception only. But it sparks our hope that it will be possible to find adequate methods to tackle the problem successfully in other areas of human behavior as well.

Appendix: Tables 1-23

Table 1. Ranking of B1 set.

		Bruno	Bert	Bodo	Bolka	Birger	Billy	Bernd	Bastian	Balduin	Balthasar	Bodmar
1a adult males	test 1 + Bodmar	1.95	2.80	3.55	4.80	5.43	5.62	6.98	8.07	9.45	10.65	6.70
1b	test 1 - Bodmar	1.88	2.70	3.32	4.53	5.12	5.25	6.45	7.38	8.65	9.72	
1c	test 2	1.54	2.67	3.02	4.82	4.19	6.00	6.56	7.30	8.93	9.96	
2 adult females		1.73	2.91	3.00	4.91	5.12	6.91	5.88	7.18	7.97	9.39	
3 boys, total		1.26	2.36	3.45	5.42	5.09	6.28	5.87	6.87	8.60	9.79	
4 girls, total		1.22	2.72	3.64	5.44	4.91	6.63	5.63	7.14	8.03	9.66	
5 boys+girls, 10-12 yrs.		1.16	2.48	3.46	5.72	5.04	6.91	5.60	6.88	8.12	9.63	
6 boys+girls, 13-16 yrs.		1.34	2.66	3.68	5.04	4.92	5.88	5.92	7.20	8.52	9.84	
7 boys, 13-16 years		1.48	2.33	3.57	4.95	5.05	5.62	6.33	6.95	8.95	9.76	
8 girls, 13-16 years		1.24	2.90	3.76	5.10	4.83	6.07	5.62	7.38	8.21	9.90	

Table 2. B1 set, test 1 (with Bodmar), adult males. n = 60 EPs; cf. Table 1/1a. ** = supernormal features. For the definition of M (=median), R (= rank number = average rank), W (=width of rank distribution), and E (= error of R = mean error of mean) cf. sec. 3.1.

Features **BF B S E F CB	Dummies	Rank distribution											M	R	W	E
1 1 1 1 1 1 1	Bruno	41	9	4	0	2	0	1	1	0	1	1	1.2	2.0	2.1	0.27
0 1 1 1 1 1 1	Bert	4	31	15	3	2	2	2	0	1	0	0	2.3	2.8	1.5	0.20
0 1 1 1 0 1 0	Bodo	10	7	22	6	4	3	3	5	0	0	0	3.1	3.5	2.0	0.26
0 0 1 1 1 0 0	Bolko	1	4	7	18	13	5	8	2	1	1	0	4.5	4.8	1.8	0.23
0 1 1 0 1 0 0	Birger	0	1	5	10	15	16	8	4	0	1	0	5.4	5.4	1.5	0.19
0 0 1 0 1 0 0	Billy	2	2	2	11	8	16	8	10	1	0	0	5.8	5.6	1.8	0.24
0 0 1 1 0 0 1	Bodmar	0	4	4	4	8	5	8	15	4	4	4	7.1	6.7	2.5	0.32
0 1 1 1 0 0 0	Bernd	1	0	1	7	5	6	12	17	5	5	1	7.3	7.0	2.0	0.26
0 1 0 1 1 0 0	Bastian	0	1	0	0	3	7	9	4	31	4	1	8.7	8.1	1.6	0.21
0 1 0 1 0 0 0	Balduin	1	1	0	0	0	0	1	1	15	37	4	9.8	9.4	1.6	0.21
0 0 0 0 0 0 0	Balthasar	0	0	0	1	0	0	0	1	2	7	49	10.9	10.6	1.0	0.14

Table 3. B1 set. Distribution test 1 (without Bodmar); adult males. n = 60 EPs; cf. Table 1/1b. ** = supernormal features.

Features **BF B S E F	Dummies	Rank distribution										Median	R	W	E
1 1 1 1 1 1	Bruno	41	10	3	0	2	0	2	0	1	1	1.2	1.9	1.9	0.25
0 1 1 1 1 1	Bert	4	31	17	1	3	3	0	1	0	0	2.3	2.7	1.3	0.17
0 1 1 1 0 1	Bodo	10	7	24	7	2	5	5	0	0	0	3.0	3.3	1.7	0.22
0 0 1 1 1 0	Bolko	1	6	6	20	11	9	5	1	1	0	4.3	4.5	1.6	0.21
0 1 1 0 1 0	Birger	0	1	5	14	16	16	7	0	1	0	5.1	5.1	1.3	0.17
0 0 1 0 1 0	Billy	2	2	3	11	13	15	11	3	0	0	5.4	5.3	1.6	0.21
0 1 1 1 0 0	Bernd	1	1	1	7	7	7	20	10	4	2	6.8	6.4	1.8	0.24
0 1 0 1 1 0	Bastian	0	1	0	0	6	5	9	34	4	1	7.8	7.4	1.3	0.17
0 1 0 1 0 0	Balduin	1	1	0	0	0	0	1	8	45	4	8.9	8.6	1.4	0.18
0 0 0 0 0 0	Balthasar	0	0	1	0	0	0	0	3	4	52	9.9	9.7	1.0	0.13

Table 4. B1 set, test 2; adult males. n=57 EPs; cf. Table 1/1c. ** = supernormal features.

Features **BF B S E F	Dummies	Rank distribution										Median	R	W	E
1 1 1 1 1 1	Bruno	41	10	2	1	1	2	0	0	0	0	1.2	1.5	1.2	0.15
0 1 1 1 1 1	Bert	1	30	20	2	2	1	1	0	0	0	2.4	2.7	1.0	0.14
0 1 1 1 0 1	Bodo	11	10	22	4	3	5	2	0	0	0	2.8	3.0	1.6	0.21
0 1 1 0 1 0	Birger	2	2	6	28	12	6	1	0	0	0	4.2	4.2	1.1	0.15
0 0 1 1 1 0	Bolko	1	3	5	11	19	13	4	1	0	0	4.9	4.8	1.4	0.18
0 0 1 0 1 0	Billy	1	1	2	2	9	19	19	3	1	0	6.2	6.0	1.4	0.19
0 1 1 1 0 0	Bernd	0	0	0	7	9	6	17	16	2	0	6.9	6.6	1.4	0.19
0 1 0 1 1 0	Bastian	0	1	0	2	2	5	13	32	1	1	7.7	7.3	1.3	0.17
0 1 0 1 0 0	Balduin	0	0	0	0	0	0	0	4	53	0	9.0	8.9	0.3	0.03
0 0 0 0 0 0	Balthasar	0	0	0	0	0	0	0	1	0	56	10.0	10.0	0.3	0.03

Table 5. B1 set. Adult females. n=33 EPs; cf. Table 1/2. ** = supernormal features.

Features **BF B S E F	Dummies	Rank distribution										Median	R	W	E
1 1 1 1 1 1	Bruno	18	9	4	1	1	0	0	0	0	0	1.4	1.7	1.0	0.17
0 1 1 1 1 1	Bert	4	11	8	7	2	0	0	1	0	0	2.7	2.9	1.4	0.25
0 1 1 1 0 1	Bodo	7	7	11	2	1	3	2	0	0	0	2.7	3.0	1.7	0.30
0 1 1 0 1 0	Birger	0	0	4	10	8	7	1	0	2	1	4.8	5.1	1.7	0.29
0 0 1 1 1 0	Bolko	3	2	1	6	10	5	1	4	1	0	4.9	4.9	2.0	0.35
0 1 1 1 0 0	Bernd	0	1	3	4	3	10	7	3	1	1	6.1	5.9	1.8	0.31
0 0 1 0 1 0	Billy	0	1	1	1	3	4	12	5	4	2	7.0	6.9	1.8	0.31
0 1 0 1 1 0	Bastian	1	0	0	1	4	3	5	13	5	1	7.7	7.2	1.8	0.31
0 1 0 1 0 0	Balduin	0	2	0	1	1	1	3	4	19	2	8.7	8.0	2.0	0.35
0 0 0 0 0 0	Balthasar	0	0	1	0	0	0	2	3	1	26	9.9	9.4	1.4	0.25

Table 6. B1 set, boys. n = 53 EPs; cf. Table 1/3.

Features **BF B S E F	Dummies	Rank distribution										Median	R	W	E
1 1 1 1 1 1	Bruno	47	3	2	0	0	0	0	1	0	0	1.1	1.3	1.0	0.14
0 1 1 1 1 1	Bert	3	41	4	1	2	0	2	0	0	0	2.1	2.4	1.2	0.16
0 1 1 1 0 1	Bodo	0	2	36	9	3	1	2	0	0	0	3.2	3.5	1.0	0.14
0 1 1 0 1 0	Birger	0	0	3	20	13	8	6	2	0	1	4.8	5.1	1.4	0.19
0 0 1 1 1 0	Bolko	2	2	2	7	13	15	6	4	2	0	5.5	5.4	1.8	0.24
0 1 1 1 0 0	Bernd	0	2	3	11	8	8	7	10	3	1	5.8	5.9	1.9	0.27
0 0 1 0 1 0	Billy	1	3	1	3	6	9	18	7	5	0	6.7	6.3	1.9	0.26
0 1 0 1 1 0	Bastian	0	0	2	2	7	9	9	18	6	0	7.2	6.9	1.6	0.21
0 1 0 1 0 0	Balduin	0	0	0	0	0	3	3	10	33	4	8.8	8.6	0.9	0.13
0 0 0 0 0 0	Balthasar	0	0	0	0	1	0	0	1	4	47	9.9	9.8	0.8	0.10

Table 7. B1 set, girls. n = 64 EPs; cf. Table 1/4.

Features **BF B S E F	Dummies	Rank distribution										Median	R	W	E
1 1 1 1 1 1	Bruno	56	5	1	1	1	0	0	0	0	0	1.1	1.2	0.7	0.09
0 1 1 1 1 1	Bert	2	35	14	8	2	3	0	0	0	0	2.4	2.7	1.1	0.14
0 1 1 1 0 1	Bodo	3	9	27	9	7	4	4	1	0	0	3.2	3.6	1.6	0.20
0 1 1 0 1 0	Birger	1	4	8	18	11	10	6	2	4	0	4.6	4.9	1.8	0.23
0 0 1 1 1 0	Bolko	2	4	2	12	13	11	9	9	2	0	5.4	5.4	1.9	0.24
0 1 1 1 0 0	Bernd	0	5	5	9	15	7	7	13	1	2	5.4	5.6	2.0	0.26
0 0 1 0 1 0	Billy	0	1	5	4	7	9	18	7	10	3	6.8	6.6	1.9	0.24
0 1 0 1 1 0	Bastian	0	1	2	1	4	11	14	20	9	2	7.4	7.1	1.6	0.20
0 1 0 1 0 0	Balduin	0	0	0	1	4	7	6	10	34	2	8.6	8.0	1.4	0.18
0 0 0 0 0 0	Balthasar	0	0	0	1	0	2	0	2	4	55	9.9	9.7	1.1	0.13

Table 8. B1 set, boys and girls 10 - 12 years old. n = 67; cf. Table 1/5.

Features **BF B S E F	Dummies	Rank distribution										Median	R	W	E
1 1 1 1 1 1	Bruno	61	3	2	0	1	0	0	0	0	0	1.0	1.2	0.6	0.07
0 1 1 1 1 1	Bert	4	44	11	4	1	1	2	0	0	0	2.2	2.5	1.2	0.14
0 1 1 1 0 1	Bodo	1	7	38	11	5	2	2	1	0	0	3.2	3.5	1.2	0.15
0 1 1 0 1 0	Birger	0	2	6	23	17	7	5	2	4	1	4.6	5.0	1.7	0.21
0 1 1 1 0 0	Bernd	0	3	5	15	13	8	8	12	1	2	5.3	5.6	1.9	0.24
0 0 1 1 1 0	Bolko	1	4	3	6	16	14	10	10	3	0	5.8	5.7	1.8	0.23
0 1 0 1 1 0	Bastian	0	1	1	2	6	17	13	19	7	1	7.0	6.9	1.5	0.19
0 0 1 0 1 0	Billy	0	3	1	4	5	8	21	10	12	3	7.1	6.9	1.9	0.23
0 1 0 1 0 0	Balduin	0	0	0	1	2	8	6	11	38	1	8.6	8.1	1.3	0.16
0 0 0 0 0 0	Balthasar	0	0	0	1	1	2	0	2	2	59	9.9	9.6	1.2	0.14

Table 9. B1 set, boys and girls 13 - 16 years old. n = 50; cf. Table 1/6.

Features **BF B S E F	Dummies	Rank distributions										Median	R	W	E
1 1 1 1 1 1	Bruno	42	5	1	1	0	0	0	1	0	0	1.1	1.3	1.1	0.16
0 1 1 1 1 1	Bert	1	32	7	5	3	2	0	0	0	0	2.3	2.7	1.1	0.16
0 1 1 1 0 1	Bodo	2	4	25	7	5	3	4	0	0	0	3.3	3.7	1.5	0.21
0 1 1 0 1 0	Birger	1	2	5	15	7	11	7	2	0	0	4.8	4.9	1.6	0.22
0 0 1 1 1 0	Bolko	3	2	1	13	10	12	5	3	1	0	5.1	5.0	1.8	0.25
0 0 1 0 1 0	Billy	1	1	5	3	8	10	15	4	3	0	6.2	5.9	1.8	0.26
0 1 1 1 0 0	Bernd	0	4	3	5	10	7	6	11	3	1	5.9	5.9	2.1	0.29
0 1 0 1 1 0	Bastian	0	0	3	1	5	3	10	19	8	1	7.7	7.2	1.7	0.23
0 1 0 1 0 0	Balduin	0	0	0	0	2	2	3	9	29	5	8.8	8.5	1.1	0.16
0 0 0 0 0 0	Balthasar	0	0	0	0	0	0	0	1	6	43	9.9	9.8	0.4	0.06

Table 10. B1 set, boys 13 - 16 years old. n = 21 EPs; cf. Table 1/7.

Features **BF B S E F	Dummies	Rank distribution										Median	R	W	E
1 1 1 1 1 1	Bruno	18	1	1	0	0	0	0	1	0	0	1.1	1.5	1.5	0.33
0 1 1 1 1 1	Bert	1	16	2	0	2	0	0	0	0	0	2.1	2.3	0.9	0.21
0 1 1 1 0 1	Bodo	0	1	12	5	2	0	1	0	0	0	3.3	3.6	1.0	0.23
0 1 1 0 1 0	Birger	0	0	2	9	1	5	3	1	0	0	4.4	5.0	1.4	0.31
0 0 1 1 1 0	Bolko	1	1	1	4	5	6	3	0	0	0	5.2	5.0	1.6	0.34
0 0 1 0 1 0	Billy	1	1	1	2	3	4	7	2	0	0	6.1	5.6	1.9	0.41
0 1 1 1 0 0	Bernd	0	1	1	1	5	3	2	5	3	0	6.3	6.3	2.0	0.43
0 1 0 1 1 0	Bastian	0	0	1	0	3	3	5	6	3	0	7.2	7.0	1.5	0.33
0 1 0 1 0 0	Balduin	0	0	0	0	0	0	0	5	12	4	9.0	9.0	0.7	0.14
0 0 0 0 0 0	Balthasar	0	0	0	0	0	0	0	1	3	17	9.9	9.8	0.5	0.11

Table 11. B1 set, girls 13 - 16 years old. n = 29 EPs; cf. Table 1/8.

Feature **BF B S E F	Dummies	Rank distribution										Median	R	W	E
1 1 1 1 1 1	Bruno	24	4	0	1	0	0	0	0	0	0	1.1	1.2	0.6	0.12
0 1 1 1 1 1	Bert	0	16	5	5	1	2	0	0	0	0	2.4	2.9	1.2	0.23
0 1 1 1 0 1	Bodo	2	3	13	2	3	3	3	0	0	0	3.2	3.8	1.7	0.31
0 1 1 0 1 0	Birger	1	2	3	6	6	6	4	1	0	0	4.9	4.8	1.7	0.31
0 0 1 1 1 0	Bolko	2	1	0	9	5	6	2	3	1	0	5.0	5.1	1.9	0.36
0 1 1 1 0 0	Bernd	0	3	2	4	5	4	4	6	0	1	5.6	5.6	2.1	0.39
0 0 1 0 1 0	Billy	0	0	4	1	5	6	8	2	3	0	6.3	6.1	1.7	0.32
0 1 0 1 1 0	Bastian	0	0	2	1	2	0	5	13	5	1	7.8	7.4	1.7	0.32
0 1 0 1 0 0	Balduin	0	0	0	0	2	2	3	4	17	1	8.7	8.2	1.3	0.24
0 0 0 0 0 0	Balthasar	0	0	0	0	0	0	0	0	3	26	9.9	9.9	0.3	0.06

Table 12. B1 set, total of all tests. EPs: n=207. For definitions of Median, R, W, E, and functions a-d, cf. sec. 3.1

Features **BF B S E F	Dummies	Rank distribution										Median	R	W	E
1 1 1 1 1 1	Bruno	162	27	9	3	3	2	0	1	0	0	1.1	1.4	1.0	0.07
0 1 1 1 1 1	Bert	10	117	46	18	8	4	3	1	0	0	2.3	2.6	1.2	0.08
0 1 1 1 0 1	Bodo	21	28	96	24	14	13	10	1	0	0	3.1	3.3	1.5	0.11
0 1 1 0 1 0	Birger	3	6	21	76	44	31	14	4	6	2	4.5	4.8	1.6	0.11
0 0 1 1 1 0	Bolko	8	11	10	36	55	44	20	18	5	0	5.2	5.2	1.8	0.12
0 0 1 0 1 0	Billy	2	6	9	10	25	41	67	22	20	5	6.7	6.4	1.8	0.13
0 1 1 1 0 0	Bernd	0	8	11	31	35	31	38	42	7	4	6.1	6.0	1.9	0.13
0 1 0 1 1 0	Bastian	1	2	4	6	17	28	41	83	21	4	7.6	7.1	1.6	0.11
0 1 0 1 0 0	Balduin	0	2	0	2	5	11	12	28	139	8	8.8	8.4	1.3	0.09
0 0 0 0 0 0	Balthasar	0	0	1	1	1	2	2	7	9	184	9.9	9.7	0.9	0.07

** = supernormal features (Bruno only)

Computed relative strengths of features (arbitrary units)

		**	BF	B	S	E	F	aver.rank resid.
rank function	a	7.0	6.6	18.4	3.9	8.5	15.5	0.34
(cf. sec. 3.1)	b	7.6	6.8	18.2	4.1	8.5	14.6	0.37
	c	5.9	7.9	20.9	1.5	7.5	16.5	0.54
	d	6.7	7.4	19.6	2.1	8.0	16.2	0.34
average		6.8	7.2	19.3	2.9	8.1	15.7	
(formal) uncertainty		.7	.6	1.2	1.3	.5	.8	

Table 13. B2 set. n = 59 EPs. For description of features, see text and Figure 13.

Features	Dummies	Rank distribution						Median	R	W	E
1 1 1 1	Bertram	34	4	11	4	3	3	1.4	2.1	1.5	0.20
0 1 1 1	Benno	8	35	11	3	2	0	2.1	2.3	0.9	0.11
0 1 0 1	Bolko	13	12	22	8	2	2	2.7	2.7	1.2	0.16
0 0 1 1	Berthold	1	2	10	31	15	0	4.0	4.0	0.8	0.11
0 0 0 1	Billy	2	5	3	13	29	7	4.7	4.4	1.2	0.16
0 0 0 0	Barthel	1	1	2	0	8	47	5.9	5.6	1.0	0.13

Table 14. Relative strengths of features in B1 set.

Group	Results of "juggling" (sec.2.2.3) BF	B	S	E	F	Computer result from Table	BF	B	S	E	F
10-12 years old	15	19	6	12	13	8	12	20	6	9	18
13-16 years old	8	28	7	15	21	9	8	30	6	14	21
females	10	28	8	12	20	5	8	27	10	9	24
males (B1/2, Table 1/1c)	9	24	6	17	22	4	8	28	2	16	24
weighted mean of all	12	25	7	15	20	12	12	26	4	13	23

Estimated uncertainty of each number +-2

Table 15. Relative strength of dummies in B3 set for different groups. a: Prediction from table 14; b: bestfit from the ranks of B3 results according to sec. 3.1; c: rank numbers in set B3. NB: Remember that in columns a and b the highest figure denotes the highest rank, in column c the lowest.

dummy	features	10-12 yrs.			13-16 yrs.			females			males		
		a	b	c	a	b	c	a	b	c	a	b	c
Bosko	BF+-+S+E+F	46	45	3.31	51	53	3.61	50	50	3.64	54	49	4.19
Bento	B+S+E+F	50	50	2.53	71	70	2.00	68	68	2.04	69	72	1.48
Basil	E+F	25	23	5.31	36	34	5.09	32	33	5,18	39	36	5.62
Beppo	BF+B+S+E	52	54	1.89	58	60	2.71	58	57	2.64	56	59	3.00
Benedikt	B+-+E+F	44	43	3.28	64	60	2.89	60	61	2.68	63	65	2.24
Bernhard	B+-+E	31	32	4.67	43	41	4.70	40	40	4.46	41	46	4.48

Table 16. C set, adult males. n = 40 EPs.

Features	Dummies	Rank distribution												M	R	W	E
BF B S E F																	
1 1 1 1 1	Christian	24	8	3	1	2	1	0	1	0	0	0	0	1.33	1.92	1.59	0.25
1 0 1 1 1	Claus	3	12	8	3	8	2	1	2	0	0	1	0	3.13	3.72	2.16	0.34
1 1 0 1 1	Carl	4	9	13	7	3	2	0	1	0	1	0	0	3.04	3.35	1.81	0.29
1 1 1 0 1	Chlodwig	1	4	6	5	7	2	6	3	4	2	0	0	5.07	5.40	2.46	0.39
0 1 1 1 1	Christobald	3	3	5	6	6	7	5	0	4	1	0	0	5.00	5.03	2.33	0.37
1 0 1 0 1	Casimir	2	0	2	5	2	8	6	6	4	4	1	0	6.67	6.57	2.41	0.38
0 0 1 1 1	Clemens	0	3	0	2	4	10	11	3	1	3	3	0	6.59	6.68	2.22	0.35
0 0 0 1 1	Chester	1	1	1	4	3	4	4	11	2	2	4	3	7.68	7.38	2.74	0.43
1 1 1 1 0	Christoph	1	0	1	6	3	2	2	2	16	4	2	1	8.69	7.57	2.58	0.41
0 1 0 1 0	Conrad	1	0	1	0	2	2	2	4	4	14	2	8	9.79	9.15	2.53	0.40
1 1 1 0 0	Constantin	0	0	0	1	0	0	1	5	3	4	15	11	10.90	10.38	1.74	0.28
1 1 0 0 0	Cecil	0	0	0	0	0	0	2	2	2	5	12	17	11.25	10.85	1.41	0.22

Computed relative strengths of features (arbitrary units)

		BF	B	S	E	F	Aver.rank residuals
rank function	a	9.7	6.4	2.4	12.3	19.2	0.62
(cf. sec.3.1)	b	9.9	6.1	2.4	12.8	18.7	0.74
	c	9.0	7.5	1.9	10.1	21.4	0.55
	d	10.3	6.0	2.4	12.4	18.8	0.27
average		9.7	6.5	2.3	11.9	19.5	
(formal) uncertainty		.5	.7	.3	1.2	1.3	

Table 17. Comparison of pairs of C and D set dummies which differ by one feature only. +% is the percentage in which the first mentioned dummy has been put before the other one.

a) pair (feature C/BF)	+% : -%
Christian:Christobald	85 : 15
Claus:Clemens	85 : 15
Average	85 : 15
Mean error (%)	4.0

b) pair (feature C/B)	+% : -%
Christian:Claus	85 : 15
Chlodwig:Casimir	72.5:27.5
Christobald:Clemens	77.5:22.5
Average	78.3:21.7
Mean error(%)	3.8

c) pair (feature C/S)	+% : -%
Christian:Carl	80 : 20
Clemens:Chester	70 : 30
Constantin:Cecil	57.5: 42.5
Average	69.2: 30.8
Mean error (%)	4.2

d) pair (feature C/E)	+% : -%
Christian:Chlodwig	90 : 10
Claus:Casimir	85 : 15
Christoph:Constantin	87.5: 12.5
Average	87.5: 12.5
Mean error (%)	3.0

e) pair (feature F)	+% : -%
Christian:Christoph	97.5: 2.5
Chlodwig:Constantin	100 : 0
Average	98.8: 1.2
Mean error (%)	1.3

f) pair (feature D/S)	+% : -%
Dagobert:Dick	71.4:28.6
David:Dieter	57.1:42.9
Diethart:Damian	61.9:38.1
Average	63.5:36.5
Mean error (%)	4.3

Table 18. D set. n = 42 EPs.

Features HM	B	S	E	F	Dummies	Rank distribution										Median	R	W	E
1	1	1	1	1	Dagobert	25	13	2	1	0	1	0	0	0	0	1.3	1.6	1.0	0.15
1	1	0	1	1	Dick	10	21	3	5	2	1	0	0	0	0	2.0	2.3	1.2	0.19
1	0	1	1	1	Donald	3	4	19	6	4	1	4	0	1	0	3.2	3.7	1.7	0.27
0	0	0	1	1	Daniel	1	2	5	14	9	5	1	5	0	0	4.4	4.7	1.7	0.26
0	1	1	1	0	David	1	0	5	5	7	8	8	7	1	0	5.9	5.7	1.8	0.28
0	1	0	1	0	Dieter	1	1	4	6	2	9	10	9	0	0	6.3	5.8	1.9	0.29
1	0	1	0	1	Dominikus	1	0	4	3	10	1	7	4	9	3	6.8	6.5	2.3	0.36
0	0	1	0	1	Dennis	0	1	0	2	3	11	5	7	0	13	7.3	7.4	2.1	0.32
1	1	1	0	0	Diethard	0	0	0	0	3	2	5	2	23	7	8.9	8.5	1.4	0.22
1	1	0	0	0	Damian	0	0	0	0	2	3	2	8	8	19	9.3	8.8	1.5	0.23

Table 19. E1 set, 13 - 16 years old. n = 38 EPs.

Features H	EY	FW	M	B	Dummies	Rank distribution								Median	R	W	E
1	1	1	1	1	Ernst	29	2	2	2	0	2	1	0	1.2	1.7	1.6	0.25
0	1	0	1	1	Edgar	4	22	9	0	0	2	1	0	2.2	2.5	1.3	0.21
1	0	0	0	1	Erich	4	5	11	3	1	4	1	9	3.4	4.4	2.5	0.40
1	1	1	0	0	Eduard	0	2	7	13	7	4	3	2	4.3	4.6	1.5	0.24
1	1	1	0	0	Erwin	1	3	5	2	11	7	5	4	5.2	5.1	1.8	0.30
1	1	0	1	0	Eugen	0	2	2	8	10	8	5	3	5.2	5.2	1.5	0.25
1	1	0	1	0	Egon	0	2	1	6	3	4	10	12	6.8	6.2	1.8	0.29
0	1	0	1	0	Elmar	0	0	1	4	6	7	12	8	6.6	6.3	1.4	0.22

Table 20. E1 set, 10 - 12 years old. n = 26 EPs.

Features H	EY	FW	M	B	Dummies	Rank distribution								Median	R	W	E
1	1	1	1	1	Ernst	22	4	0	0	0	0	0	0	1.1	1.2	0.4	0.07
0	1	0	1	1	Edgar	2	13	5	2	2	2	0	0	2.3	2.8	1.4	0.27
1	0	0	0	1	Erich	1	3	12	0	3	1	4	2	3.3	4.2	2.0	0.40
1	1	1	0	0	Eduard	0	2	1	13	6	3	1	0	4.3	4.4	1.1	0.22
1	1	1	0	0	Erwin	1	2	3	4	4	6	4	2	5.3	5.0	1.9	0.36
1	1	0	1	0	Eugen	0	1	1	4	7	4	6	3	5.5	5.6	1.5	0.30
1	1	0	1	0	Egon	0	1	2	3	2	6	5	7	6.3	6.0	1.8	0.35
0	1	0	1	0	Elmar	0	0	2	0	2	4	6	12	7.3	6.8	1.5	0.29

Table 21. Ranks in E2 and E3 sets.

	Ek(bert)	Er(hard)	Ew(ald)	En(no)	Es(so)	Eb(erhard)	El(imar)	Em(il)
E2	3.24	2.24	5.41	3.62	6.43	6.35	4.32	4.38
E3	3.13	1.42	5.87	4.00	7.42	5.95	3.84	4.37

Table 22. G set. n = 39 EPs. (N = normal width of eyebrows, W = increased width, L = increased length.)

Features N W L	Dummies	Rank distribution				Median	R	W	E
1 1 1	Gustav	34	3	1	1	1.1	1.2	0.6	0.10
1 0 1	Gert	1	22	13	3	2.3	2.5	0.7	0.11
1 0 0	Guido	1	11	24	3	2.8	2.7	0.6	0.10
0 0 0	Gernot	3	3	1	32	3.9	3.6	0.9	0.15

Table 23. GE set. n = 55 EPs.

Features (see text)	Dummies	Rank distribution				Median	R	W	E
1 1	Guido	40	13	2	0	1.2	1.3	0.5	0.07
1 0	Guenther	12	18	15	10	2.4	2.4	1.0	0.14
0 1	Gernot	3	22	18	12	2.6	2.7	0.9	0.12
0 0	Eugen	0	2	20	33	3.7	3.6	0.6	0.08

References

Baerends, G.P. – K.A. Bril – P. Bult
1965 "Versuch zur Analyse einer erlernten Reizsituation bei einem Schweinsaffen (*Macaca nemestrina*)", *Zeitschrift für Tierpsychologie* 22: 394-411.

Darwin, Charles
1872 *The expression of the emotions in man and animals.* London: Murray.
[1965] [Reprinted Chicago: University Press.]

Grzimek, Bernhard
1943 "Zum Erkennen vertrauter Menschen durch Tiere", *Zeitschrift für Tierpsychologie* 5: 431-438.

Hardy, A.
1960 "Was man more aquatic in the past?" *New Scientist* 7: 642-645.

Heinroth, Oskar
1930 "Über bestimmte Bewegungsweisen bei Wirbeltieren", *Sitzungsberichte der Gesellschaft Naturforschender Freunde* (Berlin), 333-343.

Koch, Walter A. (ed.)
1989 *The nature of culture.* Bochum: Brockmeyer.

Koehler, Otto – A. Zagarus
1937 "Beiträge zum Brutverhalten des Halsbandregenpfeifers (*Charadrius h. hiaticula*)", *Beiträge zur Fortpflanzungsbiologie der Vögel* 13: 1-9.

Kuehn, H.
1936 "Menschendarstellungen des Paläolithikums", *Zeitschrift für Rassenkunde* 4: 225-247.

Lersch, Philip
1932 *Gesicht und Seele.* München: Reinhardt.

Leyhausen, Paul
1967 "Biologie von Ausdruck und Eindruck", *Psychologische Forschung* 31: 113-227.
1973a "The biology of expression and impression",in: Konrad Lorenz – Paul Leyhausen, 272-380.
1973b "Ethological aspects of human behavior", in: *Proceedings of the International Conference "The Origin of Man".* (Accademia Nazionale dei Lincei, Roma), 219-241.
1983 "Kleidung: Schutzhülle, Selbstdarstellung, Ausdrucksmittel", in: Beat Sitter (ed.), 137-152.
1989 "The evolution of the cultural dimension", in: Walter A. Koch (ed.), 4-25.

1991 "What constitutes an aquatic mammal?", in: Machteld Roede et al. (eds.), 174-181.

Lorenz, Konrad

1935 "Der Kumpan in der Umwelt des Vogels", *Journal für Ornithologie* 83: 137-213, 289-413.

1943 "Die angeborenen Formen möglicher Erfahrung", *Zeitschrift für Tierpsychologie* 5: 235-409.

Lorenz, Konrad – Paul Leyhausen

1973 *Motivation of human and animal behavior*. New York: Van Nostrand Reinhold.

Matthaei, Rupprecht

1929 *Das Gestaltproblem*. München: Bergmann.

Roede, Machteld – Jan Wind – John M. Patrick – Vernon Reynolds (eds.)

1991 *The aquatic ape: Fact or fiction?* London: Souvenir Press.

Schmidt, Wolfgang Dietrich

1957 "Attrappenversuche zur Analyse des menschlichen Lachens", *Psychologische Beiträge* 3: 223-264. [Reprint of W.D. Schmidt, *Zur kommunikativen Funktion des menschlichen Ausdrucks. 1. Das Lachen*. Ph. D. dissertation, University of Königsberg, 1941.]

Seitz, A.

1940 "Die Paarbildung bei einigen Cichliden. I: *Astatotilapia strigigena* Pfeffer", *Zeitschrift für Tierpsychologie* 4: 40-84.

Sitter, Beat (ed.)

1983 *Menschliches Verhalten* (= 1. Kolloquium der Schweizerischen Geisteswissenschaftlichen Gesellschaft (1976). Freiburg, Schweiz: Universitätsverlag.

Tinbergen, Niko

1951 *The study of instinct*. Oxford: Clarendon Press.

Tinbergen, Niko – D.J. Kuenen

1939 "Über die auslösenden und die richtunggebenden Reizsituationen der Sperrbewegung von jungen Drosseln (*Turdus m. merula* L. und *Turdus e. ericetorum* Turton)", *Zeitschrift für Tierpsychologie* 3: 37-60.

Early social games reconsidered: Culture at play

Susan Vogel

1. The intrapersonal vs. interpersonal research paradigm

During the last thirty years, the field of Child Language Acquisition has experienced a major shift in what is considered to be the appropriate object of research. There has been a move away from separate studies of the child and his or her stages of linguistic development ("output") as well as studies of the caretaking adult and his or her adaptations to these stages ("input") to the examination of dyadic communication, i.e., studies of interaction between infants and caretakers.[1] There has also been a corresponding shift to a more holistic approach to the study of language acquisition and socialization. The child's growing mastery of language has been linked not only to the development of socio-cognitive skills but also to the "interactive structures of social activities in which the child takes part" (Corsaro – Streeck 1986: 14).

This trend has been facilitated in part by a revival of interest in the work of the Soviet psychologist and semiotician Lev Semyonovich Vygotsky (1896-1934). Two aspects of his work, as disseminated in the recent English edition (1978), have been cited most often in recent studies:[2] the well-known dictum that "an interpersonal process is transformed into an intrapersonal one"[3] and the notion of the "zone of proximal development".[4] In contrast to many developmental theories (e.g., Piaget) which explain interpersonal processes in terms of endogenous cognitive processes attributed to individuals, Vygotsky offers an alternative approach. Here, cognitive development is seen as emerging out of interpersonal processes, and the child is no longer seen as a "self-enclosed information processor" (Hickmann 1987a: 6). Although the "genetic interactionist", as opposed to the "genetic individualist" (Miller 1987), stance continues to be controversial, its adherents are growing in number, as the strengths of the interactionist position become increasingly evident.[5]

This development has far-reaching consequences for future research. It necessitates a revival of interest in naturalistic observation of dyadic or even triadic[6] communication, rich interpretation, pragmatic-oriented analysis and a developmental theory of discourse. It also raises a question that may seem somewhat trivial but is nonetheless relevant to both the study of communication and the overall subject of this volume, the origin of semiosis: what is the point of all this interaction? This essay will examine some of the recent studies of early caretaker-infant interaction from a perspective derived from Evolutionary Cultural Semiotics (cf. Koch 1986). My intention is to reveal the need for a critical discussion of the language orientation of studies of caretaker-infant interaction and to suggest that future research should at least explore the possibility of a more global, semiotic approach. Up until now, the goal of interaction studies has been to link early interaction with the acquisition of language. The next step, I propose, is to look for links with the acquisition of culture. As Bruner (1983: 103-104) remarks:

> The engine that drives the enterprise is not language acquisition per se, but the need to get on with the demands of the culture . . . Children begin to use language . . .because they need to get things done by its use. Parents assist them in a like spirit: they want to help them become "civilized" human beings, not just speakers of the language.

It may be useful to ask what we mean by "civilized" and observe how we do things not only with words but also with play.

2. Early interaction and intersubjectivity

The study of the first six months of interaction between caretakers and infants has revealed that communication during the early phases of the prelinguistic stage is highly complex and well-structured. Trevarthen (1977: 238-248) refers to interaction in the first six months as "primary intersubjectivity", which he claims has a primarily phatic function: "Both partners are fully engaged in the process of sharing and reciprocally sustaining the communication" (Trevarthen 1977: 248). Newson and Newson (1975: 280) describe this interaction as initiated by mothers, who are able to sustain the interest and attention of their babies:

> They tend to provide, quite unselfconsciously, forms of feedback which are dramatically attention-compelling because they are contingently geared - in a very delicately timed way - to actions which are spontaneously made by the infant himself . . . The resulting action sequence of the infant is therefore a combination of his own activity and *an*

> ***intelligent manipulation of that activity by the much more sophisticated adult partner.***

Primary intersubjectivity has been studied experimentally by Trevarthen by asking mothers to talk to their infants who are seated in special infant chairs. Analysis of a typical friendly exchange between a 12-week-old boy and his mother revealed that both partners are engaging in a number of comparable forms of social expression (mother: talking, moving head; infant: prespeech, arm waving) and that their interaction displays the basic turn-taking structures of (adult) conversation. Similar results have been obtained in studies of a large number of middle class mother/infant dyads in the United States and Scotland, supporting Trevarthen's (1977: 239) claim that "an exceedingly complex innate mechanism foreshadowing the co-operative intelligence of adults, and more general than the mechanism of language, was already functioning in early infancy."

A marked change in the manner of communication occurs when infants attain a workable control of grasping or prehension, at 16 to 18 weeks. Five-month-old subjects began to move beyond the world of the exclusive dyad, sometimes refusing to interact with an attentive mother and looking pointedly elsewhere. Trevarthen (1977: 254-255) interprets this development as the beginning of a new form of shared communication, one which also comes to involve objects: "secondary intersubjectivity". This sets the stage for shared experiences with objects which subsequently develop into the social games with toys that are a characteristic part of the social life of infants and their mothers in the second half of the first year.

3. Game formats and language acquisition

Bruner (1983: 45) stresses the importance of the interactive context for the development of child language. In particular, he focuses on the games of infancy that are "the staple and delight of human immaturity". These games or formats, he argues, all depend on the use and exchange of language, and they often provide the first occasion for the child's systematic use of language with an adult. Over the years, his research has been successively influenced by case grammar, speech act theory, and Hilary Putnam's historical causal theory of reference, and his analysis of the development of language acquisition illustrates the process by which mothers and infants work out a restricted and shared set of meanings and interactive sequences, which then act as referents and formats for gradually more advanced communication.

In his chapter on the growth of reference, for example, Bruner (1983: 70 ff.) begins by referring to the earliest phase of joint attention: the establishment of sustained eye-to-eye contact described by Stern (1974). The next stage sees the mother introducing objects between the child and herself as targets for joint attention, by moving the object into the child's line of vision and shaking it or "forefronting" it. She also "highlights" the object by means of a characteristic and standard vocalization illustrated in example (1) (Bruner 1983: 71):

(1) See the pretty dolly*

This pattern of vocal highlighting is typical of early interaction with infants and virtually disappears by the end of the first year. It is soon embedded in the game format "give and take", where it signals initial preparation for a bout of the game. More generally, however, Bruner relates vocal highlighting to basic techniques of managing joint attention, including the subsequent use of attentional vocatives when children are not in eye-to-eye contact and the use of a rising intonation pattern to indicate a desired shift in reference to something other than what the child is attending to. He then traces subsequent stages in the development of negotiated meaning in the context of formats, or games which impose roles, turn-taking, joint attention and a sequential structure.

Bruner (1983: 60) refers to these games as part of the "Language Acquisition Support System". Here, the mother is seen as a kind of ideal teacher, taking an established interaction sequence, highlighting and introducing a new procedure, and then gradually turning it over to the child as his or her skill increases. He terms this the "handover principle", a principle he claims is so ubiquitous that we hardly even notice it: "One sets the game, provides a scaffold to assure that the child's ineptitudes can be rescued or rectified by appropriate intervention, and then removes the scaffold part by part as the reciprocal structure can stand on its own."

Language plays an increasing role in these interactive sequences, and the mother tailors her participation to the child's apparent competence. An exchange of "book reading" with a child of 13 months, example (2) (Bruner 1983: 78), illustrates how the mother establishes the format by making all four of the moves or utterance types herself: attentional vocative, query, label, and feedback. Active babbling efforts of the young child are initially interpreted as a genuine contribution to the exchange:

(2) Mother: Look! (attentional vocative)
Child: (touches picture)
Mother: What are those? (query)
Child: (vocalizes a babble string and smiles)
Mother: Yes, they are rabbits. (feedback and label)
Child: (vocalizes, smiles, and looks up at mother)
Mother: (laughs) Yes, rabbit. (feedback and label)
Child: (vocalizes, smiles)
Mother: Yes. (laughs) (feedback)

One month later, after the appearance of some standard lexical labels, the mother begins to act as if the child were capable of words rather than mere babbles. She stiffens her requirements, so to speak, repeating her query until the child produces a shorter, lexeme-length babble. Later, only words will be accepted as fulfilling the requirements of the game.

The current phrase for this phenomenon in language acquisition is fine-tuning: "Parents speak at the level where their children can comprehend them and move ahead with remarkable sensitivity to their child's progress" (Bruner 1983: 38). If we return to the question posed earlier - what is the point of all this interaction? - it would seem that mothers are teachers, carefully scaffolding interactive sequences to promote the best possible communication with their infants. They gradually raise the standards of what they accept as a successful contribution on the part of the infant, operating on the assumption that "*something* comprehensible can be established" (Bruner 1983: 86), right from the very start. Bruner (1983: 34) regards this support system as a crucial contribution to filling in the gap between the rival views of language acquisition: empiricist associationism vs. nativism, or what he refers to as the void between "impossible" and "miraculous" explanations of child language acquisition.

4. Beyond language acquisition: Play as diversion

Bruner (1977, 1983) regards play as the primary context in which mothers support their children and negotiate with them about how to do things with words, and this restricted focus on play as the context of language acquisition can be found in a number of recent studies. Camaioni's (1986: 79) naturalistic study of mother-infant play, for example, investigates non-conventionalized vs. conventionalized games as "intersubjective procedures for constituting shared meanings". Her observations provide further evidence of a large repertoire of conventionalized games played by mothers and children (e.g., give and take, peekaboo, horsie, patacake, bye-bye, ball, build-knock down, no, point and name, put on-take off, slip on/off, open-shut, joint book-reading, question-answer, and linguistic imitation). Camaioni relates different rates of language acquisition to individual differences in social games and concludes that the extent to which a mother's discourse is matched to the child's ongoing activities affects the course and rate of early language acquisition: the more actively infants participate in conventionalized games, she suggests, the more linguistically advanced the children will be. This is a strong claim, one that deserves further investigation.

But there may be more to mother-infant interactional formats than dyadic fine-tuning in the interests of language training. The theoretical approach developed by Walter A. Koch (1986) encourages the use of evolutionary, semiotic and cultural perspectives to develop hypotheses that may enrich our understanding of the human cultural enterprise. Accordingly, the study of mother-infant interaction should be supplemented by comparative work on mother-infant interaction among primates[7] and by analysis of the rich repertoire of cultural artifacts of mother-infant interaction, e.g., the use of nursery rhymes. A detailed outline of the type of research which is called for by this approach is beyond the scope of this essay: my concern here is to point out a few basic directions.

As a prerequisite for more fruitful evolutionary research, we need to know much more about the natural occurrence, different forms, and multiple functions of play in mother-infant interaction. The usual approach is to divide mother-infant interaction into two basic categories: caretaking vs. play.[8] The next step is then to isolate particular formats that occur during play, a term which is seldom defined with any precision or subjected to further differentiation. The games documented by Camaioni (1986), for example, may not occur with equal frequency and may be fulfilling several different functions, a fact which is easily overlooked when they are all grouped together as examples of "play". The game format of "book-reading", for example, is clearly related to the growth of reference, and it occurs during non-caregiving interaction, but it has little in common with the game format of "horsie" or many of the other games documented. The question is again: What is the point of this interaction? In this case, I think a strong argument can be made that the play format of "book-reading" is used as a context for instruction, at least from the point of view of the mother. In other words, if mother-infant interaction is characterized as manipulation of activity by the more sophisticated adult partner, we should not ignore the possibility that mothers use play to manipulate in different ways.

The game of peekaboo, to take another example, seems to be quite different: it is less instructional and more diverting. It is presented by Bruner (1983: 48) as a game motivated by the infant's developing concept of object permanence and interest in achieving predictability, an approach which stresses the structural and interactional links with "book-reading" and other games. But informal observation of mothers and infants indicates that peekaboo is a game that has its beginnings in play as diversion, not play as instruction. In its earliest phases, it is a game used to entertain and distract infants, for example, during a diaper change. In an earlier discussion of the game, Bruner and Sherwood (1976: 279) describe it as a device used by a mother to "lighten" a potentially unpleasant situation, such as drying the infant after a bath. It later becomes a game which is used by mothers to divert fretting babies. Variants develop over time, as older versions prove too "boring" (Bruner 1983: 54). This indicates a need for increasingly sophisticated diversion.

This diversionary function of mother-infant play has not received much attention in discussions of human play which posit a connection between interactional formats and language acquisition. Van Lawick-Goodall's (1976: 262) description of chimpanzee mother-infant play, however, indicates how basic this function may be. Playful behavior is frequently used to distract offspring which persistently engage in an activity to which the mother objects. Again we have intelligent manipulation, but in this case the goal is diversion. Dare we posit a connection to diversion strategies or even self-diversion, entertainment?

A reconsideration of the evolutionary roots of play, in short, opens up new perspectives on the multifunctionality of playful interaction from the point of view of both participants. It may well be that the repertoire of early mother-infant games is related to many complex interactional goals and strategies, e.g., instruction, entertainment and manipulation, as well as the pleasure of the activity itself.[9]

A second set of perspectives can be derived from an analysis of cultural, in this case literary and/or musical, artifacts of mother-infant interaction. Here, what is called for is an interactional and functional analysis of situations in which highly developed texts, such as nursery rhymes and lullabies are used. At this point, evidence remains informal and anecdotal and should be followed up by naturalistic research, but a few conclusions can already be drawn from the evidence we have. The mothers I have observed use traditional texts of at least three rather different types: lullabies (3), quiet joint action rhymes (4), and rhymes building to an exciting climax and joint laughter (5):

(3) Hush-a-bye baby, upon the tree top
When the wind blows, the cradle will rock.
When the bough breaks, the cradle will fall,
And down will come baby, cradle and all.

(4) Patacake, patacake, baker's man,
Bake me a cake as fast as you can.
Pat it and prick it and mark it with a T [or child's first initial]
And put it in the oven for Tommy [or child's name] and me.

(5) Great big chicken hawk, flying in the air,
Sees a little chickie right (pause) down (dramatic pause) THERE (Aldis 1975: 249).

These texts are used in different situations: lullabies, often embedded in lengthy bedtime rituals, are used to soothe the young and lull them to sleep; joint action games constitute gentle social play and may well provide the infant with his or her first lessons in the cooperation principle; and joint laughter rhymes are typical of more vigorous social play and have been linked with early humor.[10]

Support for the special function of this third type of situation can be found in Trevarthen (1977: 255), who uses the term "play" to refer to "structured games and laughter" such as example (5). Play, according to his use of the term, develops during the phase of primary intersubjectivity and makes its first appearance well before the infant reaches the age of six months. The focus of playful interaction remains the mutual contact itself, but it differs from more gentle, protoconversational interaction: the exchanges are characterized by ritual repetition and the gradual build-up through a predictable sequence to a "twist" in the situation (Trevarthen 1977: 267) that is experienced intensely and simultaneously. The interaction - and the text - are oriented towards a goal or climax such as a tickle. The adult must remain sensitive to the infant's social signs that he or she is still willing to engage in the interaction - otherwise the game is a failure - but these vigorous games soon become beloved favorites, and there also is a marked increase in infant-initiated bouts during the last three months of the first year.[11] What is significantly different in the structure of these interactions is that they are targeted: one participant, adult or child, is the initiator, and the other participant is simultaneously the recipient and the target of the mock attack.

Preliminary analysis of even these three types of situations from the mother's point of view reveals differences in interactional goals that deserve further investigation. First, when a mother sings a lullaby to her infant, her goal is to reduce the participation of the infant to an absolute minimum (sleep). Second, when she plays a joint action game, as Bruner and others have demonstrated, the mother is providing a scaffolded introduction to turn-taking and conversational cooperation. And third, when engaging in games that lead to laughter, both participants experience the exhilaration of ritualized expectations and their violation. These three patterns, it seems to me, are all important: there is more to mother-infant interaction than fine-tuned teaching.

5. Early interaction in a cultural network

Future studies of mother-infant interaction should focus on the interactional context, goals and motivations of mothers and infants doing things with play. Nelson (1981: 181) suggested ten years ago that "how and why mothers use language may be as important for the child's pattern of acquisition as what kind of language they use." Language accompanies different kinds of play in different ways, play takes many forms, and an interactionist perspective on play may therefore increase our understanding of individual differences in language development, socialization and interactional strategies.

Cultural differences, however, may be widespread. Ochs (1987: 311) has warned that societies differ in the extent to which they engage infants as active participants in communicative activities at all. It is probable that lullabies and tickling games are universal[12] - but we need to know much more about natural mother-infant playful interaction in a wide range of cultures.

It seems, for example, that fine-tuning may be a typically Western phenomenon. Ochs (1987: 311-317) suggests that the tendency of Western middle class mothers to set up conversational activities where only one participant (the mother) is competent and then to take on all or most of the infant's communicative roles is related to a more general "philosophical orientation" of our culture: unclear thoughts or mental dispositions are considered to be suitable subjects of explicit conjecture. In other words, it is appropriate to second-guess and verbalize the thoughts, motivations and intentions of others. Mothers do this with their infants, at least in Western cultures, and we all do this as adults.

The orientation of other cultures, however, may be subtly different. Ochs (1987: 315) notes that in traditional Samoan communities, "persons are uncomfortable making explicit guesses as to what other persons could be thinking, the thoughts of others that have not been clearly expressed in language or demeanor." Infants are not usually put in the role of conversationalist until they are much older, at which time a different type of interaction occurs. In contrast to the intense face-to-face interaction of intersubjectivity, Samoan infants are held outward, facing the others in the group and directed to notice movements, remember names and repeat phrases of the caregiver. And when small children use egocentric or unintelligible speech, they are not encouraged as in Western families: they are more likely to be simply ignored.

Cultural beliefs regarding the nature of childhood and children have considerable effects on the kinds of interaction infants and young children are engaged in, and this in turn may influence early language acquisition and children's development of interactive skills and social knowledge (Cook-Gumperz et al., eds. 1986: 66). Cultural beliefs may also have effects on the forms and functions of interactional play. Multifunctional study of mother-infant interaction is only the beginning. Cross-cultural studies of entire networks of early interaction are needed to find out what we do with play.[13]

Notes

1. Cf. Schaffer (ed.) 1977, Snow 1977, Snow – Ferguson (eds.) 1977, Cook-Gumperz – Corsaro – Streeck (eds.) 1986, Hickmann (ed.) 1987.
2. For example in Budwig – Strage – Bamberg 1986, Wertsch – Hickmann 1987, McLane 1987, McNamee 1987. Recent discussions of Vygotsky's work include Bruner 1986, Lee 1987, Hickmann 1987a. For criticism of Vygotsky's approach, see Paprotté – Sinha 1987.
3. Vygotsky 1978: 57. He continues: "Every function in the child's cultural development appears twice: first on the social level, and later, on the individual level; first, between people (interpsychological), and then inside the child (intrapsychological). This applies equally to voluntary attention, to logical memory, and to the formation of concepts. All the higher functions originate as actual relations between human individuals."
4. Vygotsky (1978: 86): "The zone of proximal development . . . is the distance between the actual developmental level as determined by

independent problem solving and the level of potential development as determined through problem solving under adult guidance or in collaboration with more capable peers."

5. Piaget's philosophical standpoint is recognized, of course, even by scholars critical of his approach, and it continues to dominate research in developmental psychology. One example of how this emphasis on the individual can be modified within a Piagetian framework is found in Newson and Newson (1975 [1979]: 272): "We would not quarrel at all with his proposition that 'objective knowledge is not acquired by a mere recording of external information but has its origin in transactions between the subject and objects...'. To this statement we merely wish to add that the object with which the human infant interacts most often, and most effectively, particularly in the earliest stages of development, is almost invariably another human being." Ochs (1987: 310) discusses Piaget's emphasis on egocentrism as opposed to Vygotsky's emphasis on socio-centrism in ethnographic terms: "There may very well be cultural differences in the way in which their societies (Swiss and Russian) organize communication with infants and small children, leading them to observe different communicative capacities in the early stages of development."
6. For a discussion of the need to analyze interaction of all the members of the social network, see Lewis – Feiring – Weinraub 1981.
7. Bruner (1983: 45) also views the increased dominance of play during immaturity among Great Apes and Hominids as a preparation for the technical-social life that constitutes human culture.
8. The basic division between caregiving and play, for example, is used to assess the father's role in infancy by Parke – Tinsley (1981: 429).
9. Bruner (1983: 47) refers to Bühler's concept of Funktionslust, but only from the infant's point of view. A true interactionist should consider the emotional motivation of both participants.
10. See Vogel (1990: 294) for a discussion of different types of play and Vogel (1989: 210-245) for the development of early humor.
11. This has been documented by Bruner and Sherwood (1976) for the game of peekaboo. The change in the infant's role from passive recipient to active agent at the age of ca. 12 months is discussed in Vogel (1989: 220-221).
12. Ochs (1987: 312), for example, has observed singing and cuddling with young infants in Western Samoa, but not intense face-to-face protoconversations.
13. See Lewis – Feiring – Weinraub (1981: 263 ff.) for a discussion of the social network model.

References

Aldis, Owen
1975 *Play fighting*. New York: Academic Press.

Bruner, Jerome S.
1977 "Early social interaction and language acquisition", in: H.R. Schaffer (ed.), 271-289.
1983 *Child's talk: Learning to use language*. Oxford: University Press.
1986 *Actual minds, possible worlds*. Cambridge, Mass.: Harvard University Press.

Bruner, Jerome S. – Alison Jolly – Katherine Sylva (eds.)
1976 *Play: Its role in development and evolution*. New York: Basic Books.

Bruner, Jerome S. – Victoria Sherwood
1976 "Peekaboo and the learning of rule structures", in: Jerome Bruner – A. Jolly – K. Sylva (eds.), 277-285.

Budwig, Nancy – Amy Strage – Michael Bamberg
1986 "The construction of joint activities with an age-mate: The transition from caregiver-child to peer play", in: J. Cook-Gumperz – W.A. Corsaro – J. Streeck (eds.), 83-108.

Camaioni, Luigia
1986 "From early interaction patterns to language acquisition: Which continuity?", in: J. Cook-Gumperz – W.A. Corsaro – J. Streeck (eds.), 69-82.

Cook-Gumperz, Jenny – William A. Corsaro – Jürgen Streeck (eds.)
1986 *Children's worlds and children's language*. Berlin: Mouton de Gruyter.

Corsaro, William A. – Jürgen Streeck
1986 "Studying children's worlds: Methodological issues", in: J. Cook-Gumperz – W.A. Corsaro – J. Streeck (eds.), 13-35.

Hickmann, Maya
1987a "Introduction: Language and thought revisited", in: M. Hickmann (ed.), 1-13.
1987b "The pragmatics of reference in child language: Some issues in developmental theory", in: M. Hickmann (ed.), 165-184.

Hickmann, Maya (ed.)
1987 *Social and functional approaches to language and thought*. New York: Academic Press.

Koch, Walter A.
1986 *Evolutionary cultural semiotics: Essays on the foundation and institutionalization of integrated cultural studies*. Bochum: Brockmeyer.

Koch, Walter A. (ed.)
1990 *Geneses of language*. Bochum: Brockmeyer.

Lamb, Michael (ed.)
1981 *The role of the father in child development*. (2nd edition.) New York: Wiley.

Lawick-Goodall, Jane van
1976 "Mother chimpanzee's play with their infants", in: J. Bruner – A. Jolly – K. Sylva (eds.), 262-267.

Lee, Benjamin
1987 "Recontextualiziing Vygotsky", in: M. Hickmann (ed.), 87-104.

Lewis, Michael – Candice Feiring – Marsha Weinraub
1981 "The father as a member of the child's social network", in: M. Lamb (ed.), 259-294.

Lewis, Michael – Leonard Rosenblum (eds.)
1974 *The effect of the infant on the caregiver*. New York: Wiley.

McLane, Joan B.
1987 "Interaction, context, and the zone of proximal development", in: M. Hickmann (ed.), 267-285.

McNamee, Gillian Dowley
1987 "The social origin of narrative skills", in: M. Hickmann (ed.), 287-304.

Miller, Max
1987 "Argumentation and cognition", in: M. Hickmann (ed.), 225-249.

Nelson, Katherine
1981 "Individual differences in language development: Implications for development and language", *Developmental Psychology* 17: 170-187.

Newson, John – Elizabeth Newson
1975 "Intersubjectivity and the transmission of culture", *Bulletin of the British Psychological Society* 28: 437-446.
[1979] [Reprinted in: John Oates (ed.), *Early cognitive development*. London: Croom Helm, 1979, 271-286.]

Ochs, Elinor
1987 "Input: A socio-cultural perspective", in: M. Hickmann (ed.), 305-319.

Parke, Ross D. – Barbara Tinsley
1981 The father's role in infancy: Determinants of involvement in caregiving and play", in: M. Lamb (ed.), 429-457.

Paprotté, Wolf – Chris Sinha
1987 "A functional perspective on early language development", in: M. Hickmann (ed.), 203-222.

Schaffer, H.R. (ed.)
1977 *Studies in mother-infant interaction.* New York: Academic Press.

Snow, Catherine E.
1977 "Mother's speech research: From input to interaction", in: C. Snow – C.A. Ferguson (eds.), 31-49.

Snow, Catherine – Charles A. Ferguson (eds.)
1977 *Talking to children: Language input and acquisition.* Cambridge: University Press.

Stern, Daniel N.
1974 "Mother and infant at play: The dyadic interaction involving facial, vocal, and gaze behaviors", in: M. Lewis – L. Rosenblum (eds.), 187-213.

Trevarthen, Colwyn
1977 "Descriptive analyses of infant communicative behavior", in: H.R. Schaffer (ed.), 227-269.

Vogel, Susan
1989 *Humor: A semiogenetic approach.* Bochum: Brockmeyer.
1990 "The ontogenesis of the reality-fiction distinction: A comparative approach to play", in: W.A. Koch (ed.), 290-309.

Vygotsky, Lev S.
1978 *Mind in society: The development of higher psychological processes.* Cambridge, Mass.: Harvard University Press.

Wertsch, James V. – Maya Hickmann
1987 "Problem solving in social interaction: A microgenetic analysis", in: M. Hickmann (ed.), 251-266.

Infant semiosis[1]

Colwyn Trevarthen

Infants, by definition, cannot speak. They have no signs whatsoever by which to declare categories of experience that are good for exchange with others. They cannot compare their own experiences with others' experiences. Infants' minds are, therefore, outside the cultural imagination and practices of the persons around them. It would seem that infancy ends when symbolic communication begins, when the child begins to understand what words mean.

Nevertheless, research of the past 20 years has demonstrated that infants have motives of a protosymbolic nature that serve in the learning of symbols. Infants try to engage with the consciousness of other persons, and they offer significant messages in the ways they regulate orientation to events around them in response to other persons' interests. They make efforts to share experiences and purposes and, before many months, they take up other persons' meanings through communication with their feelings and interests, fitting into the activities of other minds by an imitative sensitivity to the expressions of their bodies. Infants are not mentally incoherent reflex organisms that produce only signals of physiological state to guide a care-provider. They are not automatic respondents to a few sign stimuli that trigger innate reactions to human movements.

In the first communication games infants play with their mothers we may observe an intelligence that has impulses to share ideas and fix meanings in a conventional code. We can trace the stages by which a mind grows that thinks in propositional narratives and in metaphorical representations, a mind that generates dynamic forms of consciousness and purpose that are fit to codification in language. The infant, we find, is even more than a self-sufficient experiencer and problem-solver, even more than a thinker. His or her cognition is endowed with the appetite to learn by picking up cultural ideas from a community of others. It exhibits an innate intersubjectivity that is adapted to cultural learning.

1. Infant semiosis and classical semiotics

The discovery of an infant intersubjectivity that is independent of either phylogenetically implanted sign stimuli or culturally acquired symbols, adds new psychological dimensions to classical semiotics, whose founders were logicians or linguists in the rational tradition (cf. Lyons 1977). They were concerned either with the organization of phenomenal experience into meaningful categories and inferences, or with the contents and structure of language. It is the emotional precocity of infants as persons in relation to other persons that is so unsettling for a rationalist or pragmatist interested in what C. S. Peirce (CP 2.227) called "scientific intelligence" (meaning one "capable of learning from experience"). How can one feel in a subtle, apparently sophisticated way without any world knowledge; above all, how can one have feelings about nothing, without any idea of cultural or logical categories?

In Peirce's Doctrine of Signs, as in Descartes "passions de l'âme" (Descartes 1649), emotions are disturbances of the mind that gain their differentiation by being attached to objects of experience. They "overflow" in bodily movement only to the extent that they are untamed by logic and because they involve the will in somewhat confused relation to experience. "An emotion is directly felt as a bodily state, or else it is only known inferentially" (Peirce, CP 1.250). Likewise, for Kant, Piaget (1962; cf. Cicchetti – Hesse 1983), and Eco (1979), emotions are peripheral to cognition, overflowing from primitive regulatory processes, inimical to reason. But in infant communication, emotions are primary and their dynamics carry thought and bring experience into understanding (Trevarthen 1993b). And they do so neither in the inferences or bodily reactions of the individual nor as by-products of cognitive work, as most of our philosophers pretend they must, but in living and direct interaction with the emotions of a partner. Infants master the two-mind problem without taking the time to discover they have a mind of their own (Trevarthen 1994). They exploit emotions in dialogue and derive the capacity for reflective awareness from forms that come to life and grow in communication.

Peirce identified the capacity for rational thinking with a matching capacity for communicating "scientific intelligence". A rhetorical process links subjects' minds through a perceived correspondence between attitudes and expressions of their bodies. Representations in expression can convey one subject's cognitive mastery of the qualities, immediate presence, and probable behavior of objects to generate

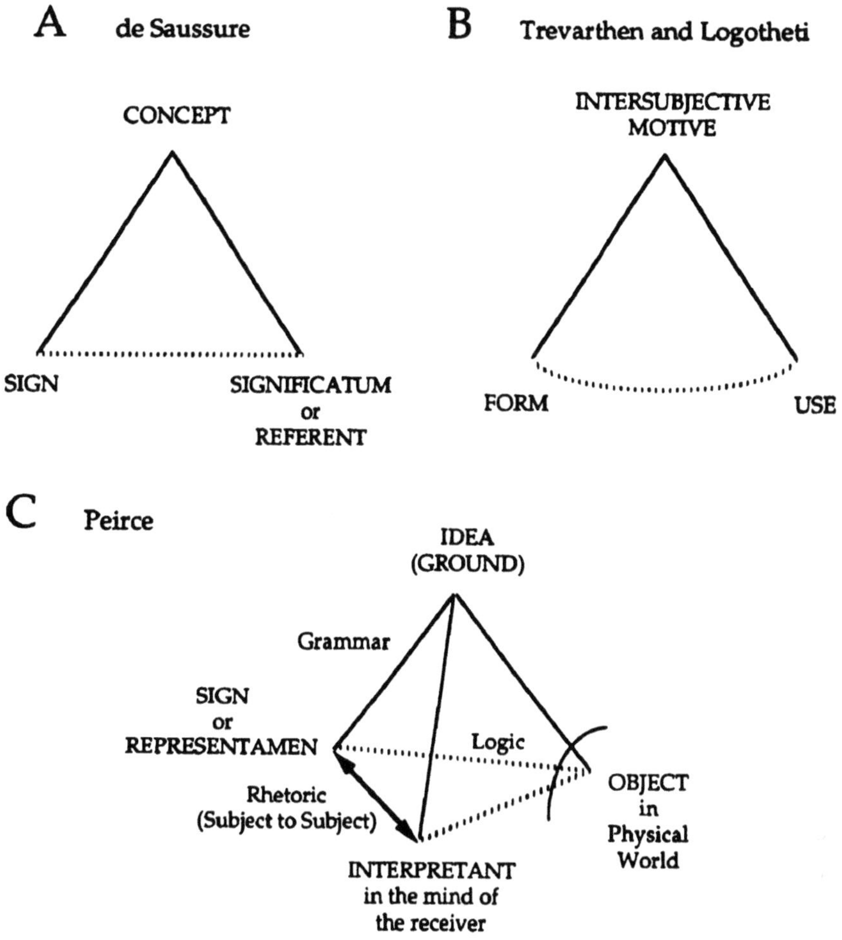

Figure 1. The rationality of symbols: A: Saussure's formulation of the relationship between a concept, a sign and the object signified (Saussure 1960; Lyons 1977). The link between the sign and the object is arbitrary or conventional. B: A diagram that emphasizes the intersubjective motive behind a symbol and the negotiation of meaning. Every symbol has an arbitrary form and an arbitrary application to an object of common interest in communication (Trevarthen – Logotheti 1987). C: Peirce's conceptualization of the relationship between a sign (representamen) that stands in imagination for the idea ("ground") of a thing (object) and an understanding (interpretant) that another may have from the sign as a message.

interpretations in somebody else's mind (see Figure 1). They create a grammar of ideas that holds information and understanding about a logic of facts or truths. A formulation of Peirce's (Figure 1C), which he calls an abstraction of sign making behavior, can be compared in diagram form with de Saussure's "signification" (Figure 1A). We have modified these classical diagrams of sign processes to emphasize the central role of the intersubjective link, which is the foundation of infant semiotics (Figures 1B, 2).

Analysis of the media of adult thinking and communicating has over centuries been the source of involved and inconclusive controversy. In modern times, the philosophy of feeling has been smothered in concern for rational cognitions and phenomena of impersonal reality. In consequence, language has been described as a system for categorizing experiences or for influencing wills. A remarkable consequence has been the flourishing of pessimistic belief that our awareness and thinking are completely overwhelmed by words and the historically constructed meaning in language; that our consciousness is overgrown with socio-historical conventions to the degree that no natural, unartificial, communication is left.

The interpersonal, intermental aspect is still deeply, artificially, mysterious in our psychology (Trevarthen 1993a). Yet we know that there are universal values in human expression, and infants appear to have the same kind of motives everywhere for gaining a place in the culture of their society, however large or small it may be. A study of infant communication can help clarify the way the individual's experience of life and the world becomes shared with others in a language. It may help us escape the fear that all we know is what we have been taught in words.

How does one individual's awareness of objects and events become transformed into signs that somebody else can understand in comparable form? Is the symbolization in thinking and recollecting of one mind for itself the primary process from which common code of language is derived by arbitration, or are the units of thinking and remembering built first in contracts of interest and purpose between intimately communicating minds? Is the thinker fated to be possessed by ideas inherent in the language he or she uses to talk and think in? We dare hope that exploration of the infant's awakening of interest in the coded concepts and practices of the parental culture will clarify these questions.

2. The unique intersubjective motives in human cultural learning

Human semiotics is evidently unique, with unique creative powers (Lotman 1979; Hurford 1989). Not even the great apes possess the essential reciprocating, narrative-spinning mentality that can look for and listen to stories and thus learn from the experience of a particular lineage of other minds (Tomasello et al. 1993). What a group of apes learns by imitation is trivial compared with what any 18-month-old human is understanding. The motives of infants for learning how to join in the human sense of the world show how their kind of communication is adapted, in the evolutionary sense, to sustain a culture (Trevarthen 1993a). Infant semiotics is the first expression of an innate motivation for cultural learning. Its precocious motivation for communication suggests that every human mind thinks conversationally about phenomena. It suggests that solitary thought is but a sociality of consciousness temporarily on its own, talking to itself.

Animals exchange signs of their emotions and motives about immediate experiences or about the immediate route to remote, remembered experiences (Marler 1984). Orienting to one another or moving in synchrony, they regulate a subtle and often complex train of social contracts and negotiate the rank of their activity in social groups. Chains of signals convey the intensity and quality of their motivations towards each other. The expressions of individuals have evolved to make intrinsic regulation of their present or future behavior accessible to other individuals. In a few species, communication of social (intersubjective) reactions has been modified to carry information to the group about what one individual perceives in the environment (e. g., specifying kinds of threat, or kinds of food, or the identity of a social attacker or rival; Gouzoules et al. 1985). But even when animal signals can, by use of coded signals, designate entities that, for the signal-receiver, are remote, as in the case of vervet monkeys "naming" a predator species to alert others in the group about the appropriate escape strategy (Cheney – Seyfarth 1985, 1992), they cannot enter into an extended speculation or a purposeful or propositional narrative of feelings, reasoning out the general abstract principle (maxim or moral) from the particular case.

The ability of all human groups to share dramatic accounts of imaginary, legendary events - recollections of the past, predictions of the future or total fantasies significant only in the joys, fears and wonder they excite - is intimately connected with the ability to categorize, evaluate, and designate. In other words, discursive, conversational intelligence is the generator of all meaning. It is also the matrix for educational instruction of the young and for the recalling of meanings in connected relationships. It is narratable memory.

3. The first signs in human communication are interpersonal

In this paper I will summarize what I understand about how an infant comes to have an imagination and interest to fix meanings in culturally approved actions and in the words of a language. Our data, from photographic evidence and microanalysis of films and video records of communication with infants, has been summarized in a number of papers over the past 20 years (Hubley – Trevarthen 1979; Murray – Trevarthen 1985, 1986; Trevarthen 1974, 1977, 1979, 1985, 1986a, b, 1987, 1990, 1993b, 1994; Trevarthen – Hubley 1978; Trevarthen – Marwick 1986). The mastery of a language begins, and remains, but part of a larger process of sharing ideas for social and pragmatic purposes, cooperatively, in a community of identified persons who know different things. A theory of intersubjective motives for cultural learning has been worked out to explain the complex communicative behavior observed in infancy (Trevarthen 1980, 1982, 1984, 1987, 1988, 1989, 1993a; Trevarthen – Logotheti 1987; Trevarthen – Marwick 1986). The above papers cite the by now very large literature on infant communication.

Infant expressions are composed as messages in conversations with friends who offer companionship in terms the infant can comprehend. It will help our analysis of these expressions and messages, and of the parenting responses that sustain them, to draw a distinction between the referential part of a sign that is aimed at some object in the world and the intersubjective or emotive part by which motives of the communicating subjects for taking interest in shared goals are established and brought to face one another in a process of conversation.

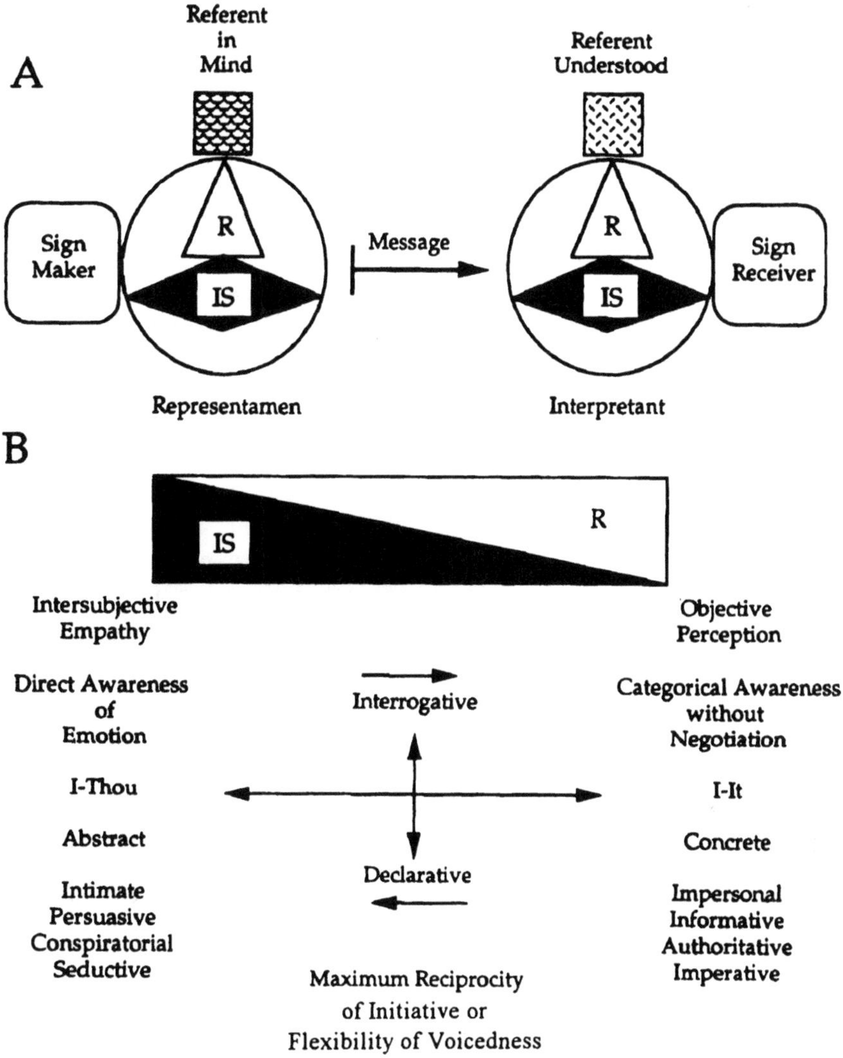

Figure 2. The communication of symbols: A: Any symbol has two components, an intersubjective motive (IS) and a reference (R). When a message is successfully passed between subjects, comparable referents with the same meaning are conceived by them. (Modified from Trevarthen 1990.)
B: Messages have different degrees of intersubjective motivation or referential application. In dialogue or conversation they are reciprocal, with differing interrogative or declarative force or "voice".

4. Neonate semiosis

Motivation to communicate and emotions that qualify communication are evident in humans from birth (Trevarthen 1993b). The classically conceived primary conditions for emotion, in bodily manifestations and in attachment to objects, fall to the background as secondary effects when we observe the interplay of feelings in a mother's communication with an alert newborn. A newborn, even one born up to two months before term (i. e., about 8 weeks before 40 weeks gestational age), will show communicative reactions to the presence of a sensitive, affectionate partner. The quality of the emotion conveyed by the way an adult moves to hold and greet the infant has a direct effect on how the infant behaves. In addition to expressions of distress or need, the baby may orient to the other with positive indications of interest, smiling in response to the feelings the other expresses. As their communication becomes established, there is a confluence of feeling and a coordination of motives into one system of emotion-change. Each expression is carried in a sequence of elementary feelings and predictable phases in the exchange. Transients of emotion become organized into complexes that are ready to give qualitative interpretation and specified identity to experiences held in common by the communicating subjects. Thus, the human contact in which meanings can be found is, under favourable conditions, within the reach of an infant a few moments after birth.

Although newborns are more advanced in their touch and audition senses and highly responsive to contact and motion stimulation, they also show awareness of persons at a distance from their bodies, and they can see. The visual orientation of a infant less than 46 weeks gestational age lacks sharp focus and foveal aim and the responses to other persons' expressions are slow, sometimes taking nearly a minute to assimilate the others' repeated signals of affectionate interest (Holmlund 1990). But some infants show remarkable precision in imitation from birth. Imitations of facial and manual expressions prove that a newborn can see another person and discriminate movements that they make to express emotions (Field et al. 1982, Maratos 1982, Meltzoff 1985, Meltzoff – Moore 1983, Kugiumutzakis 1985, 1993, Heiman 1989, Heiman – Schaller 1985).

Neonatal imitations have two characteristics that prove their communicative purpose (Kugiumutzakis 1993):

a) they are voluntary in the sense that a goal-directed effort shapes them towards the form of the "model" by successively improved approximations.
b) they are selective of forms of movement that are components of "conversational" expression and therefore can be part of a communicative exchange.

Observations of ordinary spontaneous imitations in the home show that mothers and newborns select face expressions, hand gesticulations, head nods, and vocalizations, and they generally do not imitate coughs or sneezes that can be taken as involuntary "vegetative" emissions that have no conversational value. Experimental studies confirm that newborns may imitate these expressive movements when they are artificially presented in isolation, when the infant is in a state of "quiet arousal" and optimum attention (Kugiumutzakis 1985; Heiman 1989; Heiman et al. 1989). The quantitative studies also demonstrate that there are large individual differences in the readiness of newborns to imitate. These may be indicative of future differences in social responsiveness or 'personality' (Field et al. 1982, 1985; Heiman et al. 1989).

Rapid learning of the identifying features of the mother, starting with *in utero* learning of her voice characteristics, favors a preferential response to her approach for face-to-face communication in the first days after birth. From the first emotional greetings she gives the baby as soon as it is born, the mother's expressiveness also assumes special, highly modulated forms that stimulate the attention of the infant and encourage positive expressions of communication, such as smiles, coos, hand spreading, and limb waving (Fernald et al. 1989; Papousek – Papousek 1989; Papousek et al. 1985; Stern 1974; Stern et al. 1982). These movements are more frequent when the infant is oriented to a person, and less common with any impersonal object that does not display emotional and communicative states.

Communicative interactions with newborns assume a turn-taking pattern because each partner follows an expressive turn with an alert waiting-for-response pause (Beebe et al. 1979, 1985). Turn-taking is not "taught" by the mother. It is part of the innate motivation for obtaining reciprocal communicative contact with another human being, and it is evident in the first imitative exchanges. Even a premature infant can engage in alternating vocalization on a steady beat with a sensitive partner, the periodicity of the alternating calls being close to *adagio* or 70/minute (Trevarthen 1986a, 1994). These and other features of the kinematics of neonatal communication support the theory of an innate motivation for reciprocal mutually-regulated expression of an organized set of feelings that establish a psychological context for sharing interests and purposes. The adult may be speaking a language, but the contact with the infant is by the highly predictable paralinguistic features of "baby talk", "intuitive motherese" or "infant-directed speech". The intensity of the infant's orientation and the readiness for an exchange of expressions stimulates in the adult a belief that the infant is "trying to say something" and is ready to share experiences. The impression is consolidated by the way clearly differentiated emotional expressions are coordinated in sequence with movements that seem to orient and focus the infant's eyes, fingertips, and hearing, evidently gating information uptake for perception (Trevarthen 1993b, 1994).

5. Protoconversational forms

Within two months of birth, protoconversational communication, first identified by Mary Catherine Bateson, is fully defined, the most significant developments being a marked increase in precision of eye-to-eye contact and a quickening of all responses. Holmlund (1990) reports that responses in communication have speeded up to near adult levels of proficiency by 3 months. Now the infant can be attracted to play a part in an exchange of expressions that has a reliable pattern closely resembling that of the paralinguistic components in adult face-to-face conversation. The details of this communication have been analyzed with the aid of a Double Television Communication apparatus in which mother and infant, in separate rooms, respond to one another's televised image and electronically transmitted voice (Murray – Trevarthen 1985; Trevarthen 1985, 1993b, 1994).

The attention of the infant to a partner's voice and face expression, shown by a "knit brow and jaw drop" expression of *fixed orientation*, is followed promptly by recognition of positive, affectionate elements in the partner's feelings. Periodic inviting vocalizations, touching and movements of the head, eyebrows and mouth evoke first a *smile of recognition* then an animated "utterance". At the climax of expression, the 6- to 8-week-old assumes an attitude of *declamation*, vocalizing, gesturing, and changing head posture in a coordinated assertion of communicative purpose, frequently removing orientation from the partner, who has been closely attended to until this moment. The adult, seeing this, is stimulated to give an encouraging, praising kind of *reply* that matches the level of affect of the infant's declamatory display or "utterance", interpreting it with a parallel emotional form or "affect attunement" that gives back or complements the feeling (Stern et al. 1985). Then the infant reorients to the adult and observes, before being excited again to smile and make another utterance.

As has been suggested, the face-to-face interplay of feelings and expressive impulses of protoconversation is uniquely human. No behavior like it can be seen in monkeys or apes, even though these species, and many other mammals, orient to one another and may alternate in presenting visible expressive signals and vocalizations. The fact that complex paralinguistic behaviors of human conversation have such a precocious manifestation, appearing well-developed in the first 2 months of infancy before the infant has established the stable postural base that head orienting and manipulation require, is evidence that human mental development is destined to be regulated, probably in a unique manner, by learning in communication.

6. Emotions as conversation regulators

The emotional regulation of protoconversation has been investigated by observing what happens when the human response to it fails. If a mother holds her face still in the middle of face-to-face play with her two-month-old, this causes the infant to show distress and withdrawal. A similar pattern appears when the mother presents the uncommunicative manner of simulated depression (Cohn – Tronick 1983). Real postnatal depression interferes with the infant's communication and cognition, and, if it persists, is accompanied by limited cognitive development in later months (Murray 1992).

The Double TV apparatus allows a critical experiment to be performed in which a lively, positive portion of the recording from the mother is replayed to the infant (Murray – Trevarthen 1985; Trevarthen 1985, 1993b). The distress this produces shows that a 2-month old is extremely sensitive to the contingent responsiveness of the live mother's expressions, which is lost when a physical recording is offered. If, conversely, a portion of the recording of a communicating infant is replayed to the mother, she experiences an uneasy loss of contact and may feel the infant is avoiding her or that she is somehow giving the wrong signals (Murray – Trevarthen 1986).

7. The dynamic motives in protoconversation that represent a conversational "other"

All the above observations of the patterns in protoconversation and their intersubjective controls support the theory that the infant is born with dynamic motives that both generate and respond to behaviors with particular temporal (kinematic) and anatomic (physiognomic) features that are also regulated in their intensity or force (energetics) (Trevarthen 1986a). In spite of the immaturity of the infant's nervous system, human standards of expression and emotional response ensure that adult and infant can meet on a common ground of affect signals and set up a cooperative dialogue. These findings have led to the conclusion that the infant's brain is born with a live "virtual other" in it, ready for assimilating the particular quality of expressions from an actual partner whose identifying features are rapidly learned (Bråten 1988). The newborn has a complete field of emotions that are most differentiated and most active in the presence of an affectionately responsive partner (Trevarthen 1993b). The dynamic patterns of feeling in protoconversation in which the infant follows and joins in rapidly transforming expressive sequences, give the clearest evidence that each human mind is innately organized for intersubjective participation with the interests and feelings of another human mind.

8. Body-sense and play: Developing metacommunication

Developments beyond three months show the infant moving towards negotiations of expression with increasing initiative. The baby starts to lead the mother's attention out of the conversational confrontation towards items in the shared environment.

From three months, a baby is increasingly curious about events of an impersonal or physical kind that offer affordances for exploration by sensory tracking and manipulation. This change in cognitive appetite accompanies growth of the muscles, receptors, and motor coordinations of the developing body and brain that confer a power to overcome the pull of gravitation so the baby can support the head in free rotation and project the arms out to grasp objects. Reaching and grasping brings new substances, forms and textures into the infant's control and the baby experiments with systematic transformation of experiences in sight, sound, touch, and taste. The intrinsic identifying properties of objects are discovered at an accelerating pace by the combined application of hands, eyes, ears, and mouth to pick up useful information.

Because the baby often shifts attention to an object, the partner who seeks communication with the baby experiences a loss of contact. Attraction of the baby to communicate demands a playful quick-changing intensity in expression that a younger infant would find confusing. The baby now inhabits a more lively field of movement and sensation that another can share. Soon play routines are discovered that facilitate lively and enthusiastic participation with the infant (Stern 1974; Bruner – Sherwood 1975; Trevarthen 1977, 1986a, 1987, 1990; Trevarthen – Hubley 1978; Trevarthen – Marwick 1987). Games of vocalization and dance are practiced in which the force and rhythms of movement get stronger. Patterns of expression are created that are more complex than any interactions infant can have with an object.

The dynamic sequence of play with a 3- to 6-month old is progressively more dramatic and sustained through more elaborate syntactical forms than in protoconversation. The baby can now anticipate a progression to a climax of teasing over several seconds. This interest in a "narrative sequence" of feelings and expressions that has a beginning, a development, a climax and a resolution or dénouement, the last being carefully paced to be startling and provocative of laughter, is manifest when the infant is responding to classical baby songs and action chants. In many languages, these songs and rhymes have the same elaborate features that testify to the complexity of the infant's communicative motives at this age. Communication appears to be activating universal motive templates that will later serve for transmission of symbolic meanings in communication.

The core element of a baby song is a four line stanza lasting 12 to 15 seconds with a base pulse of *andante*, a tripping rhythm, simple pitch shifts and rhyming syllables at specific points and variations in the beat to regulate excitement in the last two lines (Trevarthen 1986a, 1987, 1990). The controlled complexity of temporal patterning in these vocal action games gives evidence for a "grammar of expression" that the infant can share more than a year before words are learned. Infants quickly learn to recognize songs and eagerly join in when invited, recognizing the name or initial action forms for a song when a familiar partner offers them. This behavior is not merely immediately imitative; it soon becomes part of the child's own spontaneous and "deferred" expression. Furthermore, a baby of six months who has been encouraged to play in song may express self-satisfaction or joy by "singing" when alone (Papousek – Papousek 1981).

9. Evolving games with increasing meaning

As the infant gains in skill at manipulating objects, the parents, siblings, and other playmates use objects as vehicles for teasing play with the baby. "Person-person games" become transformed into "person-person-object games" and the baby is given reactive "toys" that make noises or roll or bounce when handled, shaken, or thrown (Trevarthen – Hubley 1978). Again, the incorporation of such object-exploring behaviors into communication games or negotiated "formats" (Bruner – Sherwood 1975; Ratner – Bruner 1978) is peculiarly human. The narrative "plots" of such play have no equal in the fighting and chasing games of young mammals, even though some animal play may incorporate objects that are chased or fought over.

In the teasing and joking games and musical entertainments of infants approaching the middle of the first year, communication is complex and highly productive. But, the topic remains the communication itself, i.e., it remains "metacommunicative". The baby anticipates and responds with increasing skill in forms of dynamic syntactical cohesion that convey entailment, qualifications, contrast, repetition with changed emphasis, subordination, opposition, release, and so on. The infant can learn patterns of different poetic or dramatic form and enjoy their repeated presentation. This provides the "text" for awareness of causal relationships between the many objects that now can be discriminated, identified, and recognized in the infant's awareness.

The infant can gain interest in experiences through the *presentations* that others make to change these experiences in momentarily surprising but quasi-predictable and often repeated playful ways. Identities and relationships or operations are "dramatized" by the emotions coupled to their demonstration (Trevarthen – Hubley 1978). The infant's imitative awareness appreciates the excitement of a partner's attempts to "attune with" what the infant is doing and the changes of action and experience that the infant is showing (Stern et al. 1985). Soon the infant is expert at seeking for these feelings of the other and for the habitual forms of display. By "emotional referencing", when the infant looks toward a caretaker upon meeting a new situation or object, or after performing some deliberately demonstrative "act", the infant's learning and skill is evaluated or "tagged" by the other person's emotions (Klinnert et al. 1983).

Moreover, the infant about 7 or 8 months old, about to gain new mobility through crawling, can share a space of orientations with another person, taking up their direction of gaze or their pointing (Scaife

– Bruner 1975; Butterworth – Grover 1988). This means that by that age, at least, the other person's awareness is convergent with and can be linked to the infant's awareness in a common space of experience. Such joint awareness is recognized as a key element in communication that leads to cultural learning (Tomasello et al. 1993).

10. The wordless social "Me" expressed in protosigns, teasing, and fear of strangers

Around six months after birth, an infant can be observed imitating signal gestures and mannerisms (Uzgiris 1981, 1984) and the baby can be easily enticed by enthusiastic praise from family members to perform a "trick" that impresses other persons (Trevarthen 1987, 1990). An arbitrary expressive form can be learned and recalled later in another context to be presented as a "significant message", the "reference" of which is, however, trapped in the family game in which it was first learned. When, as may occur, the "sign" is given to a stranger, apparently in an attempt to "break the ice" of an awkward, fearful confrontation, it is puzzling to the stranger and may be mocked, which distresses the infant (Trevarthen 1986b, 1990). Reactions to a stranger may become coquettish or challenging but they remain ambivalent. The baby begins to laugh easily, and this is combined with a sensitivity to who it is that laughs at or with the infant. The laughter of a stranger may instill fear. Fear of strangers seems to be linked with a tendency to supercilious expressions and clowning when the baby is with a partner who is either not supportive - for example, when the mother keeps a still face - or too pointedly observant of the infant's self, as when a stranger is trying to "make friends" (Trevarthen 1990).

Thus self-referred, other-sensitive emotions become stronger and clearer in the second six months. The baby begins to recognize his or her self in a mirror and in experiments it has been shown that toward the end of the first year infants have a preference for seeing pictures of infants of the same sex as him or her self. When looking in a mirror the baby makes "experiments" with babbling, face grimaces and hand gestures, and repeats imitations of the exaggerated expressions others offer in play.

We may interpret these complex developments in the infant's behavior as showing Bateson's metacommunication or "communication about communication" (Bateson 1956), creating Winnicott's potential space between the infant's "me" (subject) and "not-me" (object) (Winnicott

1960). It would appear to be preparatory to learning in an active "zone of proximal development" in which the partner can give cognitive, logistical, and practical support (Vygotsky 1962), and the child can begin to learn through "guided participation" (Rogoff 1990). The infant is manifestly gaining a more sensitive insight into the other's states of mind, and can be said to be expressing a "sense of intersubjective self" (Stern 1985)

The new motives affect the way the baby plays with objects and brings attention to them into communication with another person. The baby picks up comical ways of handling things and starts showing objects for others, seeking congratulation, holding them up as a joke with a look, a gesture, a grimace, or a vocalization that can become a coded act that is transmitted between the child and the "audience" of the other (Hubley – Trevarthen 1979; Trevarthen 1986b, 1990; Trevarthen – Hubley 1978; Trevarthen – Marwick 1987). The object is given a communal meaning by the form of gesture and expression (Halliday 1975).

These behaviors and the significant gestures that are picked out from games have the characteristics of "protosigns". They have been arbitrarily defined in the loyal and affectionate communication of the family or with familiar playmates. The do not have useful currency with strangers who are perceived by the infant as somehow threatening and destructive of meaning (Trevarthen 1990). This is the age at which infants have been observed to exhibit increased "fear of strangers". Such social sensitivity is the emotional obverse of attachment to familiar persons. The fact that it increases as the infant gains in self-awareness and "showing off" with familiars, suggests that the beginning of coded communication is part of a motivation for defining a cooperative group in which meaning is consolidated by daily practices that must be protected against "strange" customs or beliefs (Trevarthen 1988, 1989, 1993a). Subsequent developments into language confirm this idea.

Around nine months it may be observed that an infant is taking a new, much more definite initiative in games, attempting to direct a partner to behave in a certain way or to repeat a playful act (Hubley – Trevarthen 1979). The infant may show considerable skill in teasing an adult, as if knowing how to manipulate feelings and predict reactions (Reddy 1990). Such behaviors appear to be part of the transition to a new constructive sharing of interest in things and tasks.

11. Sharing experiences, purposes, and feelings in a pragmatic context

At nine or ten months a change occurs in the way an infant perceives what a partner in communication intends to do, or wishes to share. Before this, the baby will happily manipulate objects for his or her own interests alone, unless the object is made part of a game. The new awareness makes the infant want to meet the other person's point of view, to share manipulation of objects and cooperate. The infant now perceives the partner as a source of new ideas concerning objects. They can develop joint conscious control in the performance of a task that involves combining objects in some invented, arbitrary way. Instead of transforming awareness of the affordances of objects entirely by self-produced movements, learning alone, the infant observes the gestures and orientations of the partner and takes note of the accompanying vocal expressions and their references to the child and to the situation.

The infant is now aware of where a parent is looking and follows pointing gestures (Scaife – Bruner 1975). He or she starts to point to direct other's attention with a vocal expression of enquiry in face and voice, is responsive to tone of voice and uses inflections of voice to mark enquiry, request, instruction, refusal, etc. (Halliday 1975; Dore 1983). In previous months toys were presented in an animated way with a "gloss" of excitement, and movements of the baby were reflected back with "affect attunement" by the adult partner to entertain the infant (Stern et al. 1985). Now these feelings are projected into objects along common lines of attention in the shared field of interest (Butterworth – Grover 1988). Thus, objects can gain a coded interpersonal meaning.

In this prelinguistic phase of communication, as in the paralinguistics of the conversational communication of adults, the different forms of expression of various parts of the body are synchronized and equivalent (Trevarthen 1986b, 1994). When adults are chatting, they express their feelings and their ideas about events and objects in words, in facial expressions, in hand gestures and in changes of the posture of their body, simultaneously (Kendon 1980). The central motivation for all these expressions has a narrative organization that McNeill (1992) calls "inner speech". When speech becomes difficult, hand gestures are even more explicit.

Infants express themselves with their whole body from the first protoconversations (Trevarthen 1986a, b). At the end of the first year, when they are beginning to make "acts of meaning" in "protolanguage" (Halliday 1975), they express their communicative purpose with combinations of vocalization and word that show great versatility and that are influenced by imitation of how other persons express themselves. This is the stage at which, in a deaf family that uses hand signs for daily communication, a deaf infant begin to make the first coded word-like expressions in the form of imitated hand movements. At the same age, a hearing infant may be starting to make imitative word-like vocalizations. The deaf signing child will subsequently vocalize less, and the hearing infant will use speech more than gestures (Volterra – Caselli 1985). This is the beginning of the narrowing of possibilities towards the maternal language. Gradually, the intuitive grammar of emotions and interpersonal control of motives gives way in the second year to the lexico-grammar coded in words (Halliday 1975, 1978).

12. Taking up cultural tools and roles: Negotiating coded play with peers

Conventional use of tools, roles, and rituals of performance is mastered in the second year, beginning before speech, preparing the way for rapid learning of the maternal language. The change from manipulating for private gain to imitation of others' directives, indications, and evaluations leads the toddler to ideas that are already coded in words in the communication of older members of the community round the child (Tomasello et al. 1993).

The cooperative learning of language needs flexibility of imagination which is expressed in metaphorical or pretend play (Trevarthen – Logotheti 1987). Objects and actions become assimilated into shared purposes that can lead them to change identity. When the desired objects or events are absent and no substitute presents itself, they may be created entirely in imagination to satisfy the motive for play and communication. The child can invent play actions alone, too. Both solitary and shared play, if motivated by pretence, are creative of meanings that are ready to be shared. The development of the child's imagination and future learning are dependent on the ability to exchange points of view and imitated ideas with a companion, an ability that is tragically deficient in an autistic child.

When toddlers under two years of age are together, they show transitions of cooperation in play, swapping roles and imitating (Nadel-Bulfert – Baudonnière 1982; Stambak et al. 1986). Infants can attend to one another's interests and give and take objects from the first year. In the second year, they imitate invented actions, making synchronized, pretend play with similar objects a form of joke. They like to share meanings that have been made up in the play, codifying their actions by choosing the same object or the same way of acting. As in the relations of the younger infant with family members, toddlers in play groups form friendships that confirm and extend the possibilities for shared pretence. Reciprocal imitation permits negotiation of play themes and leadership in the creation of games. The children learn from each other as they trade initiatives. It is clear from the imaginative play of toddlers with their mothers or with peers that motivation for creating meanings and for coding objects and experiences in conventional categories, or symbolization, are part of the child's thinking before words are mastered (Trevarthen 1990; Trevarthen –Logotheti 1987)

Around the middle of the second year, at 15 to 20 months, a child has a fragile social identity that, like that of the 7 to 8 month old, is acutely aware of the potential difficulties of communication with strangers (Kagan 1982). It would appear that the imagination that is reaching out to learn how other persons categorize their experiences is sensitive to the risks of imitating without understanding. The withdrawal of a shy child into a private fantasy world may have much to teach us about the pathology of symbolic thought. Developments in preschool years show how mastery of thinking is dependent on a free and flexible regulation of contact with other minds by emotions.

There are impressive gains in social awareness as children begin to speak in their second year (Bretherton et al. 1981; Trevarthen 1986, 1987, 1993a; Trevarthen – Marwick 1986). They have fluent inventive or creative fantasy, recognizing objects of technical, industrial or artistic importance, roles and postures, socio-dramatic performances with pretend emotions, moral positions. These ideas and skills are imitated from others and spontaneously displayed to portray the child's personality as observed by others.

13. Conclusions: The primary motives for signs are interpersonal "forms of social life"

The way infants gain entry to language brings out the primacy of interpersonal cognitions and their emotional regulation. Emotions are clear in infants; they govern formation relationships to persons who share the infant's consciousness and extend the infant's power of action. They enable transfer of evaluations and reinforcements to the infant and the orientation of the infant to present circumstances. They enable the infant to learn by being taught.

Communicable ideas are built by combination of ideas generated by both intersubjective and pragmatic motives. Study of infant communication has lead some investigators to concentrate on pragmatics, the activities and intentional basis for cooperation in the prevailing circumstances, the initiatives of infants and their responses to maternal messages being interpreted in terms of the philosophy of Speech Acts (Bates et al. 1975). We need to enrich this insight with appreciation of the dynamic interpersonal motives that allow human minds to sense one another's impulses and preferences directly (Bretherton et al. 1981), as well as by way of orientations to a shared reality.

Infants show a shared grammar of action (Bruner 1975, 1983). But they do more. They demonstrate a person-awareness and obey what may be called a grammar of interpersonal motives, or, better, a grammar of emotions. The emotional syntax of the games that parents play with younger infants show the beginnings of a narrative organization that has many fundamental functions of language long before the infant can learn a word, regulating such terms as interpersonal contact, declarative/imperative tone, voice or personal attitude, indexical or deictic directive, quality of the substantive goal, identity of the substantive goal, locus of initiative, state of comprehension (wonder, surprise, recognition, impatience, anger, delight), and rational entailment.

Negotiation of the interpersonal values of motives for cognition and action lays the groundwork for cooperative task performance. In the second year, when the infant is proficient in cooperation and active in protolanguage, the motives of communication are ready for the learning of words. As words are learned, they attach to the known persons in communication, to their actions or to the objects that are indicated in their shared interest and actions. Different toddlers with differing parental support and differing temperaments may show preference for one or other of these kinds of topic in their first speech (Dore 1983; Nelson 1973; Barrett 1981).

The sharing of point of view and of continuous evaluation of events allows the imitated vocalizations and gestures to be caught and fixed in memory. The infant has to have motives to imitate and to display. Reciprocity of motivation is deficient in the developmental disorder of autism, which confirms that intrinsically regulated brain developments are behind the normal process of symbol learning in the second and third years (Trevarthen 1989b; Rogers – Pennington 1991).

The way infants enter into the language of their parents may help us develop a theory of how human cultural intelligence and its linguistic elaboration evolved (Trevarthen 1989a, 1993a). Comparison with apes shows that the developments around the end of the first year bring out forms of imitation and cooperative awareness that are innately different in humans (Tomasello 1988; Tomasello et al. 1993). This is a time when conspicuous developments are beginning in the cerebral hemispheres of the human infant that can be related to language and cultural learning.

The "theory of mind" approach to thinking about symbolic communication has given new vitality to investigations into how language and the concepts codified in it are responsible for enriching a communicable consciousness in the child. Metarepresentation, "thinking about thinking" is taken to be something that has to develop or be constructed by complication of a Piagetian kind of cognition that makes representations (Leslie 1987).

Both emotions and "minds of others" can be theorized in language as a luxury of philosophizing or to facilitate the negotiating of differences in viewpoint or attitude, but emotions and representations of others' minds begin their work in infancy with no theory, simply as natural correspondences that humans in contact with one another generate and partake of. Infants, though unsophisticated, are proficient in both emotional correspondence and imitation of the ideas incipient in the attentions and actions of another person. They communicate well and seek cooperative awareness before they can articulate, or be tested on, any "theory of mind".

It is evident from the way a child enters language that "thinking" and "communicating" are closely interdependent, and that symbols potentiate both. But there has been a philosophical bias in our psychology that favors attention to intrasubjective thinking and conceiving over the intersubjective communicating. Sociologists and linguists such as Malinowsky (1923), Halliday (1978), Habermas (1979), and Rommetveit (1987) have stressed that language is a medium for social cooperation in a culture, that it is not merely a representational system for cognitive processes in the individual that has a conventional code applied to it to permit transfer of thoughts. But, the theory of signs is classically rationalistic.

Peirce (CP 2.297) concludes that the word "symbol", from the Greek, means a thing 'thrown together', which is to say it is something made by persons agreeing together or making a contract. It is a meaning shared; it is even 'any expression of sentiment': "A symbol, once in being, spreads among the people. In use and in experience, its meaning grows." But he quickly collapses this insight into a cognitive interpretation for thinking in the mind of one subject. A word as a symbol is, he says, (CP 2.297-2.302):

> applicable to whatever may be found to realize the idea connected with the word. . . . The symbol is connected with its object by virtue of the idea of the symbol-using mind, without which no such connection would exist. . . . Symbols grow. They come into being by development out of other signs, particularly from icons, or from mixed signs partaking of the nature of icons and symbols. We think only in signs. These mental signs are of mixed nature; the symbol-parts of them are called concepts. If a man makes a new symbol, it is by thoughts involving concepts. So it is only out of symbols that a new symbol can grow.

Piaget took symbols only in their representational form as elements of deferred conception in one mind. For him, the essence of a concept is that it is an experience that has been made abstract by logical operations of comparison, class inclusion, ordering and so on. The communicable aspect is secondary and arrived at by mutual adjustment of significance between conceivers wanting common rules for thinking and knowing, to overcome conflict of egocentric ideas.

According to Peirce, the emotions expressed give communicative value or potential for the "interpreting subject" (the one who receives) to the cognitions and adjustments of purpose in the "representing subject" (the one who emits). Emotive and directive changes guide attentions and intentions of the interpreter into cooperative correspondence. The timing of expressions acts as a frame for organization of the functions of motivation and for construction of the intersubjective correspondence.

If no concept of a symbolic kind is possible until there has been an argument between conceiving minds and a negotiation of a convention of meaning, then every idea that has a *name* or a *form* or a *use* that is agreed has these in an agreed *meaning*. What an individual can conceive with no communication to others cannot be symbolic. In Peirce's terms, it can be merely an "icon" perceived in recollection of its object, or, at most, a reasoned out set of "indices" that recalls a set of experiences driven by and guiding action in space and time. This would be called a "procedural memory" now. Once a human subject is a member of a language community with its world of symbols and metaphors, then the individual can spin new thoughts and create narratives of imagining and pretending, in "declarative memory".

Peirce's Firstness, Secondness, and Thirdness are cognitive states of an individual introspective but active Subject. He is certainly interested in how these are made available to a second subject by communication (from representing subject to interpreting subject), but he does not describe the emotional qualities (paralinguistic expressions) by which the subjects enter into one another's mind states. His classification of signs, which has been founded on differences in the supposed mental representations or experiences of phenomena, can be matched by a classification that distinguishes forms of intersubjective communication, ways that express from one person to another attitudes or ideas about each other or their shared world (Table 1). These different messages, as units of a language, are symbolized by different terms and they are combined in meaningful narratives by the same transforming processes that motivate the changes and exchanges of expression in preverbal or paraverbal communication. In other words, the syntax of verbal expression in speech or text is derivative of, or built upon, a nonreferential process that regulates the changes and exchanges of motivation and feeling between subjects in all communication where cooperative awareness is being created. This is the level of semiotic process at which infants communicate.

Table 1. Peirce's typology of signs matched by a typology of intersubjective communication.

	Phenomenal Attribute	Communicative Sign	Motivation
Firstness	Quality	Icon	Emotion about Object *in Subject*
Secondness	Actuality	Index	Direction, Place and Form for Object in Intended Action *of Subject*
Thirdness	Probability by Agreement	Symbol	Cooperative Identity and Value *between Subjects*

Note

1. My interest in and knowledge of semiotics are largely the consequence of communication with Walter Koch. Professor Koch invited me to his International Symposium on "The Nature of Culture" in Bochum in October, 1986, and there I met experts in the field for the first time. Contacts I made at that stimulating meeting have been continued with much profit to me. The present paper was written in the home of David and Anna Smillie, Sarasota. Their kindness and David's library and scholarly interest were essential to its completion.

References

Aaronson, D. – Robert W. Rieber (eds.)
1975 *Developmental psycholinguistics and communication disorders.* (Annals of the New York Academy of Sciences 263.) New York: Academy of Sciences.

Bakeman, Roger – Lauren B. Adamson
1984 "Coordinating attention to people and objects in mother-infant and peer-infant interaction", *Child Development* 55: 1278-1289.

Barrett, Martyn D.
1981 "The communicative functions of early child language", *Linguistics* 19: 273-305.

Bates, Elizabeth – L. Camaioni – Virginia Volterra
1975 "The acquistition of performatives prior to speech", *Merrill Palmer Quarterly* 21: 205-226.

Bateson, Gregory
1956 "The message 'This is play'", in: B. Schaffer (ed.), 145-242.

Bateson, Mary C.
1975 "Mother-infant exchanges: The epigenesis of conversational interaction", in: D. Aaronson – Robert W. Rieber (eds.).

Beebe, Beatrice – D. Stern – Jerome Jaffe
1979 "The kinesic rhythm of mother-infant interactions", in: Aron W. Siegman – Stanley Feldstein (eds.).

Beebe, Beatrice – Jerome Jaffe – Stanley Feldstein – K. Mays – D. Alson
1985 "Inter-personal timing: The application of an adult dialogue model to mother-infant vocal and kinesic interactions", in: Tiffany M. Field – Nathan Fox (eds).

Bever, Thomas G. (ed.)
1982 *Regressions in mental development: Basic phenomena and theories.* Hillsdale, N.J.: Erlbaum.

Bråten, Stein
1988 "Dialogic mind: The infant and adult in protoconversation", in: Marc E. Carvallo (ed.), 187-205.
Bretherton, Inge – S. McNew – Majorie Beeghly-Smith
1981 "Early person knowledge as expressed in gestural and verbal communication: When do infants acquire a 'theory of mind'?", in: Michael E. Lamb – L.R. Sherrod (eds.).
Bruner, Jerome S.
1975 "The ontogenesis of speech acts", *Journal of Child Language* 2: 1-19.
1983 *Child's talk: Learning to use language*. New York: Norton.
Bruner, Jerome S. – Alison Jolly – Katherine Sylva (eds.)
1975 *Play: Its role in evolution and development*. Harmondsworth: Penguin.
Bruner, Jerome S. – Victoria Sherwood
1975 "Early rule structure: The case of peekaboo", in: Jerome S. Bruner – Alison Jolly – Katherine Sylva (eds.), 277-285.
Buck, Ross
1984 *The communication of emotion*. New York: Guilford Press.
Bullowa, Margaret (ed.)
1979 *Before speech: The beginning of human communication*. Cambridge: University Press.
Butterworth, George – Grover, Lawrence M.
1988 "The origins of referential communication in human infancy", in: Laurence Weiskrantz (ed.).
Butterworth, George – Paul Light (eds.)
1982 *Social cognition: Studies of the development of understanding*. Brighton: Harvester Press.
Carvallo, Marc E. (ed.)
1988 *Nature, cognition, and system*. Dordrecht: Kluwer
Cheney, Dorothy L. – Robert M. Seyfarth
1985 "Social and nonsocial knowledge in vervet monkeys", in: Laurence Weiskrantz (ed.).
1992 "Précis of 'How monkeys see the world'", *Behavioral and Brain Sciences* 15: 135-182.
Cicchetti, Dante – P. Hesse
1983 "Affect and intellect: Piaget's contributions to the study of infant emotional development", in: Robert Plutchik – H. Kellerman (eds.).
Cohn, Jeffrey F. – Edward Z. Tronick
1983 "Three-month old infants' reaction to simulated maternal depression", *Child Development* 54: 185-193.

DeCasper, Anthony J. – A.A. Carstens
1981 "Contingencies of stimulation: Effects on learning and emotion in neonates", *Infant Behavior and Development* 4: 19-35.

Descartes, René
1649 *Les passions de l'âme*. Amsterdam: Louis Elzevier.
[1911] [English translation by E.S. Haldane – G.R.T. Ross. Cambridge: University Press.]

Dore, John
1983 "Feeling form and intention in the baby's transition to language", in: Roberta Golinkoff (ed.).

Duncan, Starkey – Donald W. Fiske
1977 *Face-to-face interaction*. Hillsdale, N.J.: Erlbaum.

Eco, Umberto
1979 *A theory of semiotics*. Bloomington: Indiana University Press.

Emde Robert N.– Zeynep Biringen – R.B. Clyman – D. Oppenheim
1991 "The moral self in infancy: Affective core and procedural knowledge", *Developmental Review* 11: 251-270.

Fernald, Anne – Traute Taeschner – Judy Dunn – Mechthild Papousek – Benedict de Boysson-Bardies – Ikuko Fukui
1989 "A cross-language study of prosodic modifications in mothers' and fathers' speech to preverbal infants", *Journal of Child Language* 16: 477-501.

Field, Tiffany M.
1985 "Neonatal perception of people: Maturational and individual differences", in: Tiffany M. Field – Nathan Fox (eds.), 177-197.

Field, Tiffany M. – R. Woodson – R. Greenberg – D. Cohen
1982 "Discrimination and imitation of facial expressions by neonates", *Science* 218: 179-181.

Field, Tiffany M. – Nathan Fox (eds.)
1985 *Social perception in infants*. Norwood, N.J.: Ablex.

Golinkoff, Roberta (ed.)
1983 *The transition from pre-linguistic communication*. Hillsdale, N.J.: Erlbaum.

Gouzoules, Harold – Sarah Gouzoules – Peter Marler
1985 "External reference and affective signalling in mammalian vocal communication", in: Gail Zivin (ed.).

Habermas, Jürgen
1979 *Communication and the evolution of society*. London: Heinemann.

Halliday, Michael A. K.
1975 *Learning how to mean*. London: Arnold.

1978 *Language as social semiotic*. London: Arnold.

Heimann, Mikael

1989 "Neonatal imitation, gaze aversion, and mother-infant interaction", *Infant Behavior and Development* 12: 495-505.

Heimann, Mikael – Joseph Schaller

1985 "Imitative reactions among 14-21 days old infants", *Infant Mental Health Journal* 6: 31-39.

Heimann, Mikael – Katherine E. Nelson – Joseph Schaller

1989 "Neonatal imitation of tongue protrusion and mouth opening: Methodological aspects and evidence of early individual differences", *Scandanavian Journal of Psychology* 90: 90-101.

Hinde, Robert A. – A.N. Perret-Clermont – Joan Stevenson-Hinde (eds.)

1986 *Social relationships and cognitive development*. Oxford: Clarendon.

Holmlund, C.

1990 Development of turn-taking as a sensorimotor process in the first 3 months: A sequential analysis. [Paper presented at the 5th International Congress of Child Language, Budapest.]

Hubley, Penelope – Colwyn Trevarthen

1979 "Sharing a task in infancy", in Ina C. Uzgiris (ed.), 4: 57-80.

Hurford, James R.

1989 "Biological evolution of the Saussurean sign as a component of the language acquisition device", *Lingua* 77: 187-222.

Jahoda, Gustav – I.M. Lewis (eds.)

1988 *Acquiring culture: Cross cultural studies in child development*. Beckenham, Kent: Croom Helm.

Kagan, Jerome

1982 "The emergence of self", *Journal of Child Psychology and Psychiatry* 23: 363-381.

Kendon, Adam

1980 "Gesticulation and speech: Two aspects of the process of utterance", in Mary Ritchie Key (ed.), 207-227.

Key, Mary Ritchie (ed.)

1980 *The relationship of verbal and nonverbal communication*. New York: Mouton.

Klinnert, M.D. – Joseph J. Campos – James F. Sorce – Robert N. Emde – Marilyn Svejda

1983 "Emotions as behavior regulators: Social referencing in infancy", in: Robert Plutchik – Henry Kellerman (eds.).

Koch, Walter A. (ed.)

1989 *The nature of culture*. Bochum: Brockmeyer.

Kugiumutzakis, John E.
1985 *The origins, development, and function of early infant imitation.* (Acta Universitatis Uppsaliensis, 35.) Uppsala University: Ph. D. dissertation.
1993 "Intersubjective vocal imitation in early mother-infant interaction", in: Jacqueline Nadel – L. Camaioni (eds.).
Lamb, Michael E. – L.R. Sherrod (eds.)
1981 *Infant social cognition: Empirical and theoretical considerations.* Hillsdale, N.J.: Erlbaum.
Lelord, G. – J.P. Muh – M. Petit (eds.)
1989 *Autisme et troubles du development global de l'enfant.* Paris: L'Expansion Scientifique Française.
Lenneberg, E.H. (ed.)
1974 *Language and brain: Developmental aspects.* (Neurosciences Research Program Bulletin, 12.) Boston: Neurosciences Research Program.
Leslie, Allan
1987 "Pretense and representation: The origins of 'Theory of Mind'", *Psychological Review* 94: 412-26.
Lewis, Michael – L. Rosenblum (eds.)
1974 *The effect of the infant on its caregiver.* New York: John Wiley.
Lindblom, Bjorn – Rolf Zetterstrom (eds.)
1986 *Precursors of early speech.* Basingstoke: Macmillan.
Lipsitt, Lewis P. – Carolyn K. Rovee-Collier (eds.)
1981 *Advances in infancy research.* Vol. 1. Norwood, N.J.: Ablex.
Lock, Andrew (ed.)
1978 *Action: Gesture and symbol: The emergence of language.* London: Academic Press.
Lotman, Jurij
1979 "Culture as collective intellect and problems of artificial intelligence", in: L.M. O'Toole – Anne Shukman (eds.), 84-96.
Lyons, John
1977 *Semantics.* Cambridge: University Press.
Malinowski, Bronislaw
1923 "The problem of meaning in primitive language", Supplement to Charles K. Ogden – Ivor A. Richards, 296-336.
Mantanegro, Jacques – Anastasia Tryphon – Sylvain Dionnet (eds.)
1987 *Symbolisme et connaissance. Symbolism and knowledge.* (Cahier No. 8, Jean Piaget Archives Fondation.) Geneva: Jean Piaget Archives Fondation.

Maratos, Olga

1982 "Trends in development of imitation in early infancy", in: Thomas G. Bever (ed.), 81-101.

Marler, Peter

1984 "Animal communication: Affect or cognition", in: Klaus R. Scherer – Paul Ekman (eds.), 345-365.

McNeill, David

1992 *Hand and mind: What gestures reveal about thought.* Chicago: University Press.

Meltzoff, Andrew N.

1985 "The roots of social and cognitive development: Models of man's original nature", in: Tiffany M. Field – Nathan A. Fox (eds.).

Meltzoff, Andrew N. – M. Keith Moore

1983 "Newborn infants imitate adult facial gestures", *Child Development* 54: 702-709.

Murray, Lynne

1992 "The impact of postnatal depression on infant development", *Journal of Child Psychology and Psychiatry* 33 : 543-561.

Murray, Lynne – Colwyn Trevarthen

1985 "Emotional regulation of interactions between two-month olds and their mothers", in: Tiffany M. Field – Nathan Fox (eds.).

1986 "The infant's role in mother-infant communication", *Journal of Child Language* 13: 15-29.

Nadel, Jacqueline

1986 *Imitation et communication entre jeunes enfants.* Paris: Presses Universitaires de France.

Nadel, Jacqueline – L. Camaioni (eds.)

1993 *New perspectives in early communicative development.* London: Routledge.

Nadel-Bulfert, Jacqueline – Pierre-Marie Baudonnière

1982 "The social function of reciprocal imitation in 2-year old peers", *International Journal of Behavioural Development* 5: 95-109.

Neisser, Ulric (ed.)

1994 *Ecological and interpersonal knowledge of the self.* Cambridge: University Press. [in press]

Nelson, Katherine

1973 *Structure and strategy in learning to talk.* (Monographs of the Society for Child Development, Serial No. 149, 38 (1-2).) Chicago: University Press.

Nespoulous, Jean-Luc – Paul Perron – André Roch Lecours (eds.)
1986 *The biological foundation of gestures: Motor and semiotic aspects*. Hillsdale, N.J.: Erlbaum.
Ogden, Charles K. – Ivor A. Richards
1923 *The meaning of meaning*. London: Routledge and Kegan Paul.
Olson, David (ed.)
1980 *The social foundations of language and thought: Essays in honor of J. S. Bruner*. New York: Norton.
O'Toole, Lawrence M. – Anne Shukman (eds.)
1979 *Dramatic structure: Poetic and cognitive semantics*. (Russian poetics in translation 6.) Oxford: Holden.
Papousek, Hanus
1967 "Experimental studies of appetitional behavior in human newborns and infants", in: Harold W. Stevenson – E.H. Hess – H.L. Rheingold (eds.).
Papousek, Mechthild – Hanus Papousek
1981 "Musical elements in the infant's vocalization: Their significance for communication, cognition, and creativity", in: Lewis P. Lipsitt – Carolyn K. Rovee-Collier (eds.), 163-224.
1989 "Forms and functions of vocal matching in interactions between mothers and their precanonical infants", *First Language* 9: 137-158.
Papousek, Mechthild – Hanus Papousek – Marc H. Bornstein
1985 "The naturalistic vocal environment of young infants: On the significance of homogeneity and variability in parental speech", in: Tiffany M. Field – Nathan Fox (eds).
Peirce, Charles Sanders
1931-1958 *Collected Papers of Charles Sanders Peirce*, vols. 1-6 ed. Charles Hartshorne and Peter Weiss, vols. 7-8 ed. Arthur W. Burks. Cambridge, Mass.: Havard University Press.
Perlmutter, Marion (ed.)
1984 *Parent-child interaction and parent-child relations in child development*. (The Minnesota symposia on child psychology 17.) Hillsdale, N.J.: Erlbaum.
Piaget, Jean
1962 *Play, dreams, and imitation in childhood*. London: Routledge and Kegan Paul.
Plutchik, Robert – Henry Kellerman (eds.)
1983 *Emotion, theory, research, and experience. Vol. 2: Emotions in early development*. New York: Academic Press.
Ratner, Nan B. – Jerome S. Bruner
1978 "Games, social exchange, and the acquisition of language", *Journal of Child Language* 5: 391-400.

Reddy, Vasudevi
1990 "Playing with others' expectations: Teasing and mucking about in the first year", in: A. Whiten (ed.), 143-158.
Rogers, Sally J. – Bruce F. Pennington
1991 "A theoretical approach to the deficits in infantile autism", *Development and Psychopathology* 3: 137-162.
Rogoff, B.
1990 *Apprenticeship in thinking: Cognitive development in social context*. Oxford: University Press.
Rommetveit, Ragnar
1987 "Meaning, context, and control: Convergent trends and controversial issues in current social-scientific research on human cognition and communication", *Inquiry* 30: 77-99.
Saussure, Ferdinand de
1960 *Course in general linguistics*. London: Owen.
Scaife, Michael – Jerome S. Bruner
1975 "The capacity for joint visual attention in the infant", *Nature* 253: 265-266.
Schaffer, B. (ed.)
1956 *Group processes*. New York: Josiah Macy Foundation.
Schaffer, H.R. (ed.)
1977 *Studies in mother-infant interaction: The Loch Lomond symposium*. London: Academic Press.
Scherer, Klaus R. – Paul Ekman (eds.)
1984 *Approaches to emotion*. Hillsdale, N.J.: Erlbaum.
Sebeok, Thomas A. – Jean Umiker-Sebeok (eds.)
1990 *The semiotic web 1989*. Berlin: Mouton de Gruyter.
Siegman, Aron W. – Stanley Feldstein (eds.)
1979 *Of speech and time: Temporal speech patterns in interpersonal contexts*. Hillsdale, N.J.: Erlbaum.
Stambak, Mira – M. Ballion – M. Breauté – S. Rayna
1986 "Pretend play and interaction in young children", in: Robert A. Hinde – A. N. Perret-Clermont – Joan Stevenson-Hinde (eds.).
Steele, Ross – Terry Threadgold (eds.)
1987 *Language topics: Essays in honour of Michael Halliday*. Vol. 1. Amsterdam: John Benjamins.
Stern, Daniel N.
1974 "Mother and infant at play: The dyadic interaction involving facial, vocal, and gaze behaviors", in: Michael Lewis – L. Rosenblum. (eds.).
1985 *The interpersonal world of the infant: View from psychoanalysis and development psychology*. New York: Basic Books.

Stern, Daniel N. – L. Hofer – Wendy L. Haft – John Dore
1985 "Affect attunement: The sharing of feeling states between mother and infant by means of intermodal fluency", in: Tiffany M. Field – Nathan Fox (eds.).

Stern, Daniel N. – S. Spieker – K. MacKain
1982 "Intonation contours as signals in maternal speech to pre-linguistic infants", *Developmental Psychology* 18: 727-735.

Stevenson, H.W. – E.H. Hess – H.L. Rheingold (eds.)
1967 *Early behavior: Comparative and developmental approaches*. New York: Wiley.

Stokoe, William C. Jr. – Virginia Volterra (eds.)
1985 *Sign language studies*. Silver Spring, M.D.: Linstock.

Tomasello, Michael
1988 "The role of joint attentional processes in early language development", *Language Sciences* 10: 69-88.

Tomasello Michael – A. C. Kruger – H. H. Ratner
1993 "Cultural learning", *Brain Behavior and Evolution* 16(3): 495-552.

Trevarthen, Colwyn
1974 "The psychobiology of speech development", in: E.H. Lenneberg (ed.), 570-585.

1977 "Descriptive analyses of infant communication behavior", in: H. R. Schaffer (ed.), 227-270.

1979 "Communication and cooperation in early infancy: A description of primary intersubjectivity", in: Margaret Bullowa (ed.), 321-347.

1980 "The foundations of intersubjectivity: Development of interpersonal and cooperative understanding in infants", in: David Olson (ed.), 316-342.

1982 "The primary motives for cooperative understanding", in: George Butterworth – Paul Light (eds.), 77-109.

1984 "Emotions in infancy: Regulators of contacts and relationships with persons", in: Klaus Scherer – Paul Ekman (eds.), 129-157.

1985 "Facial expressions of emotion in mother-infant interaction", *Human Neurobiology* 4: 21-32.

1986a "Development of intersubjective motor control in infants", in: Michael G. Wade – H.T.A. Whiting (eds.), 209-261.

1986b "Form, significance and psychological potential of hand gestures of infants", in: Jean-Luc Nespoulous – Paul Perron – André Roch Lecours (eds.),149-202.

1987 "Sharing makes sense: Intersubjectivity and the making of an infant's meaning", in: Ross Steele – Terry Threadgold (eds.), 177-199.

1988 "Universal cooperative motives: How infants begin to know the language and skills of the culture of their parents", in: Gustav Jahoda – I.M. Lewis (eds.).

1989a "Motives for culture in young children - their natural development through communication", in: Walter A. Koch (ed.), 80-119.

1989b "Les relations entre autisme et développement socio-culturel normal: Arguments en faveur d'un trouble primaire de la régulation du développement cognitif par les émotions", in: G. Lelord – J. P. Muh – M. Petit – D. Sauvage (eds.), 56-80.

1990 "Signs before speech", in: Thomas A. Sebeok – Jean Umiker-Sebeok (eds.), 689-755.

1993a "An infant's motives for speaking and thinking in the culture", in: A. H. Wold (ed.).

1993b "The function of emotions in early infant communication and development", in: Jacqueline Nadel – L. Camaioni (eds.).

1994 "The self born in intersubjectivity: An infant communicating", in: Ulric Neisser (ed.).

Trevarthen, Colwyn – Penelope Hubley

1978 "Secondary intersubjectivity: Confidence, confiding, and acts of meaning in the first year", in: Andrew Lock (ed.), 183-229.

Trevarthen, Colwyn – Helen Marwick

1986 "Signs of motivation for speech in infants, and the nature of a mother's support for development of language", in: Bjorn Lindblom – Rolf Zetterstrom (eds.), 279-308.

Trevarthen, Colwyn – Katerina Logotheti

1987 "First symbols and the nature of human knowledge", in: Jacques Montanegro – Anastasia Tryphon – Sylvain Dionnet (eds.), 65-92.

Uzgiris, Ina C.

1981 "Two functions of imitation during infancy", *International Journal of Behavioral Development* 4,1: 1-12.

1984 "Imitation in infancy: Its interpersonal aspects", in: Marion Perlmutter (ed.).

Uzgiris, Ina C. (ed.)

1979 *Social interaction during infancy: New directions for child development.* Vol. 4. San Francisco: Jossey-Bass.

Volterra, Virginia
1981 "Gestures, signs, and words at two years: When does communication become language?", *Sign Language Studies* 33: 351-362.
Volterra, Virginia – M.C.C. Caselli
1985 "From gestures and vocalizations to signs and words", in: William C. Stokoe – Virginia Volterra (eds.).
Vygotsky, Lev S.
1962 *Thought and language*. Cambridge, Mass.: M.I.T. Press.
Wade, Michael G. – H.T.A. Whiting (eds.)
1986 *Motor development in children: Aspects of coordination and control*. Dordrecht: Martinus Nijhof.
Watson John S.
1972 "Smiling, cooing, and the game", *Merrill-Palmer Quarterly* 18: 323-339.
Weiskrantz, Laurence (ed.)
1985 *Animal intelligence*. Oxford: University Press.
1988 *Thought without language*. Oxford: Clarendon.
Werner, Heinz – Bernard Kaplan
1963 *Symbol formation*. New York: Wiley.
Whiten, A. (ed.)
1990 *Natural theories of mind: Evolution, development, and simulation of everyday mindreading*. Oxford: Blackwell.
Winnicott, Donald W.
1960 "The theory of the parent-infant relationship, *International Journal of Psychoanalysis* 41: 585-595.
[1990] [Republished in Donald W. Winnicott, *The maturational process and the facilitating environment*. London: Karnak Books.]
Wold, Astri H. (ed.)
1993 *The dialogical alternative: Festschrift for Ragnar Rommetveit*. Oxford: University Press.
Zivin, Gail (ed.)
1985 *The development of expressive behavior: Biology-environment interactions*. Orlando: Academic Press.

PART III

Glottogenesis: Phylogeny, ontogeny, and actogeny

Language and the origin of semiosis

Robin Allott

Language is the type of semiosis which has been most closely examined and which has served as a model for considering other forms of semiosis. Semiotics has been based, certainly in the case of language, very much on the proposition of Saussure that the sign is arbitrary and that the sign is conventional or social. This paper seeks to trace out the implications for semiotics of a very different account from Saussure's of the origin, development and functioning of language (cf. Allott 1989), leaving it open whether one should conclude, in the light of this, that language does not constitute a paradigm or model for a general science of semiotics (and is not a typical or useful example of a semiotic system) or that language should be treated as the paradigm, but a totally changed paradigm, so that the new view of language will require a restructuring of semiotics and lead to a much more biological and indeed neurological approach to the science of signs.

1. Language: Superordinate or subordinate?

It is still very much a live issue whether language is the prototypical semiotic system, from which all other semiotic systems (at any rate human semiotic systems) derive and on which they should be modelled or whether a science of semiotics should go much wider than linguistics and in effect prescribe to linguistics how language should be approached and analysed as only one among many semiotic systems. Followers of Peirce and Saussure would classify semiotics as superordinate and language as subordinate. Others, e.g., Barthes and the Tartu school, would assert the primacy of language, whilst others would occupy various middle positions. It is notable, however, that those who treat language as the semiotic paradigm are for the most part far from accepting the complex, formal development of linguistics as the model for other sign systems.

Perhaps the problem can be traced back to Locke. Locke's remarks at the end of Book IV of the *Essay* (21.4) virtually amount to a specification of the program which Peirce proposed for himself but never completed. Locke's influence, though indirect, can also be seen in Saussure. A typically representationalist view of the sign was accepted by Saussure and underlies his account of the relation between language and the study of sign systems more generally. It was against this background that he outlined the specific character of language as a system of signs in which the only essential thing was the union of meaning and sound images, comparable to "writing-systems, the signs of deaf-mutes, symbolic rites, polite formulas, military signals, etc." (Saussure [1916]: 16). Succeeding writers on semiotics have carried the argument to one or the other extreme. Kristeva (1969: 58) sees that for Saussure linguistics is only part of a general science of signs and proposes that linguistics in due course will merge in a universal science or, more colorfully: "La sémiotique se construira alors à partie du cadavre de la linguistique, d'une mort . . . à laquelle la linguistique se résignerait après avoir préparé le terrain à la sémiotique." At the other extreme, as already noted, Barthes asserts the unquestioned primacy of language from which all other sign systems must be derived.

The most sustained recent attack on the Saussurean account of the relation between language and semiotics is that by Holdcroft (1991: 156, 160):

> [If one followed Saussure] the way to begin the study of language is to develop the theory of signs, to describe the principles that hold for all signs and sign systems, and then to derive a description of linguistic sign systems from the theory and the principles . . . However, . . . it is surely a questionable strategy [because] to begin with, it takes it for granted that different sign systems have something interesting in common, in spite of the fact that the examples given are very diverse . . [More fundamentally,] the conception of a sign that Saussure employs is . . . not a plausible candidate for one of signs in general.

What is one to conclude? To relate the structures and processes of language to those of other sign systems, the first step surely is to examine what the structures and processes of language are. The science, or technique, that is meant to do this is linguistics. Linguistics as a science or craft has passed through many stages since Saussure's *Cours* was published. The stages have produced radically different accounts of the functioning of language; the Chomskyan system has been in process of nearly continuous change for more than 30 years. In either its earlier Transformational-Generative form or in its more recent Government-Binding form it is difficult to see how general coding and decoding principles, general sign system principles can be extracted which would be readily applicable to other aggregates or systems of signs as they have been identified by semioticians. And, of course, beyond and apart from Chomsky there are many other competing and inconsistent accounts of language. How can one say that language is the semiotic prototype if there is absolutely no consensus about the proper way to characterize and analyze language?

Equal confusion reigns about the principles, characteristics, scope of any general semiotic science, and about the description and analysis of other sign systems identified as such by semioticians. There is uncertainty and disagreement at every level, from the lowest: the identification of the sign, to the highest concepts of underlying structure in sign systems. How can one say that language is an exemplar, a specimen of a sign system, subordinate rather than superordinate, if the criteria by which one identifies signs and the methodology for analyzing sign systems are apparently completely lacking? A semiotic science has to be founded on recognition of commonalties between diverse sign systems, but there is no systematic account of what language shares with other sign systems.

2. The "science of signs" ?

Rather gloomy current assessments of semiology or semiotics, of the prospects of the "science of signs" can be set against the hopes of the originators. I have already referred to Locke's prospectus for *semeiotikė* and Saussure's equally optimistic write-up of the then nonexistent science of semiology. More recent statements of what semiotics might be or aim to be can be contrasted with the far-reaching ambitions of Peirce and Morris. Peirce saw semiotics as the key to human knowledge and, following Locke, as to be identified or at least to be closely linked to a new science of logic. His aspiration was "to erect a philosophical edifice that shall outlast the vicissitudes of time" (Peirce, CP 1.1). The ambitions of semiotics have been large; the achievements remarkably small. Why should this be? One answer may be that there was unclarity at the very beginning in the concept of the sign, the atomic element in a hypothetical system. However, some scholars have foreshadowed a different and perhaps more promising line of attack for semiotics. Thom has expressed the view that semiotics would profit from closer contact with biology or ethology. In the set of replies to Koch's valuable questionnaire, Pinxten suggested that the borders of semiotics are the neurobiological sciences where "absolutely amazing advances are being made today . . . in neurophilosophy and the general model of connectionism" (Pinxten 1989: 34).

What can usefully be said about the fundamental problem in semiotics, the nature of the sign? Ever since Locke made the relation between signs (words) and ideas a central feature of human understanding, there has been debate and divergence about the nature of the sign. "Sign" is a word used in ordinary speech; if it is to be given a more precise meaning in semiotics, it is from the ordinary meanings of "sign" that the narrowing down must start. The range of meanings and applications of the word "sign" to be found in any dictionary shows that one cannot use the word as though what it means is self-evident. Nor is there agreement about the classification of signs. Peirce's three broad classes of sign: index, icon and symbol, are still in active use but, though he later went on to elaborate his classification, the baroque, or Gaudiesque, proliferation of sign classes which he outlined has not been useful (any more than the rather similar elaboration by Morris).

There is no greater clarity about the "sign process", by which is meant the sequence of events or operations extending from the origination of the sign (typically by a human source), through its reception by the human destination and then on to its internal effects on the receiver. Peirce analyzed the process into three elements, object, sign and interpretant. This sequence of events can be described in mentalist or behaviorist terms. Morris translated the sequence into stimulus and response; characteristic sign-using behavior, on this view, is exemplified in putting on a coat at the sight of rain clouds rather than in reading a book.

There seem to be logical as well as psychological difficulties with the sign process as it has been described. Peirce said that nothing is a sign which is not interpreted, so on this view signs have no objective existence; whether a sign is a sign depends upon the receiver of the sign. Then the interpretant of the sign, which is what constitutes the sign as a sign, may be an idea or it may also be, or be described as, a habit, a behavioral disposition, a tendency to action. Yet the interpretant also may be a sign, or a more developed sign as Peirce described it, which within the receiver in its turn may give rise to or be associated with other interpretants, which also would be signs, in what could be, in Peirce's terms, part of an infinite series of signs. This mentalist account seems unclear, even bewildering. The behaviorist alternative has been discarded by most linguists, psychologists and neurologists.

3. Language origin and function: The motor theory

Language is the capacity of one individual to alter, through structured sound emission, the mental organization of another individual. Language is more than speech just as perception is more than the structure and functioning of the eye. In both cases we have also to be concerned with the neural organization underlying the functions of speech and visual perception. The theory is that language was constructed on the basis of a previously existing complex system, the neural motor system. The programs and procedures which evolved for the construction and execution of simple and sequential motor movements formed the basis of the programs and procedures going to form language.

A principal theme is the mosaic evolution of language, the fitting together of a whole array of elements, anatomical, neural and behavioral. Two important behavioral elements for language are imitation and the categorical perception of speech sound, both abilities found in some animals. Imitation, of speech or other sound or bodily movement, involves a remarkable and complex linking of perception and motor organization. The capacity to discriminate categorically between human speech sounds has, surprisingly, also been found in a variety of animals and in extremely young human infants. These and other behavioral prerequisites for language depend on the intimate involvement of the motor control system and the existence of crossmodal processes. Development of the language capacity has resulted from the progressive establishment of new crossmodal or transfunctional neural linkages. Cerebral reorganization in the sense that the interconnectedness of different brain regions concerned with what are usually considered distinct functions has substantially increased. This extensive relation between language and the motor system is what one might reasonably expect, given the central role of the motor system in all behavior and the essentially motor character of speech production. The motor system forms the indispensable mediator between language and perception.

The essential additional hypothesis is that the motor system, prior to the development of language, was built up from a limited number of primitive elements - units of motor action - which could be formed into more extended motor programs. If this is so, then one can look for a direct correspondence between the primitive motor elements and the fundamental elements of spoken language, the phonemic system. The processes of word formation and syntactic rules for constructing word sequences would then be derived from the neural rules governing the union of motor elements into simple and more complex actions. If language is in this way derived from the motor system, there is no reason to believe that any aspect of language - sound elements, words or syntactic structure - is necessarily arbitrary.

The relation between motor programming and speech programming can be examined at the phonemic, lexical and syntactic levels. For phonemes, this leads to the idea of an invariant program for each phoneme, or "auditory targeting", a motor alphabet underlying speech, related to the elementary motor patterns underlying other forms of action. On the motor theory, the categorization of speech sounds is derived from organization prior to language, and specifically from the categorization of motor programs used in constructing and executing all forms of bodily action. What the rhesus monkey, or the chinchilla , in their capacities for the categorical perception of speech sounds, share with the young human infant is very similar skeletal and muscular organization. The specificity of the phoneme is the accidental result of the application of the different elementary motor subprograms to the muscles which went to form the articulatory system. The theory suggests that the phonemes of all languages are likely to be selected from a limited set of possible distinctive speech sounds.

The link between the motor system and the formation of words follows. The hierarchical structure of the motor system is built on the basis of a limited set of motor elements. These are combined in an unlimited number of ways (motor words). Words in speech are a read-out of neural structures in the same way as actions or facial expressions. A word, as a neural structure, can be formed from the coactivation of the motor subprograms for phonemes which are then melded or shingled together to form a distinct neural program for the whole word. Experimental approaches with the creation of artificial words have suggested that there can be a lawful relation between speech sounds and auditory or visual percepts. Research into sound symbolism suggests that there is an isomorphism at the motor level between speech and the contents of perception. The object seen produces a motor pattern which is readily transferable as a motor program to the articulatory system and so becomes the associated word for the thing. The neuromuscular sequences which are the immediate motor programs underlying words are derived from the integration of the neural structures underlying perception in all its forms (visual, auditory, tactile etc.) and motor organization.

The phonemes which go to form any word manifest pre-existing motor programs for the general control of bodily activity, and if, in the simple case, the words relate to actions or familiar external objects and percepts, then inherent in each action represented by a word is its own overall motor program, its own combination of motor control elements. Motor programs are also of essential importance in perception, in vision (eye movements in scanning and in other adjustments), and in perception through touch. The motor theory suggests that the selection of the structure of phonemes for a particular action or object derives from the motor aspects of the action or perception to which the word refers.

There must, on the theory, be a close relation between the structuring of motor activity, motor syntax, and the organization of language, speech syntax. The motor theory suggests that syntax also has a physiological origin, an origin in cerebral organization for action, for motor control. If one treats syntax, at its simplest, as the putting together of motor-derived words, there is a clear analogy between the formation in speech of a sentence or a phrase (or more extended discourse) on the one hand and the formation of an extended bodily movement, an action sequence, a complex neural motor routine on the other hand. The specifically syntactic aspects, i.e. ordering regularities, are in the motor theory derived from ordering regularities in the neural sub-structure for motor control and perception (ordering regularities in visual perception and in the execution of bodily actions). For each of the three main components in syntax (word classes, ordering regularities and function words), the relation to the motor theory may take the form of a relation directly with the organization of action - the grammar of action, or a relation directly with the organization of perception - the grammar of vision. Vision, like action, is substantially motor-based. Scanning of a visual scene by the eye is a serial process, just as spoken language is a serial process. The iconicity of syntax in terms of word order derives from this.

4. Motor theory and other semiotic systems

From the summary of the motor theory in the previous section, one can pick out a number of distinct aspects which may be helpful in looking at other semiotic systems.

Animal (vertebrate) precursor features. In other semiotic systems one ought to consider what the evolutionary history might have been and whether anything in the system has parallels in animal organization and communication.

Mental organization change. The role of language is to produce changes in the mind (brain) of the hearer which structurally resemble those in the mind (brain) of the speaker. This presumably would be a feature of all semiotic systems where there is a human source and destination. It may even have relevance where the communication is mediated in space or time e.g. processes similar to secondary forms, such as written language.

Neural basis. All semiotic systems and "messages" ultimately will be represented by changes in neural patterning: changes in synaptic strengths, establishment of new excitatory or inhibitory connections, dendritic growth (as hypothesized in PDP or connectionist approaches to neural modelling).

Crossmodal processes. The role of any semiotic system may be to produce specific action from perception, whether the perception is of the behavior of a human being or of changes otherwise in the external environment. This will depend on the existence of crossmodal links between different brain functions. What these may be can be examined in relation to particular sign systems.

Motor system centrality. All behavior in evolutionary terms is a matter of motor primacy, and one would expect the motor primacy to be reflected in the construction and functioning of other semiotic systems.

Primitive motor elements (programs). Because of the limitations of the human (and other) brains - there cannot be preformed neural programs to provide for every possible sentence or every possible action - and the demands of an unpredictable environment, semiotic systems have to be open-ended. The parsimonious way is to rely on a limited set of primitive elements which can be combined in many ways to meet the needs of any situation. Other sign systems may use the primitive motor elements underlying language or other primitive elements.

Categorical perception. Production and perception of natural sign systems must have evolved together for them to operate as communication. In perception, there must be provision for extracting primitive elements or combinations of the primitive elements reliably from the incoming "message".

Non-arbitrariness. Insofar as the motor elements are the product of evolution of neural organization, language rests ultimately on a non-arbitrary basis. It is for examination how far other systems have a (perhaps covert) non-arbitrary base.

Hierarchical pattern of system. Motor control is a hierarchical process, with relevant parameters being fed in at the appropriate level. This is a necessary organizational economy, and one can look for a comparable hierarchic structure in other sign systems. The primitive elements in any sign system must be capable of being melded together to form higher-level structures, to provide a "lexicon" to match the multiplicity of objects or actions in the field to which the sign system applies.

Derivation of second-level patterns from the structure of perception. In the case of language, the particular structure of words is derived from the structure of the perception or action to which the words refer. Whether something like this happens in other sign systems can be examined.

Extended sequences. In the case of language, syntax is based on motor syntax. Motor syntax may also be a source for the organization of extended sequences in other sign systems, or if not, one can examine exactly what rules govern these sequences. Besides forms relating to actions or percepts, there will also be schemes for organizing sequences, both in terms of order and through functional devices (function words in language).

Diversity. An obvious feature of language is geographic and community diversity. Can one identify similar diversity in other systems? The existence of geographic or community diversity strongly suggests an original population-genetic basis, manifested physiologically or neurologically, affecting preferences in the choice of the sign set by the sign-using group.

Genetic/cultural development. Sign systems are dependent on the existence of shared abilities for production and reception of the signs between members of the community. The current forms of a language are the product of a mixed genetic/cultural development. In other sign systems where geographical or other diversity is found, one would expect a similar mixed genetic/cultural pattern of development of the system.

In a very tentative way, one can comment on how far these aspects of the motor theory of language may apply to other forms of communication, to other sign systems. Given the primacy of cerebral motor control in the theory, the most likely group of sign systems to which the theory may be extended are those with a prominent motor element. This includes:

Gesture. Gesture can take various forms. The most familiar is gesture directly accompanying speech. McNeill has in many papers and books defended the thesis that such gesture in some way appears to be a direct derivative of the speech process, illustrating the concepts with which the utterance deals. Gesture of this kind certainly could be analyzed very much along the lines of the motor theory.

American Sign Language and other sign languages (BSL). These have been extensively researched and analysed by Stokoe, Bellugi and others to demonstrate their rightful claim to be treated as languages though not direct derivatives of spoken language. These sign languages contain iconic elements as well as conventionalized elements; they have morphological and syntactic structure though understandably different from the morphology and syntax of spoken language. Again an obvious candidate for treatment in terms of the motor theory.

Facial expression. Ekman for many years has undertaken research which shows the universal and natural character of facial expression and the scope for analysis of facial expressions in terms of primitive elements of muscular action which combine to form complex expressions. Neurological research, for example by Ojemann and Mateer, has demonstrated a close cortical relation between speech areas and areas controlling facial expression.

Postural expression. All bodily actions, including articulation and gesture, can be seen as modifications of bodily posture. Posture is not something static but a dynamic, constantly modified state. The motor elements which form part of the motor theory of language are modifications of posture which result in particular bodily movements.

All these motor-based sources of expression are important elements in making possible nonverbal communication and the empathetic perception by one individual of another's intentions and emotional state. Less obviously, the motor theory or aspects of it may be relevant also for:

Music. The performance of music is a highly motoric activity and perceived music is powerful in producing responsive movement in the listener. In language also there are aspects with some relation to music, tone and pitch, rhythm, the melodic line of the sentence. In song, language and music join to produce their effect on the listener. Macdonald Critchley (1977) and others have written perceptively about music and the brain. It would be worth considering whether features of the motor theory might be helpful in analyzing musical production and musical perception.

Dance. There have been attempts to prepare an alphabet of choreography in terms of primitive elements from which more elaborate dance patterns are constructed. Dance is a motor response to music, whether the dance is classical ballet or disco-dancing.

Insofar as the motor theory proposes and requires a neurological/physiological matching between motor action and perception (particularly visual perception), there is some prospect that features of the motor theory might be relevant even for:

Figurative Art. Drawing and painting are motor activities and one could treat both the creation of art and the response to it in motor terms.

It is interesting that the categorical perception of colour in many ways resembles the categorical perception of speech sounds.

Alphabetic systems or hieroglyphs. These are specific examples of visual patterning. Koehler's MALUMA/TAKETE experiment shows the possibility that visual patterns such as these may reliably be linked to word patterns.

Obviously a much closer examination has to be undertaken of each of the sign systems before one could reach firm conclusions about the relevance of the motor theory or of aspects of the motor theory. Nevertheless, the motor basis of the sign process looks a good candidate to serve as a unifying concept in semiotics.

5. Semiosis, language, and perception

The confusion about what constitutes a sign makes one wonder whether there is something fundamentally wrong. Is the whole idea of the atomic sign a mistake? Are not "signs" only meaningful in the context of the total organization of the individual's "knowledge-structure" and of the total relevant environment as perceived by the individual - in much the same way as the meaning of a word depends upon the whole context in which it is spoken? The key questions then would be: How is the sign marked off, distinguished from the totality, that is, how does the subject decide where to focus his attention, whether in regarding the external world or regarding his own interior world? What makes for the salience of the sign? How does the "sign" modify the subject's internal mental [neural] organization, that is, the mode of interpretation of the sign ? This brings one very close to issues about the nature and process of perception.

A sign plays a role in perception. The sign must, in some sense, be perceived before it can function as a sign by being matched with its hypothetical interpretant (a preexisting idea or concept, tendency to action or whatever an interpretant is supposed to be). Perception thus is essential to, and prior to, the sign process. Peirce and Morris touched very briefly on this: Morris simply noted "the difficult question as to whether or not perception is to be regarded as a sign process" (Morris 1946: 34), but Peirce (CP 4.424) more categorically said: "The immediate object of all knowledge and all thought is, in the last analysis, the Percept." Thus, semiosis in some sense is perception. From survey of current views of psychologists and neurologists on the nature of perception, one can draw the following points:

1. Perception cannot readily be analysed sequentially from the object, through the sensory channel to the "mental" pattern of the concept, and interpretation of the concept. The perceiver brings a prior structure to each act of perception, and the act of perception is directed (as an aspect of attention), not simply a mechanical pick up of whatever information there happens to be.

2. Perception is not intermittent but continuous, producing continuous change in the perceiver. The prior structure which the perceiver brings to each act of perception is an elaborate neural structure which in some way constitutes a model of the expected environment as well as a model of the individual himself, body and mind. Each new perception modifies the total model.

3. The creation of these models is closely linked, perhaps directly dependent on the representation in the brain of the perceiver's own body. The body image serves as an organizing construct from which all other perception derives. The structure of oneself is the means by which we order our experience of the world (the structure of oneself including the permanent awareness of oneself as body plus accumulated experience and attitudes).

4. Perception is associated with action in many different ways. The act of perception is itself a motor performance, requiring direction of the head, the eyes, even the bodily posture. And perception is closely involved with the determination of patterns of action following perception. Not all perception leads immediately to action: reading a text may modify the reader's "ideas", produce neural changes, without resulting in any overt motor action. But, in evolutionary terms, perception developed to serve the need for action, for response to the environment, and the structuring of perception and of bodily action are likely to be very similar, complementary and interlocked.

5. Both the structures of action and the structures of perception are something that we have inherited, that evolved long before Homo sapiens. It is likely that our organization for perception and for action are the same as or very similar to those of other primates, and probably much further down the evolutionary scale. Signification for us and for animals must be a very similar process.

What bearing has this view of perception on semiotics, the science of signs, language, communication? What distinguishes perception from communication? Here we encounter problems with the vagueness of terms: if we identify communication with all sign processes, messages from the environment, then perception and communication become identical. The cloud communicates information about rain to us. This, however, does not seem a very helpful use of "communication". The narrowing down of the concept seems to require that we should call communication only the transfer of meaning from one animate creature to another. Our awareness of the other extends not only to awareness of them as a particular body but also of aspects of their self, perceived through language and through interpretation of the indicia of the other's attitudes, structure, emotions, character etc. If this is so then the concept of the "sign" takes on a quite different property: it will be because one human and another, one animal and another, share similar behavioral organization, perceptual organization, language organization, ultimately neural organization, that an action of one can serve as a sign conveying meaning to the other. The meaning of "sign" then has to be deepened to recognize it as the product of a complex shared structure.

Perhaps we should seek to construct a theory of signification (the extraction of meaning) rather than a theory of signs. We should not conceive of the neural processes as successive processes of coding and decoding but rather as transformations which result from propagation of the input through a complex system (the total network of neurons). Rather than serial processes, something much more like the ideas in parallel distributed processing will be relevant. Connectionism leads away from the atomic sign, the atomic idea, the atomic concept, to the complex global modifiable network, producing expectancies which control perception, that is, how changes in the environment are processed from moment to moment.

References

Allott, Robin
1989 *The motor theory of language origin*. Lewes: Book Guild.

Holdcroft, David
1991 *Saussure: Signs, system and arbitrariness*. Cambridge: Cambridge University Press.

Koch, Walter A. (ed.)
1989 *Culture and semiotics*. Bochum: Brockmeyer.

Kristeva, Julia
1969 *Semeiotiké: Recherches pour une sémanalyse*. Paris: Seuil.

Locke, John
1706 [1972] *An essay concerning human understanding*. (5th edition.) 2 vols. London: Dent.

Morris, Charles
1946 *Signs, language and behavior*. New York: Prentice-Hall.

Peirce, Charles S.
1931-1958 *Collected Papers of Charles Sanders Peirce.* 8 vols. Eds. Charles Hartshorne, Paul Weiss, and Arthur W. Burks. Cambridge, Mass.: Harvard University Press.
1982 *Writings. Vol. I (1857-1865).* Ed. Max H. Fisch et al. Bloomington: Indiana University Press.

Pinxten, Rik
1989 "Culture and semiotics", in: Walter A. Koch (ed.), 34-35.

Saussure, Ferdinand de
1916 [1966] *Course in general linguistics.* [Trans. Wade Baskin.] New York: McGraw-Hill.

Language evolution: A Darwinian process[1]

Bernard H. Bichakjian

1. The foundations of linguistics

Unlike biology, where science was able to prevail upon metaphysics as early as the last century, linguistics still lives in the cozy world where the postulated distinction between body and mind provides a rationale for thinking that evolution must be confined to bones and other such physical features. Mental constructs and cultural carriers such as language are, of course, placed high above the reach of such a mechanical process, but like all edifices, this beautiful dream house is as strong as the foundation that supports it. In the present case, the entire construction rests on an act of faith, because the underlying dichotomy between body and mind is not as mutually exclusive as it is assumed. After all, if humans are capable of greater mental activities, it is not because an extraordinary agent has endowed them with a special superimposed faculty, but because our species has developed the necessary somatic apparatus.

The latter approach is the one that will be used here. After a brief survey of the state of the art needed to explain why earlier attempts to introduce an evolutionary perspective into linguistics met with such little success and the presentation of empirical data showing that sounds and syntactic strategies do display a developmental history, this paper will argue that the key to understanding language evolution must be sought in the investigation of the biological underpinning of linguistic features - the biochemical events that ensure their acquisition, the genetic mechanisms that can modify them by modifying their acquisition, and the selective pressures that guide their course.

2. The state of the art

It started with Schlegel (1772-1829), who at the dawn of the nineteenth century coined the term *comparative grammar* and expressed the conviction that "die vergleichende Grammatik . . . uns ganz neue Aufschlüsse über die Genealogie der Sprachen auf ähnliche Weise geben wird, wie die vergleichende Anatomie über die höhere Naturgeschichte Licht verbreitet hat" (1808: 28). The task was indeed to organize the kindred languages spoken from here to William Jones's faraway India into a comprehensive family tree and to reconstruct the common ancestral language. This second endeavor led to the formulation of phonological laws relating the sounds of each of the derived languages with rigid regularity to their original form in the protolanguage. Used inferentially, these laws produced the reconstruction that was aimed at, and, deductively, the derivational history of the modern languages.

The use of the term *law* to label the observed correspondences between two sets of sounds may seem excessive, but this concept was used in biology, and in a way comparative grammar was modeled after comparative anatomy, and it harbored scientific ambitions. Unfortunately, the scientific criterion was not fully met, because, in spite of their rigor, these laws remained descriptive, and no attempt was made to explain the observed changes as a logical result of a higher epistemological principle (cf., e.g., the hierarchical organization of biological, chemical, physical, and mathematical laws). Ease of articulation (cf., e.g., the pronunciation of *cupboard* as against that of *cupcake*) phonetic attrition (cf., e.g., the substandard *fis'* instead of the prescribed *fist*) were adduced here and there, but, as Hermann Paul (1920: 68) emphatically stated, the goal was purely descriptive: "Das Lautgesetz sagt nicht aus, was unter gewissen allgemeinen Bedingungen immer wieder eintreten muss, sondern es konstatiert nur die Gleichmässigkeit innerhalb einer Gruppe bestimmter historischer Erscheinungen."

2.1 Schleicher: A Darwinist with a cultural handicap

The German Indo-Europeanist and amateur botanist August Schleicher (1821-1868) was in a sense an exception. He became fascinated by Darwin's *On the Origins of Species* and ardently pleaded that linguists apply to languages the methods used in the natural sciences (Schleicher 1863: 6), but his own application was such a failure that he achieved the opposite effect. Since Indo-European was rich in inflectional markers, whereas modern languages are conspicuously poor in this respect (cf., e.g., the Lat. *scrip-s-isse-t* and its English counterpart *(s)he might have written*, where personal pronouns and temporal and modal auxiliaries are used instead), Schleicher (1852: 14-30) concluded that Indo-European developed from its origin until the dawn of history and then reversed its course, with the derived languages undergoing a process of decay (German *Verfall*) that was said to be still active today.

Schleicher's failure to understand how evolution works is patent. Would it have been thinkable for a naturalist to claim that, from the origin of life until, say, the heyday of the reptilians, species evolved by developing advantageous features and that since then evolution has been an accumulation of deleterious changes? The pattern would be in direct conflict with the principle of natural selection. That such an untenable claim was made by someone who was an admirer of Darwin and who urged that the methods used in the natural sciences be applied to languages may come as a surprise, but it is all the more significant since it clearly shows that even a pro-Darwinian nineteenth-century linguist was unable to place language change in a truly evolutionary perspective.

Actually, Schleicher was a brilliant linguist, who otherwise made important contributions to comparative grammar, but he had a handicap. Just as naturalists were hampered in their evolutionary work by the church's opposition, so were linguists by the prevailing cultural bias they had acquired during their education. The nineteenth-century scholars who happened to be linguists were foremost learned men. They had studied Greek and Latin and learned to consider the golden ages of Athens and Rome to have been all-time highs in the history of Western civilization. Since language, literature, and other forms of artistic and intellectual expression were seen as an integrated object of adulation, the mere thought that languages could have developed advantageous features after the days of Demosthenes and Cicero would have been sheer heresy. Given the *ne plus ultra* status granted to Sanskrit, Greek, and Latin and their intricate inflections, which were considered to be the pinnacle of grammatical organization, a straight line of development from the origin of speech to present-day languages was simply inconceivable. That is why Schleicher, who could not fit the loss of morphological markers into a Darwinian scheme, turned to philosophy and found in Hegel's *Phenomenology of the Mind* the dialectic he needed. Since Hegel had claimed that everything in the universe had developed in a seesaw pattern swinging repeatedly from thesis to antithesis and thence to a higher-level synthesis, which in turn would become a new thesis triggering its own antithesis (see Figure 1 overleaf), Schleicher had a model that could accommodate the rise and fall and the presumed future rebirth of morphological markers. Unfortunately, Hegel's model, perhaps well-suited for rhetorical exercises, did not prove to be a viable evolutionary theory, and Schleicher's application to the Indo-European languages, though meant to serve as an example, became instead a deterrent against the evolutionary approach.

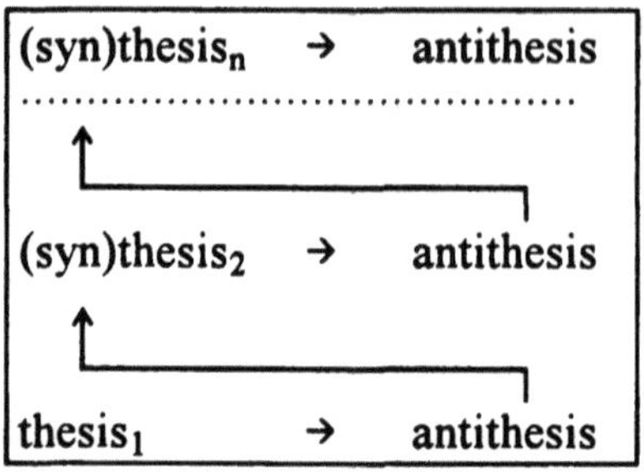

Figure 1. Hegel's dialectic

The century that had started with Schlegel's expecting grammar to apply the comparative method used successfully in anatomy ended with linguistics and biology far apart from each other - the naturalists could clearly observe and explain a unidirectional progression from the origin of life to humans, while the linguists could only organize languages into families and describe the changes that had taken place. Unlike Darwin and his followers, they had failed to produce a theory that would enable them to place their data in an evolutionary perspective and understand the changes that they had observed.

2.2 The twentieth century: An insightful beginning

At the beginning of the present century, some insightful observations were made about language evolution, but unfortunately these ideas did not become influential. The French Indo-Europeanist Antoine Meillet (1866-1936) pointed out that languages essentially proceed in one general direction (1918/1965a: 65). Soon afterwards, perhaps independently of him, the American linguist and anthropologist Edward Sapir (1884-1939) argued that languages "drift" and "that linguistic drift has direction" (1921/1949: 155). The Danish linguist Otto Jespersen (1860-1943) went further. As a true evolutionist, he argued unequivocally that languages are becoming evermore efficient (1922/1964: 364; but also as early as 1891 and again in 1941). Advocating any kind of superiority for the modern languages of Western Europe, Greece, and India over Latin, Classical Greek, and Sanskrit, respectively, was obviously a dangerous proposition: not only did it go against the prevailing cultural dogma, but increased linguistic efficiency was difficult to prove. Today, Meillet is read primarily by Indo-Europeanists, Sapir's "drift" is generally given a very loose interpretation, and Jespersen is remembered essentially for his other work, namely his *Modern English Grammar*. While his merits as an Anglicist are indisputable, the failure to stress his insight into language evolution could be equated with remembering Beethoven as a great pianist and "forgetting" to mention that he had also written symphonies.

2.3 Getting off-course with Saussure

While the ideas of Jespersen and like-minded scholars were going unheeded, linguistics was reaching a turning point in its history. Whereas up till then languages had been seen as sums of features, each of which had to be traced back to the protolanguage, the Swiss Indo-Europeanist Ferdinand de Saussure (1857-1913), who is considered to be the father of modern linguistics, stressed the importance of recognizing the systemic nature of languages (Saussure 1916/1931: 124). The static notion of "system" later gave way to that of "set of rules", as the well-known American linguist Noam Chomsky (1957: 49) introduced his generative conception of grammar, but throughout the twentieth century the emphasis has been on the coherence of the internal organization of languages. In such a perspective, it is not just the features that change, but the system or set of rules that is supplanted by another (cf. Jakobson 1931/1964: 316, Martinet 1962: 2, and Postal 1968: 270).

In itself, the holistic approach was an important contribution to linguistics - just as humans are not apes with a number of new attributes, but different organisms altogether, so is French an integrated system of its own, not a form of Latin with some isolated features added here, others taken away there. But, whereas biologists have been able to combine their recognition of the unity of an organism with the need to trace the developmental sequence of the constituting features - humans have a more developed central nervous system than other living primates - linguists only described the internal organization of languages and the changes that convert one system into another. Since outgoing and incoming items have not been comparatively assessed, and indeed often peremptorily assumed to be equally gratuitous, though the markedness theory was a modest step in the right direction, linguistic features have not been hierarchically ordered, much less organized in a developmental history. It is known, for instance, that Latin had declensions and that the Romance languages gradually eliminated this feature, but the relative merits of the two grammatical devices have not been properly discussed and certainly not determined. Lacking such a developmental scale and apparently unwilling to build one, Saussure and his successors, from

Jakobson (1896-1982) to Chomsky, unlike the biologists, who recognize in fossil and living organisms the successive steps of the evolution of species, have failed to discern in the typological gamut of extinct and extant vernaculars the differential stages of language evolution. Instead of seeing them as specific products of an evolutionary process, they have viewed linguistic systems as cases of a preexisting entity - unique or diverse, innate or acquired - that the contingencies of life can modify aimlessly (cf. Saussure 1916/1931: 105, 111-112).

To explain these modifications of linguistic systems, an abundance of factors have been adduced. It has been successively argued that changes are triggered by the conflict between the intellectual, i.e., matter of fact, and the expressive functions of language (Jakobson 1931/1964: 333-336), the opposing needs of linguistic communication - maximum clarity and minimum effort (Martinet 1955), added sociological motives - being understood while registering one's social affiliation (Labov 1972), the contradiction between juvenile logic and adult artifice (Halle 1962), and the fickle nature of humans (Postal 1968). These days, we seem to have succumbed to the exotic charm of creolization (cf. for a discussion Polomé 1980), which no doubt will also wane. While the popularity of such convenient solutions is indeed short-lived, cases can certainly be found to illustrate the actions of each of these factors (cf. also Kiparsky 1988), but such explanations remain nevertheless *ad hoc*, and their advocates cannot be credited with having offered a comprehensive theory of language evolution.

As the twentieth century approaches its end, the art seems to be in the following state: formal representations of the organization of languages and equally formal descriptions of shifts from one organization to another are in the center of linguistic preoccupations, but the relative merits of outgoing and incoming items are not assessed, the developmental history of linguistic features is not observed, and consequently the very notion of language evolution is absent.

2.4 The importance of an evolutionary approach to linguistics

The failure to recognize the existence of evolutionary processes in language is not just a lacuna found in an isolated area of research. It affects the entire field of linguistics because we cannot understand the true nature of language unless we uncover the course of its development and the mechanisms that produce it. Indeed, without the proper understanding of the developmental processes, linguists are just as handicapped as a biologist who would not know that species have evolved.

3. Finding inspiration in biology

The evolutionary framework is therefore indispensable in linguistics, and the developmental processes can be easily recognized, especially if one turns to evolutionary biology for inspiration. It is generally admitted that humans have evolved not so much by acquiring entirely new features, but through a neotenous process, whereby the growth of the somatic features that appeared early during the ancestor's ontogeny was spread over a longer period (cf., e.g., Gould 1977). As the genetic clock ticked more slowly, more and more juvenile features were preserved into adulthood, in expanded and/or better adapted forms, while those that used to supplant them during the ancestor's maturity were concomitantly pushed out of the descendant's ontogeny. On the basis of the evidence extrapolated from the ontogeny of living apes, it is concluded that humans have on the one hand preserved the favorable brain/body ratio and the cerebral plasticity, which their ancestors enjoyed only during their foetal and juvenile periods, and on the other eliminated the protruding jaws and the sloping forehead, that used to appear later in their ancestors' lives. These are indeed just a few examples.

Neoteny should not be confused with paedogenesis. The latter is the acceleration of sexual maturity, which aborts the growth of somatic features and usually leads to degeneration. Since natural selection can hardly favor such a process, paedogenesis has not played a major role in evolution (cf. de Beer 1958: 63-64 and 90-91). Neoteny, on the other hand, is a very important process - both common and beneficial - because the slower growth of somatic features can foster developments that confer selective advantages.

Neoteny is of course a biological process that applies to anatomical features and the attendant physiological functions, but languages are not biological organisms, nor are linguistic features biological organs (cf. Bichakjian 1989 for a discussion of the innateness theory). How this genetically based evolutionary process could perhaps explain the development of linguistic features will be discussed later, but first data will be presented to show (1) that developmental changes exist in languages, (2) that these changes are also produced by a process whereby early acquired features supplant late acquired ones, and (3) that a shift to ever-earlier acquired features confers on a language and its speakers a whole range of selective advantages.

4. Developmental changes in the Indo-European languages

While *ad hoc* references are often made to individual developmental changes in linguistics (cf., e.g., Kuryłowicz 1964: 130), a systematic account of the development of linguistic features is still wanting. The data that will be presented here are taken from the Indo-European languages, because of all the language families this is the one whose ancestral form and evolution, which spans some six or seven millennia, have been the most intensely investigated. Moreover, many of the developments that took place in the Indo-European languages can also be observed in the history of Hamito-Semitic, Finno-Ugric, Dravidian and Altaic languages (cf. Bichakjian 1991a), while some of the changes that can be inferred in the surmised history of African and American Indian languages, though outwardly different, probably belong to the same process (cf. Bichakjian 1991b). The situation in linguistics can be compared with the one in biology. Birds and mammals may have developed different features, but the evolutionary process inferred in one class applies to the other as well.

4.1 The development of phonological features

In the Indo-European languages the following developmental changes have taken place. The sound system of the protolanguage included very few vowels. While *i* and *u* were heard under certain circumstances - almost by accident - and *a* in certain types of words, *e* was the only regular vowel, but this limited inventory was supplemented with a series of *h*-like sounds called laryngeals. Gradually the laryngeals disappeared, leaving behind the vowels *i*, *e*, *a*, *o*, *u*, organized in two sets: one long, one short. In a subsequent change, the quantitative distinction between long and short segments disappeared or began to recede in favor of additional qualitative distinctions (cf., e.g., the Lat. *legit* 'he reads' and *lēgit* 'he read' and their French reflexes *(il/elle) lit* and *(il/elle) lut*, respectively).

In the consonantal part of the sound system, the shift has been from a set made almost exclusively of complex stops (cf. Gamkrelidze – Ivanov 1973, Gamkrelidze 1987) to sets of simple consonants evenly or nearly evenly distributed among stops and fricatives.[2] The general direction can be clearly observed in a comparison between the Indo-European and Modern French obstruents.

$p^?$	$t^?$	$k^?$	$k^{w?}$		p		t		k
p^h	t^h	k^h	k^{kw}		b		d		g
b^h	d^h	g^h	g^{wh}			f	s	š	
	s					v	z	ž	
Indo-European					*Modern French*				

When the ancestral obstruents are compared with those of another modern Indo-European language, the contrast may not prove to be equally sharp, but the general direction is unmistakably the same.

4.2 The development of morphological features

The morphology of the protolanguage has undergone two sets of changes: one concerning the marking of grammatical distinctions, the other the grammatical distinctions themselves. In the ancestral language, grammatical features were for the most represented with suffixes. For instance, the Latin *leget* conveyed, in addition to the lexical meaning of 'reading' expressed by *leg-*, the grammatical notion of futurity derived from an earlier notion of volition - the original value of the subjunctive - represented by *-e-* and the equally grammatical notion of third person singular, indicated by *-t*. While the Latin form illustrates the ancestral morphological structure, the modern one can be observed in its English equivalent *(s)he will read*, where all the grammatical notions are expressed with independent particles. Languages thus developed pronouns, auxiliaries, and, in the case of nouns, articles and prepositions that supplanted unending declensional and conjugational paradigms.

The most significant changes produced by the evolution of the grammatical distinctions took place in the linguistic representation of action. Whereas the protolanguage stressed the modality and the aspect of an action, today, the focus is on temporal distinctions. The ancestral language had, in addition to the imperative, an indicative, a subjunctive, and an optative mood, expressing respectively an occurrence, a will, and a wish. In the Latin benediction *Pax Domini sit semper vobiscum*, 'May the peace of the Lord be with you always', *sit*, 'may . . . be', displays both the ancestral optative marker and the original grammatical notion of 'wish.' The subjunctive and the optative soon merged into one mood and fostered the parallel development of a future tense. Today, the hybrid mood, often called *subjunctive*, is receding or has disappeared altogether.

The other verbal distinction that changed was aspect. Since grammar books often use the label *tense* instead of *aspect*, speakers of languages such as English are not always aware of aspectual distinctions. The difference, for instance, between *I saw* and *I have seen* is not one of tense but of aspect. Both actions belong to the past, but *I have seen* conveys a resultative meaning, which *I saw* does not. The protolanguage had already begun making a present-past distinction, but this opposition was limited to a specific domain, while the dominant distinction was one of aspect. Verbs were put in the imperfective, perfective, or resultative aspects depending upon whether the subject was considered to be, respectively, in the process of performing the action, in the act of completing it, or in a state resulting from it. These were aspectual distinctions, but they gradually became supplanted by temporal ones in most of the derived languages.

4.3 The development of syntactic features

Syntax has also been the theater of great changes. How the cognitive notion of agent evolved into the grammatical function of subject is a fascinating matter, but it will not be presented here because the shift is not immediately observable and the inferred process would require a considerable amount of extrapolated data. Instead, the focus will be placed on the development of subordination and on the reorganization of structural elements.

One of the common misconceptions about earlier forms of syntax is based on the assumption that, before languages developed conjunctions and relative pronouns, which make subordination possible, utterances were made of strings of short sentences, each one containing a single verb and that verb being in a finite form. According to this assumption, earlier speakers would have produced sequences such as *Paris ran off with Helen; the Greeks attacked Troy.* The empirical data make it abundantly clear that nothing could be further from the truth. Complex sentences are no doubt recent developments, but subordination existed long ago, albeit in a different form. Whereas modern languages have become equipped with the necessary structural devices to subordinate one sentence to another - grammar books speak of main and subordinate clauses - the protolanguage used participial and other such nonfinite forms to express subordinate actions: *Paris having run off with Helen, the Greeks attacked Troy* (cf. also the following Sanskrit example cited by Meillet [1964: 374]: *tásyāvanénijānasya mátsyaḥ pāṇí á pede* 'while he was washing himself a fish came into his hands,' but lit. 'of him washing himself a fish into the hands came'.) Today, verbals are still used in certain circumstances, but on the whole participial and other such constructions have been supplanted by subordinate clauses, especially when the process is used recursively. It is quite common to produce sentences, such as *The Greeks attacked Troy because Paris had run off with Helen while Menelaus was away*, but *Menelaus being away, Paris having run off with Helen, the Greeks attacked Troy* would be more difficult to process and less precise since the distinction between temporal *while* and causal *because* cannot be expressed. The development of the complex sentence is therefore a shift from "subordinate" **phrases** to subordinate **clauses**.

The other area of syntax where a major change has taken place is word order, which not only has become more rigid as subject and direct object have lost their characteristic case markers, but also reversed its pattern. When grammatical structures are analyzed in terms of *heads* and *modifiers* (cf. Jackendoff 1977 for the original proposal, and Lefebvre – Massam 1988 for necessary revisions), it immediately becomes apparent that Indo-European, and indeed all protolanguages, were predominantly head-last (Givón 1979), while the historical data clearly reveal the action of an ongoing shift from head-last to head-first structures. The reversal can easily be observed when a characteristic Latin sentence, such as the following taken from Plautus (ca. 254-184 BC), is compared with its English equivalent:

haec	*me*	*vid-isse*	*cert-o*	*sci*	*-o*	(Mil. 299)
these things	me	seen have	sure-ly	know	I	
1	2	3 4	5 6	7	8	

I	*know*	*for*	*sure that*	*I*	*have*	*seen*	*these things*
8	7	6	5	2	4	3	1

Forms and phrases, such as *sure-ly* or *(I) certainly know*, do reveal the existence of head-last structures in English, but there can be no doubt that Latin, presented here as an illustration of the protolanguage, featured predominantly head-last structures, while English, representing the extant languages, generally displays head-first combinations.

The rate of the sweeping change from the ancestral to the modern order and the sequence in which the various grammatical structures have shifted or are shifting from a head-last to a head-first model may of course differ from one language to another, but the propensity for head-first structures is manifest everywhere.

4.4 Unity and significance of developmental changes

The above changes outline the development of the main phonological, morphological, and syntactic features in the Indo-European languages (cf. Bichakjian 1988 for a less sketchy presentation). While other developments or other phases of the above developments may have taken place in other language families (cf. Bichakjian 1991a and 1991b), the Indo-European data clearly show that developmental changes do take place in linguistics.

While each of the above-indicated changes may not strike experienced Indo-Europeanists as a dazzling revelation, the importance of the foregoing lies in the realization that the history of the Indo-European languages is not a string of *ad hoc* changes, but the history of the development of linguistic features. The failure to recognize until now the evolutionary nature of these changes is in a sense understandable: these developments do not seem to have anything in common (cf., e.g., new sets of vowels supplanting long/short distinctions, tense prevailing upon aspect, and structures becoming head-first), nor are the advantages of the new features immediately obvious. But this is where neoteny can prove to be useful.

5. The paedomorphic[3] evolution of languages

If the developmental changes outlined above are projected on an ontogenetic scale, it becomes strikingly obvious (1) that the phonological, morphological, and syntactic features of the ancestral language were all acquired late by native children, and (2) that the sequence of changes that has taken place has consisted of replacing late acquired items with ever-earlier acquired alternatives. These conclusions are based on the evidence extrapolated from extant languages where near equivalents of the original features can be found, though not all of them in the same language.

5.1 The evidence from language acquisition

Indeed, systems of different sounding vowels are learned before systems opposing long and short ones, which in turn are mastered before vowels combined with laryngeals. Plain stops and fricatives are acquired before complex stops. Prepositions, auxiliaries, and similar grammatical particles are used properly before complete declensional and conjugational paradigms are committed to memory. The systematic use of temporal distinctions is already established when aspectual and modal ones appear. Subordinate clauses emerge surprisingly early, while participial constructions are latecomers. Finally, head-first structures, especially when they reach a certain length, become part of the children's competence markedly before the reverse model. (For a discussion of the psycholinguistic data and their bibliographic sources, see Bichakjian 1988).

This is not to say that exceptions should absolutely be ruled out. Of course, a special context could be found where a given consonant of the early acquired set may appear after one of the later group. Some case markers in some languages may become used before the first preposition has made its entry, and the English *red ball* will not require more time than the French *ballon rouge*. But such comparisons are idle, because what matters is not the acquisition of an isolated item, but the mastery of an entire system as opposed to another. Regular accusatives may indeed appear very early, but the significant fact is that children who learn to speak languages that rely on the use of prepositions are done with this task by the age of four, while children who must absorb the declensions of their native languages are still at it at the age of eight. When the true alternatives are compared, the evidence is unambiguous and overwhelming - *languages have evolved by continuously replacing their features with ever-earlier acquired equivalents*. The nature of language evolution is now revealed: it is a shift to ever-earlier acquired features, which applies to all linguistic components (phonology, morphology, and syntax) and to all languages.

It is opportune to recall at this point that the biological theory of neoteny does not imply in any way that today's men and women are immature apes because they have preserved into adulthood some of the foetal or juvenile features of their ancestors. Likewise, the early acquired linguistic features that have become part of a given language are no less mature than the late acquired ones that they have supplanted. English is not Indo-European baby talk any more than today's humans are infantile australopithecines. English and Indo-European are equally mature languages, but with the important difference that English speaking children acquire their native language considerably earlier (cf. *mutatis mutandis* Slobin 1966).

5.2 The selective advantages of early acquired features

The early acquisition of a given language is indeed very important. It is a known fact that language plasticity like other types of cerebral plasticity is not a potential held constant during a given period of time, but a decreasing disposition during a specific ontogenetic span. Therefore, the sooner the neural pathway of a given linguistic feature is laid out, the stronger it is (cf. *mutatis mutandis* LeVay – Wiesel – Hubel 1980: 2). Better established neural pathways are not only less impervious to negative somatic and psychological conditions (e.g., fatigue or insecurity) and pathological processes (degenerative or accidental aphasia), but also more economical in their regular use, since their activation requires less energy. Given the importance of early established neural pathways, the shift to earlier acquired linguistic features now becomes a development towards more advantageous alternatives.

The advantages of earlier established neural pathways are not only cybernetic. The faster a child becomes proficient in language, the faster (s)he can acquire social skills, and, as in the case of linguistic features, the earlier the neural basis of social behavior is laid out, the better will the child function socially throughout his or her life time. This argument has been used by Messer (forthcoming, p. 17 of the ms.) to point out the advantages of acquiring one's native language during childhood instead of adulthood. It applies also within the period of childhood - the lower the age of linguistic proficiency the greater the advantage for the individual.

Finally, a faster mastery of linguistic skills provides a child with an earlier access to the kind of knowledge that can be acquired only through linguistic communication. This early access to knowledge is highly profitable, because the child who has become proficient in language sooner is further ahead in his or her mental development than the other members of his or her age group and possesses a more solid knowledge, since the neural networks that represent it are stronger. Above all, the child commands a greater wealth of data and mental skills, since the time, energy, and neural potential that were freed by earlier acquired linguistic features can be applied to other fields of learning.

The ever earlier achievement of linguistic proficiency ties in with the "early" birth characteristic of our species. Instead of the expected twenty-one, the human gestation period has been "reduced" to nine months, in part, at least, under the selective pressures exerted by the advantages of an earlier start in life. The sooner the infant leaves the uterine confine, the sooner his or her social and mental development can be intensified. This advantage would be partly neutralized if language acquisition would instead be postponed more and more. Instead, an increasingly early proficiency strengthens and expands the advantages of an "early" birth.

The projection of the historical data on an ontogenetic scale has revealed that linguistic features have developed towards ever-earlier alternatives. The advantages provided by these alternatives in the establishment of early specialized neural pathways, the development of social behavior, and the accumulation of knowledge now suggest that natural selection could indeed have guided language evolution along the above-defined direction. Like its biological counterpart, language evolution is therefore a shift to more advantageous alternatives, i.e., a Darwinian process.

5.3 The biological correlates

The parallel between language evolution and neoteny is now obvious, but is it just a parallel or is language evolution also a neotenous process? In the present state of the art, biology cannot provide all the necessary facts, but it does present helpful indications. It is known that genomes include two types of information - structural and regulatory - and that the growth of a biological feature is produced by the regulated expression of the corresponding DNA sequence. Neoteny, then, takes place when the timing devices slow down a given protein synthesis, spreading thereby the growth of an organ over a longer period (King – Wilson 1975).

Like many other genes, the DNA sequences coding for our potential to acquire and use human languages have not been identified, but it is obvious that this potential is a species-specific feature, and, as such, it may be surmised to be part of our genome. Clinical evidence from speech pathology supports this logical conclusion and suggests the existence of a language factor on the X chromosome (Lehrke 1974; Renier 1983: 47) and perhaps language regulating genes on chromosome 15 (Gardner forthcoming). Moreover, the difference between the deficient speech developments of identical twins on the one hand and fraternal ones on the other - the patterns concord far better in the first case than in the second, where one of the siblings may not even display any speech deficiency at all (Luchsinger – Arnold 1965: 373; Benton 1975) - clearly supports not only the hypothesis that our potential to speak is genetically coded, but also the attendant view that the ontogeny of speech, which manifests itself in the form of language acquisition, is genetically regulated.

In addition to the clinical observations, there is a body of solid experimental data suggesting that language acquisition is biologically regulated. Intensive studies of the visual cortex have led to the conclusion that plasticity is probably produced by the phosphorylation of proteins that determine the shape of neurons (Aoki – Siekevitz 1988: 39-42). As phosphate groups are added, the cortical neurons grow longer and more active dendrites, ready to assemble into pathways when outside stimulation is provided. Conversely, when phosphorylation decreases, the neurons grow more rigid, and their dendrites eventually become static, making neuronal interconnections increasingly difficult. The plasticity period is then over.

While these experiments and observations were conducted on the visual cortex, the proteins that determine the shape of neurons are not a special feature of this region. They are found throughout the brain (Aoki – Siekevitz 1988: 40), suggesting therefore that cerebral plasticity, in general, and linguistic plasticity, in particular, are produced by essentially similar biochemical processes.

Since language acquisition is achieved through a linguistic stimulation of the appropriate regions of the brain during their period of plasticity, the phosphorylation of the neurons in Broca's and Wernicke's areas and in the attendant parts of the brain, which determines the shape and behavior of dendrites, and the myelination of connecting fibers, which improves the signaling efficiency of the impulse, in fact, set the terms and the rate of language acquisition. These biochemical processes can of course be influenced by contingent factors, but primarily they are regulated at the DNA level, thereby making language acquisition also a genetically regulated process.

If, as the evidence strongly suggests, our potential to acquire linguistic features is genetically regulated, it is only logical to conclude that, as in the case of somatic features, a change in the timing of the process would bring about a change in the linguistic features that are eventually acquired, and by so doing would make languages evolve. This correlation *mutatis mutandis* has been convincingly established by Konrad Lorenz (1971: 180), who argued that "for anybody who has correctly grasped the fundamental unity and conceptual inseparability of form and function, it will be immediately obvious that the persistence of juvenile behavioral characteristics in man is intimately correlated with that of morphological characters." And he concluded: "*The constitutive character of man - the maintenance of active, creative interaction with the environment - is a neotenous phenomenon*" (Lorenz 1971: 180). It would be consistent with the Viennese ethologist's view to add that the paedomorphic evolution of languages is part of that "neotenous phenomenon". The genetic regulation of the ontogeny of speech, and the consequences that a modification of this regulation can have on language acquisition and hence on language evolution must, of course, remain putative for the time being, but as Lenneberg (1967: 240) pointed out: "Although we can only speculate on this point, our speculations with regards to language are no more daring than with regard to most other structural or functional features".

Though such methodological support is helpful, it is even more encouraging to observe that substantial collateral indications have been coming from the biological sciences. In a recent paper, Russell Gardner of the Texas Medical School has come to the conclusion "that we must find genetic blueprints underlying language production, that is, DNA sequences that allow human children to learn language and then to foster its continued use. We need to search for DNA sequences that when modified in turn modify language production" (forthcoming, p. 2 of the ms.). Given the rapid developments taking place in genetics (cf. Chumakov et al. 1992), Gardner's molecular research program may be carried out before long, but already Robert Sokal and his associates (1992: 7671), who have compared genetic and linguistic distance, have come to the conclusion that "the remaining significant partial correlation between language and genetics, after geography is held constant, indicates a relation between these two variables above and beyond that caused by their common spatial differentiation". Of course, Gardner's call and the findings of Sokal et al. do not provide the required molecular evidence, but, with the biological observations complementing the linguistic evidence, they do lend support to the acceptability of a genetic explanation of language evolution.

It is appropriate to bring this discussion to a close with a quote from Walter A. Koch (1991: 32), who, with his encyclopaedic knowledge of evolutionary matters, has rightly pointed out that "no one expects genes and languages to be unequivocally correlated but then no one should be surprised if there is *some* correlation". That correlation will become more and more apparent as our knowledge of gene maps unfolds. In the present state of the art, as I stipulated above, not all the necessary facts have been uncovered, but the evidence available and the existing indications do suggest that language evolution is a genetically controlled process whereby features are continuously replaced with earlier acquired and, therefore, more advantageous alternatives.

6. Understanding apparent exceptions

The foregoing was a presentation of language evolution: its existence, its nature, and its putative explanation. The emphasis was intentionally laid on developmental changes. This is not to say that all other possible sources of language change are excluded. Population movements can contribute to a more or less significant remodeling of linguistic systems, grammarians and other such pundits love to tamper with languages, and fads and sociocultural concerns are reflected in the words we choose. All these factors do play a role at one time or another, but they do not constitute a counterargument against a theory of language evolution.

Hybridization takes place in biology, and language contact in linguistics; neither one raises doubts about the existence of developmental processes, nor do they present a threat to their respective theories of evolution. Like abdominal delivery, which fosters the survival of a child with a handicap, or genetic engineering, the interventions of grammarians may initiate movements that are independent of, and even contrary to, the natural evolutionary processes (cf. Bichakjian 1986), but the injunctions of grammarians do not belie the validity of a theory of language evolution any more than the actions of obstetricians and molecular biologists invalidate the principles of the Synthetic Theory. The faddish or sociocultural innovations have at least one thing in common with developmental changes - they also bring in advantageous features, but in their case the advantage lies simply in the novelty of the item or in the social value that is attributed to it. These advantages are transient, however, and so are the features that carry them. Their erratic behavior does not conflict with the existence of developmental changes and the theory of linguistic evolution that accounts for them.

Bringing the developmental sequences into focus and advocating the existence of evolutionary processes in linguistics cannot imply, of course, that the changes have been the same everywhere or that they have occurred at the same rate. The phylogenies of fishes and humans may not have traveled the same distance, but fishes and humans have evolved according to the same principles, and these very principles also explain in part at least the extinction of fossil species. The disparities that exist in the biological world have their counterparts in linguistics, neither set conflicts with the corresponding evolutionary theory, instead it is the theory that explains them.

7. A Darwinian theory of language evolution

The data presented in this paper, their interpretation, and their discussion suggest that languages evolve by developing evermore advantageous features. It was argued that early acquired features confer on a language and, thereby, on its users definite selective advantages. Since the chronology of the acquisition of linguistic features is on record, and the greater strength of earlier established neural pathways has been independently demonstrated, the arguments that were used have tangible support.

One could also argue that the incoming items have been more advantageous because they are more functional than the outgoing ones. Plain stops and fricatives do not have the distributional constraints of complex consonants, independent particles combine more freely that inflectional markers, embedded sentences have a greater syntactic potential than participial phrases, but these functional advantages, however real and important, have not been stressed here because a universal gage of functionality remains wanting.

Finally, it is not implied that no other paradigm could be used to assess the relative advantages of linguistic features. Neural network simulators (cf., e.g., Jones – Hoskins 1987) may be able to show that a given feature can be produced and processed with a more economical algorithm than another (e.g., the use of prepositions vs. declensions). Perhaps there will also be an algorithmic explanation for the fact that the maximum level of acceptance is much higher for the head-first structures than it is for the opposite ones. New possibilities of assessing the advantages of linguistic features will certainly become available, but the role of language acquisition will remain central, and, given a putative correlation between the scales of algorithmic complexity and their sequences of mastery, the new evidence may be expected to corroborate the paedomorphic explanation of linguistic evolution.

There can be, of course, no certainty about the evidence that future research will provide, but neither can there be doubt about the empirical data that are available today: (1) linguistic features display developmental sequences that are produced by continuously replacing the existing item with an earlier acquired alternative; (2) earlier acquired features and, thereby, earlier linguistic proficiency, provide speakers with definite advantages in the layout of neural pathways and in the development of social and mental skills. These sequences and their advantages call for a Darwinian explanation of evolution for languages, and the theory of linguistic paedomorphosis is an attempt to meet that need.

Notes

1. An earlier version of this paper was published in Russian in *Voprosy Jazykoznanija* 2 (1992): 123-34.
2. Stops are sounds produced with a complete interruption of the flow of air at one point of the vocal tract (e.g., *p*, *t*, or *d*), while fricatives are the result of the turbulence created by partial closures along the vocal tract (e.g., *f*, *s*, or *z*); complex sounds are produced with at least two simultaneous articulations (e.g., the initial consonant of Engl. *time*, noted *th*, which combines the blockage of the flow of air by an apico-alveolar closure [the tip of the tongue against the teethridge] with the release of subglottal pressure).
3. Whereas biologists today prefer the word neoteny, which refers to the preservation of juvenile and foetal features, paedomorphosis, which is a near synonym, seems more appropriate in linguistics, since it applies only to the former.

References

Aoki, Chiye – Philip Siekevitz
1988 "Plasticity in brain development." *Scientific American* 259, 6: 34-42.

Bahner, Werner – Joachim Schildt – Dieter Viehweger (eds.)
1991 *Proceedings of the Fourteenth International Conference of Linguistics, Berlin 1987*. Berlin: Akademie-Verlag.

Beer, Gavin de
1958 *Embryos and ancestors*. (3rd edition.) Oxford: Clarendon Press.

Benton, A.
1975 "Developmental dyslexia: Neurological aspects", in: W.J. Friedlander (ed.).

Bichakjian, Bernard H.
1986 "When do lengthened vowels become long? Evidence from Latin and French, and a paedomorphic explanation", in: W. Leo Wetzels – Engin Sezer (eds.): 11-36.
1988 *Evolution in language*. Ann Arbor, Mich.: Karoma.
1989 "Language innateness and speech pathology", in: Jan Wind – Edwin G. Pulleyblank – Eric de Grolier – Bernard H. Bichakjian (eds.): 209-32.
1991a "Evolutionary patterns in linguistics", in: Walburga von Raffler-Engel – Jan Wind (eds.): 187-224.

1991b "From family tree to phylogeny", in: Werner Bahner – Joachim Schildt – Dieter Viehweger (eds.): 2446-2452.

Callary, Edward (ed.)
in press *Studies in language origins IV.* Amsterdam: Benjamins.

Chomsky, Noam
1957 *Syntactic structures.* The Hague: Mouton.

Chumakov, Ilya et al.
1992 "The continuum of overlapping clones spanning the entire human chromosome 21q", *Nature* 359: 380-86. [This paper is co-signed by 36 co-authors, including Daniel Cohen.]

Fisiak, Jacek (ed.)
in press *Proceedings of the International Conference on Biology of Language: Essentialist vs. evolutionist in the nature of language.*

Friedlander, W.J. (ed.)
1975 *Advances in neurology.* New York: Raven.

Gamkrelidze, Thomas V.
1987 "The Indo-European glottalic theory: A new paradigm in Indo-European comparative linguistics", *Journal of Indo-European Studies* 15: 47-59.

Gamkrelidze, Thomas V. – Vjaceslav V. Ivanov
1973 "Sprachtypologie und die Rekonstruktion der gemeinindogermanischen Verschlüsse", *Phonetica* 27: 150-156.

Gardner, Russell, Jr.
in press "Are there language-regulating genes on chromosome 15?" in: Edward Callary (ed.).

Givón, Talmy
1979 *On understanding grammar.* New York: Academic Press.

Gould, Stephen J.
1977 *Ontogeny and phylogeny.* Cambridge, Mass.: Belknap Press of Harvard University Press.

Halle, Morris
1962 "Phonology in generative grammar", *Word* 18: 54-72.

Jackendoff, Ray
1977 *X Syntax: A study of phrase structure.* Cambridge, Mass.: The MIT Press.

Jakobson, Roman
1964 "Principes de phonologie historique", in: N.S. Troubetzkoy, 314-341 [A rev. and tr. version of "Prinzipien der historischen Phonologie", *Travaux du Cercle Linguistique de Prague* 4(1931): 247-67].

Jespersen, Otto
1891 *Studier over engelske kasus.* [Ph.D. dissertation, University of Copenhagen.]

1941 *Efficiency in language*. Copenhagen: Munksgaard.
1922 *Language: Its nature, development, and origin*. London: Allen and Unwin.
[1964] [Reprinted New York: Norton.]

Jones, William P. – Josiah Hoskins
1987 "Back-propagation: A generalized delta learning rule", *Byte* October: 155ff.

King, Mary-Claire – A.C. Wilson
1975 "Evolution at two levels in humans and chimpanzees", *Science* 188: 107-116.

Kiparsky, Paul
1988 "Phonological change", in: Frederick J. Newmeyer (ed.), 363-415.

Koch, Walter A.
1991 *Language in the upper pleistocene*. Bochum: Brockmeyer.

Kuryłowicz, Jerzy
1964 *The inflectional categories of Indo-European*. Heidelberg: Winter.

Labov, William
1972 *Sociolinguistic patterns*. Philadelphia: University of Pennsylvania Press.

Lefebvre, Claire – Diane Massam
1988 "Haitian Creole syntax: A case for DET as head", *Journal of Pidgin and Creole Languages* 3: 213-43.

Lehrke, R.G.
1974 *X-linked mental retardation and verbal disability*. New York: International Book Corporation.

Lenneberg, Erich
1967 *Biological foundations of language*. New York: Wiley.

LeVay, Simon – Torsten N. Wiesel – David H. Hubel
1980 "The development of ocular dominance columns in normal and visually deprived monkeys", *Journal of Comparative Neurology* 191: 1-51.

Lorenz, Konrad
1971 *Studies in animal and human behaviour*. Vol. 2. [Tr. Robert Martin.] London: Methuen.

Luchsinger, Richard – Godfrey E. Arnold
1965 *Voice - speech - language: Clinical communicology: Its physiology and pathology*. (2nd Engl. edition.) Belmont, CA: Wadsworth.

Martinet, André
1955 *Économie des changements phonétiques. Traité de phonologie diachronique*. Bern: Francke.
1962 *A functional view of language*. Oxford: Clarendon Press.

Meillet, Antoine
1937 *Introduction à l'étude comparative des langues indo-europennes.* (8th edition.) Paris: Hachette.
[1964] [Reprinted Alabama: University of Alabama Press.]
1965a "Convergence des développements linguistiques", in: A. Meillet 1965b, 61-75. [First published in *Revue philosophique* 85 (1918).]
1965b *Linguistique historique et linguistique générale.* Vol. 1. (2nd. edition.) Paris: Champion, Vol 1. [1st ed. 1918.]

Messer, David, J.
in press "Language acquisition and the essentialist-evolutionist debate", in: Jacek Fisiak.

Newmeyer, Frederick J. (ed.)
1988 *Linguistic theory: Foundations* (The Cambridge survey of linguistics, vol. 1). Cambridge: University Press.

Paul, Hermann
1920 *Prinzipien der Sprachgeschichte.* (5th edition.) Halle: Niemeyer.

Polomé, Edgar C.
1980 "Creolization process and diachronic linguistics", in: Albert Valdman (ed.), 185-202.

Postal, Paul
1968 *Aspects of phonological theory.* New York: Harper and Row.

Raffler-Engel, Walburga von – Jan Wind (eds.)
1991 *Studies in language origins II.* Amsterdam: Benjamins.

Renier, Willy O.
1983 *X-linked mental retardation: A clinical study of X-linked syndromes with mental retardation.* [Unpublished Ph.D. dissertation, University of Nijmegen.]

Sapir, Edward
1949 *Language: An introduction to the study of speech.* New York: Harcourt. [First published in 1921.]

Saussure, Ferdinand de
1931 *Cours de linguistique générale.* (3rd edition.) Paris: Payot. [First published Lausanne, 1916.]

Schlegel, Friedrich
1808 *Über die Sprache und Weisheit der Indier: Ein Beitrag zur Begründung der Alterthumskunde.* Heidelberg: Mohr.

Schleicher, August
1852 *Les langues de l'Europe moderne.* [Tr. Hermann Ewerbeck.] Paris: Ladrange. [German original: *Die Sprachen Europas in systematischer Übersicht.* Bonn: König, 1850]

1873 *Die Darwinsche Theorie und die Sprachwissenschaft.* (2nd. edition.) Weimar: Böhlau. [First ed. 1863.]

Slobin, Dan I.
1966 "The acquisition of Russian as a native language", in Frank Smith – George A. Miller (eds), 129-148.

Smith, Frank – George A. Miller (eds.)
1966 *The genesis of language.* Cambridge, Mass: The MIT Press.

Sokal, Robert R. – Neal L. Oden – Barbara A. Thomson
1992 "Origins of the Indo-Europeans: Genetic evidence", *Proceedings of the National Academy of Sciences of the United States of America* 89: 7669-7673.

Troubetzkoy, Nikolaj Sergeevitsch
1964 *Principes de phonologie.* [Tr. Jean Cantineau.] Paris: Klincksiek.

Valdman, Albert (ed.)
1980 *Theoretical orientations in Creole studies.* New York: Academic Press.

Wetzels, W. Leo – Engin Sezer (eds.)
1986 *Studies in compensatory lengthening.* Dordrecht: Foris.

Wind, Jan – Edwin G. Pulleyblank – Eric de Grolier – Bernard H. Bichakjian
1989 *Studies in Language Origins I.* Amsterdam: Benjamins.

Deep reconstruction of languages and semantics

Vitaly Shevoroshkin

1. Linguistic, archaeological, and genetic evidence of evolution

In order to reconstruct, in our imagination, the remote past of humans, scholars study, first of all, material remainders of previous times: human bones, tools, paintings, etc. These studies have advanced dramatically over recent years - both, thanks to new findings and to the new interpretations of old findings. Only a few scholars know, though, that, in many cases, linguists, and not anthropologists and archaeologists, were the first who came to correct conclusions on the basis of linguistic reconstructions. Their conclusions were confirmed by archaeologists somewhat later and quite independently.

Such was the conclusion that Near Eastern cultures, of the Natufian type, were much older and much more developed than previously thought. Linguists not only reconstructed the lexicon of Natufians, which contained words designating some very sophisticated tools and processes. They also dated it by the 12th millennium BP (= before present). And they showed that Natufians spoke the Afro-Asiatic language just before this language started to disintegrate into daughter languages (Semitic, Cushitic, Chadic, et al.).

Linguists were the first to show that Indo-Europeans inhabited some territory in the Near East (later known as Anatolia) some 8 millennia ago and that they were strongly influenced by both Semites, the descendants of the above mentioned Afro-Asian, and by Hurri-Caucasians. Indo-Europeans borrowed from them many cultural words, designations of various domesticated animals and plants, tools, religious terms, and so on.

Linguists were the first to conclude that the process of the domestication of wolves, i.e., of the origin of dogs, should be dated by the 15th to 14th millennium BP, the time when the Nostratic language was spoken. The Nostratic word *k'üjnA or *kujnA (where k' stands for a "glottalized" k and A stands for "a or ä") designated both 'wolf' and 'dog', which should indicate the beginning of the above process of domestication. And the reconstruction of a language as old as Nostratic is, in itself, a major achievement of the modern humanities, as some linguists have put it.

Linguists were the first to indicate that the massive peopling of America was a relatively late process. It happened less than twelve thousand years ago. And linguists correctly identified three major waves of migration from Eastern Siberia to North America: the first wave consisted of Amerind people (the ancestors of most American Indian languages); the next wave consisted of the ancestors of American Indians speaking Na-Dene languages (Athapascan, Eyak, and a few more); the last wave (or several smaller waves) consisted of Eskimo-Aleuts.

Anthropologists do not know, as a rule, about these conclusions made by linguists on the basis of solid linguistic facts. Much more attentive to the linguists and their results are the geneticists who have provided some excellent contributions to the study of our remote past, mainly on the basis of the study of "gene distances" between different populations. And in the current argument between scholars about the genesis of Homo sapiens sapiens, the linguists support those archaeologists and geneticists who conclude that modern man appeared first in Africa and then spread to the Near East (about 100,000 years ago), from where humans spread, much later, to other parts of the world. This conclusion is based on the data showing a genetic relationship between Khoisan and all other known languages of the world.

2. The search for a primordial language

Scholars are rather fond of speculation about languages spoken long ago, above all, about the proto-language of humans, the "Mother Tongue" or "Proto-World" (or the like). Since recent research shows that the human vocal tract underwent, between 200,000 and 100,000 years ago, some prominent changes making it able to produce quite different sounds and sound combinations, we may conclude that humans spoke a rather developed language (or languages) before they came from Africa to the Near East over 90,000 years ago. On the other hand, scholars who study the development of tools and cave paintings conclude that the thinking abilities underwent a very slow development over millennia until they reached a level which is identical to the modern one. This happened, as it seems, somewhat less than 35,000 years ago, just at the time of the beginning of the first great expansion of humans from the Near East. This expansion, as it seems, "covered" most areas of previous population waves, including those of hominids.

It is quite possible that the most ancient languages we are able to reconstruct, are some "lucky" ones. They managed to survive, whereas many of their sisters were less lucky: they were "covered" by the survivor languages or simply died out. Even when we reconstruct some very ancient languages (14,000 years old and much older), we notice that some words do not "fit" well the pattern of the language under reconstruction. Such residual words might be borrowings from other languages, including those "covered" by the language we reconstruct. And when languages are not real languages yet, but mutually understandable dialects, their bearers can borrow from another dialect practically any words, even the most stable ones. It is only when languages become mutually non-understandable that the most stable words cease to be subject to borrowing.

Apparently, the ancestors of Khoisan people in the southern part of Africa never left this continent whereas the bearers of Afro-Asiatic languages went (relatively recently) "back to Africa", the continent which their ancestors had left so many millennia ago.

Geneticists also include Congo-Saharan peoples as well as Ethiopians in the African population groups. It is clear that Ethiopians do not belong here linguistically: Ethiopic language belongs, alongside South Arabic, to a subgroup of Semitic languages. Apparently, long ago the ancestors of the Ethiopians spoke a non-Semitic language before adopting Semitic (or Afro-Asiatic). It is not excluded that the Congo-Saharan peoples have adopted Congo-Saharan languages relatively early: before that time, their ancestors spoke Khoisan-like languages. Indeed, according to the geneticists, the Congo-Saharan population is African, and thus very different from all non-African population groups (including Afro-Asians, i.e., people speaking Afro-Asiatic languages).

Linguists, in their preliminary comparisons of the most stable words in Khoisan, Congo-Saharan, and Afro-Asiatic languages find impressive evidence of possible genetic relationship among all of them.

Further below, I will speak in detail about the existing, very precise, reconstructions of two ancient languages (Nostratic, of which Afro-Asiatic is a daughter, and Sino-Caucasian), of their remote, but clearly existing, genetic relationship, as well as of a probable genetic relationship of all known languages shown by comparison of reconstructed roots. This research allows us to provide a preliminary reconstruction of the "Proto-World" language, the "lucky" language which was spoken some 90,000 to 100,000 years ago, and which was the only one to "survive": all known languages seem to be its offspring.

But what can we say about the sound shape and semantics of those words which were spoken by our ancestors so many millennia ago? First of all, we should not trust unfounded speculations by scholars who look for "primordial" features in such languages as Adyghan or Ubykh (part of the North Caucasian family), Aranda (Australian), or reconstructed Indo-European.

It is wrong to conclude that the "Proto-World" language had certain vowels simply because the above languages had these vowels, etc. Some North Caucasian languages have only two (a, ə) or three vowels, which does not mean that their parent language (Proto-North-Caucasian) had only this many vowels. It certainly had more, as both Starostin (1984) and Nikolaev (1987) have shown in their recent reconstruction of North Caucasian. Even more inappropriate is the conclusion that the proto-proto-language had just one vowel or none. This conclusion is based on the belief that there are languages with one vowel. All known languages have two or more vowels, which can be explained by a feature present in all languages, namely, the existence of oppositions such as consonants vs. vowels, low vowel(s) vs. high vowel(s), obstruent consonants vs. sonorants, verbs vs. nouns, "full-meaning" words vs. "grammatical" words, etc. There should be at least two vowels in a language to enable a phonological distinction between them: a word of the type ta is semantically different from tə; na is different from nə; tan is different from tən, etc.

All this does not exclude the possibility that, at some very old and "primative" stage, human language(s) used vowels as mere "supporters" of consonants. We simply do not know anything about it. But it is inappropriate to draw such conclusions on the basis of wrongly interpreted situations in existing or reconstructed languages. It is wrong to suggest that the Indo-European language had only one vowel, or only two vowels, e and o, originating from ə and a. Indo-European had at least five vowels, namely, a, e, i, o, u. This reconstruction supported by many Indo-Europeanists is confirmed by Nostratic data. The Nostratic proto-language, i.e., the parent of Indo-European alongside several other languages, had all these vowels plus ä (and maybe ü). In most cases, Indo-European shows the vowels e and o, which does not mean that these were its only vowels. There are no languages with only e and o as vowels. The mere fact that a language has the vowels e and o presupposes that this language has also a, i, and u. This is one of the universal rules.

Artificial archaization of languages is a sad fact, well known from the history of linguistics. One would speak about Aranda as a language which conserved such primordial features as just one vowel (a) lack of generalizing words such as "snake", "tree", etc. In reality, any Arandic language has at least two vowels, a and ə. This leads us to another universal stating that if a language has only two vowels, they are a and ə. These two vowels have replaced three Proto-Australian vowels. As for the "lack of generalizations", this idea originates from the lack of elementary knowledge of Australian languages by many people who used to write about these languages, as well as about aborigine Australians.

Finally, it is wrong to consider the "primordial" human language as having very short words, just a consonant, or a combination of one consonant and one vowel, just because words and roots of North Caucasian languages frequently have this shape. North Caucasian languages developed many consonants which allows for differentiation between many stems of the type C-, CV-, VC- (C = a consonant; V = a vowel). This fact contributed to the shortening of many words. But in the Proto-North-Caucasian language, there were less consonants, and the words were relatively long.

It is quite appropriate to use data from child language and its development when studying the origin of human language(s). But the most "primative" features of the early language of children show rather what human language was like before the human vocal tract became sufficiently developed (see above). At this, very early, stage of human language (or, rather, pre-language) there might have been no distinction between vowels, between words of verbal and those of nominal nature, etc. There might be only a few consonants, the most contrastive ones, such as m, t, k (or perhaps we should speak about "syllabophonemes" mV, tV, etc.). All these suggestions are educated guesses and nothing more. Yet, they still might be of interest to different kinds of scholars, including anatomists and philosophers.

3. Reconstruction of Proto-World words

What we can build upon are two series of data: conclusions about the most stable words (which might indicate what the primordial lexicon looked like), and about the shape of human words of the proto-proto-language. We are now speaking about the "first language" as a language, not as a pre-language or the like. The first language, or languages, apparently had many consonants, several vowels, different kinds of words, and different types of sentences. First, I would like to discuss the shape of such a language.

In Blažek's (1989) reconstruction of the "Proto-World" (see below), there are only three vowels (a, i, u, a rather "popular" vowel system in world languages, a very contrastive one). Apparently, additional vowels, such as e, o, ä, ü, appeared much later.

There were a fair amount of consonants, Including glottalized consonants of the k', t', p', lateral obstruents λ', λ, ɫ etc. All these consonants were present in such archaic languages as Nostratic, Dene-Caucasian, Amerind, partly in Khoisan and other languages.

This is how far "back in time" we are able to go, and still not indulge ourselves in speculation. Participation of Khoisan languages in the reconstructed "Proto-World" words belonging to the most stable part of the basic lexicon (words which do not undergo borrowings and are seldom, if ever, replaced by other words) seems to indicate that words of this kind are some 90,000 years old. Words which we still consider as "Proto-World" but which are lacking in Khoisan (provided that they were never present) might be much younger.

As soon as we leave the reconstruction of the three vowels and the various consonants on the basis of words present in most ancient reconstructed languages, we are doomed to speculation. But what is not speculative is the conclusion about a frequent appearance in the "Proto-World" of word variants. On a lesser scale, such variants can be found in even much younger languages, such as Nostratic. Let us take a few examples:

[A] Nostratic *ńiĺa 'slippery and moist (about the inner layer of rind, skin)' and (apparently related, according to Illich-Svitych) *ńülV 'tear out, scrape off' (as in Uralic *ńül-ke 'to skin, tear out hair', Dravidian *nuḷ- 'pinch, pinch off, tear off', Altaic *ńü/öl- 'strip, tear out, scrape off'). Beside the variation of the type i:u, note the following: (1) Illich-Svitych considers the word *ńila "descriptive" (the sound-shape of the word is motivated, as in onomatopoeia). (2) The meaning is both complex and concrete ("primitivism"). (3) There was apparently no early differentiation between adjectival/nominal/verbal aspects of meaning, though, in general, the meaning is predicative, not subjective. Note that ń is a soft n; ĺ is a soft l.

[B] Nostratic *Himi 'suck, swallow' and *ʔEmV 'seize, take' (ʔ is a glottal stop) might be related and are descriptive, according to Illich-Svitych (1990) (H might be ʔ, E means "e or i"). It seems, the original word is connected with a snapping sound made by the mouth. It is important to note that the amount of descriptive words in Illich-Svitych's Nostratic dictionary is rather high and that, as a rule, they are phonetically very stable.

Illich-Svitych (1990) considers as descriptive not only a few dozen onomatopoetic words, but also such words as Nostratic *biĆa 'small', *bongä 'fat, swell', *didV 'big', *ʔebu 'breast', *k'ab/p'a 'seize' (note variants), *ʕama 'knead, mash, soften', *nämV 'soft', etc. Some scholars consider words designating round objects as well as words designating processes of the type 'eat/chew/enjoy food' (they are frequently clearly onomatopoetic), 'live, vital force' (< 'breath'), 'bend/bent/protuberant', 'hot/burn/live coals' etc. But there is not always clear evidence of the descriptive character of such words. It is also well known that many languages distinguish between adverbs or pronouns of the type 'this, here' vs. 'that, there' (etc.) by adding different vowels to the root, usually front vowels in the former case, and non-front in the latter.

"Proto-World" words reconstructed by Blažek (1989) also show a descriptive character in some cases. Many words (especially those for body parts) have variants, cf., e.g., *gini/nigi 'tooth', *t'ali/dali 'tongue', *t'ari/t'uri 'leg', *matu/mati 'forehead', etc. (see Blažek 1989; for other data on deep reconstruction see also Shevoroshkin 1989a, 1990, 1992).

4. Lack of categorial differentiation in archaic words

We may assume that both an onomatopoetic/descriptive character of such stable words as those for body parts (etc.) and a rather frequent sound alternation in such words may, indeed, indicate that we deal here with a very archaic feature. Note also that these kind of words tend not to possess a clear differentiation between categories, such as "nominal" vs. "verbal", etc. I would like to add a few examples from the "Proto-World" list:

[C] *kuni 'elbow/knee' (rather: 'joint') and a corresponding verb meaning 'to bend' (present in Khoisan ['elbow'], Nilo-Saharan ['knee'], Nostratic [*küjñA 'bend at the joints; joint'], Austro-Asiatic ['elbow'], Australian ['elbow'] and other languages). The apparent indivisibility of the underlying notions 'elbow', 'knee', 'bend (as elbow/knee)' makes it clear that both elbow and knee were named after their most prominent function/quality. This was just one simple word.

This conclusion may be drawn as far as many other ancient words are concerned, namely, those for body parts (such as Proto-World [=PW] words for 'eye/see' [*ʔaku/ʔuKV; note the variants; Nostratic *ʔak'u/ʔuk'V 'eye/see'; Nilo-Saharan *ʔak^{w}i id.; Amerind *ʔaku 'eye'; only as remnants in Khoisan and Australian]; 'nose/smell'; ear/hear', etc.), for animals [*bo´ra 'grey-brown animals such as a bear or a beaver'], insects [*k'VmV 'biting insect and to bite as it does'] etc.). In all of these cases, something is named in accordance with an important quality and function, but it is only from our point of view that a complex description of a corresponding meaning is valid. For the ancient people it was a simple and regular phenomenon of their everyday life. They would use some "descriptive" sound complex to designate a certain phenomenon under certain circumstances (say, when hunting, making fire, performing a ritual, etc.), and this word would be used from generation to generation to designate some simple notion from their point of view.

We still have, in some cases, both grammatically and semantically undifferentiated words even in modern languages, describing rather regular phenomena. Consider the use of one word to attract attention to the whole situation: "Rain", "Fire", "Cold", "Dark", etc. Having a somewhat more complex sentence type, e.g., Subject + Predicate, we would be able to describe an enormous amount of situations quite adequately, provided we have enough words to fit this pattern: "I [am] cold", "John [is] hungry", "He [is] here", "[The] Wolf [has] escape(d)", "[The] Bear [is] dead" (which is the same, of course, as "...[has] died") etc. As subjects, "typical" nouns can be used as well as pronouns. As predicates, "typical" verbs can be used as well as adverbs/adjectives and nouns. Again, all of this is from our point of view. Semantic analysis even of short sentences of our languages is exceedingly complex. Our ancestors were much less complex.

I hope that I have made my thesis clear. At some very ancient stage, there was no differentiation of the type noun vs. verb vs. adverb, etc., but there was a clear differentiation of the type subject vs. predicate. And there was much less differentiation in the meaning of words. There were much less words, which is the reason as to why alternations were frequent. This is a trace of some very archaic period of language which had not yet been "standardized". But let us substantiate the thesis of "no differentiation" between many meanings in past times in detail.

In many cases, Nostratic, as well as other languages, such as Sino-Caucasian and Amerind, show no differentiation between "nouns" and "verbs" in the above sense (e.g., 'joint' = 'bend'; 'nose' = 'smell'; 'tongue' = speak', etc.). We may add such examples as 'leaf/grow (about plants)'; 'fire/burn'; 'luck/to favor'; 'ford/to ford'; 'cold/freeze'; 'sore/hurt'; 'flood/flow'; 'wound/to wound'; 'worm, snake/creep, crawl'; 'sharp edge/to bore'; 'licking organ(s) [tongue, lip]/lick'; 'flame/ blaze, 'sun/shine'; etc.

Other cases show no differentiation between "adverbs/adjectives" and "verbs", e.g., 'high/(a)rise'; 'wet/soak'; 'round/rotate, roll'; 'secretly/

steal'; 'hot, warm/burn' [and 'fire': see above]; 'afraid/frighten' [and 'fear']; 'thick/swell'; 'broad/stretch, extend'; 'bright/shine' [and 'daylight']; 'cold/freeze'; 'dead/die' (or: 'ill, sick/die'); 'sharp/cut'; 'strenuous/strain oneself'; etc.

The following cases show no differentiation between "adverbs/ adjectives" and "nouns": 'dark/night'; 'old/year'; 'young/young animal'; 'wet, slippery/mud'; 'low/low place, lowland'; 'bright, shiny/sun' [and 'to shine', 'daylight', etc.]; 'big, plenty, good/abundance'; 'hot/fire' [cf. above]; 'humid, moist/moisture'; 'dark, dirty/spot, stain'; 'empty, hollow, light/hollow of a tree'; 'round/ball'; 'alive/vital force' [also 'to live']; 'dead/spirit of the dead' [different from 'dead/die' or 'ill, sick/ die'; cf. above].

We have already cited such undifferentiated words (belonging, from our point of view, to one and the same category) as 'wet' and 'slippery'; 'big', 'plenty' and 'good'; 'dark' and 'dirty'; 'empty', 'hollow' and 'light'; 'worm' and 'snake' etc. We can add many other examples.

Nondifferentiated "nouns": 'wound/pain' , 'male/man'; 'bosom/udder'; 'end/edge/brim'; 'chest/heart'; 'rind/hide/skin'; 'sunlight/dawn'; 'day-light/morning'; 'ache/damage'; 'kin/root'; 'head/brains'; 'belly/guts'; 'sinew/thread'; 'lip/jaw'; 'nail/claw'; 'paw/foot'; 'fingernail/finger'; 'half/side'; 'spell/song' [also possible: 'spell/name']; 'stone/flint'; 'fire/ heat'; 'fog/mist/cloud'; 'piece/part'; 'wolf/dog'; 'clay/glue'; 'heap/ group'; 'mountain/hill'; 'sap/resin'; 'trunk/tree/wood'; 'thorny branch/ thorn'; 'sickness/grief'; 'fence/wall'.

Nondifferentiated "adjectives/adverbs": 'soft/weak'; 'loose/slack, not tight, not strained'; 'enough/too much'; 'fat/healthy'; 'smooth/shiny'; 'dark/black'; 'sleek/bald'; 'strong/mighty/old'; 'bent/crooked/dislocat-ed'; 'angry/jealous'; 'long/spread/broad'.

Nondifferentiated "verbs": 'grab/procure'; 'split/cut/divide'; 'grow/ appear'; 'dwell/ sit/live/ stay'; 'be silent/be quiet'; 'carry/drag'; 'search/ find'; 'suck(le) / swallow'; 'see/understand/know' [in some cases, 'see' is original]; 'steal/hide'; 'swim/float'; 'fear/shake'; 'call/ask'; 'protect/ help'; 'trace/chase'; 'grow/appear'; 'need/lack'; 'drip/ ooze', etc.

There are at least two substantial objections against such interpretation of reconstructed words from 14,000-year old languages: at times, we might not be sure that we are reconstructing the meaning(s) of certain words in a very precise way, and indeed, even if we are, there is absolutely no guarantee that we deal with the traces of the most ancient phenomenon (in many cases, we can easily find exact parallels in contemporary languages, or languages which are not very old). But I see a clear pattern emerging from the solid studies in Nostratic and Sino-

Caucasian, including their comparison, on the one hand, and the tentative reconstruction of Proto-World (by Blažek 1989 and others) on the other hand. We find here exactly what we would expect to find: sound symbolism (= our "alternations"), a rather developed phonetic system but a massive nondifferentiation of meanings, sound motivation of many words, clear "inner form" of many words, primitive "complexity", concreteness of many meanings, etc.

The above groupings of meaning (to which a fair amount of other pairs and triplets of meanings can be added) might be considered as a rough classification on which one can base further search, and further specifications. For instance, our pair of adjectives 'fat/healthy' is based on Nostratic comparisons and reconstruction: Nostratic *kojHa 'fat, healthy' > IE *gʷeiH- 'be healthy, live', Uralic *kōja 'fat' (both noun and adjective). Illich-Svitych (1990) remarks that IE shows a development 'be fat > be healthy > live'. Uralic (as it very frequently is the case) seems to have conserved the original meaning. Starostin (1984) has compared the Nostratic word with Sino-Caucasian (= Dene-Caucasian) *GwVjV 'thick' (his G = "back g") represented by North Caucasian and Yeniseian (words meaning 'thick'). This confirms Illich-Svitych's assumption that 'fat' was the original meaning. Thus, we have here a semantic evolution in IE (as it frequently is the case) of a Nostratic meaning 'fat' which was practically identical to the Sino-Caucasian 'thick'. We should rather list the "primordial" nondifferentiated 'fat/thick' with a relatively late development 'fat> healthy'. Since the meanings 'fat', 'thick' have not disappeared, we may assume that a principally new meaning ('healthy') has evolved thanks to the evolution of thought. But the opposition 'fat/thick' (≈ 'good'): 'thin' (≈ 'bad') seems rather old.

In many cases, the same meaning is reconstructed for several words in some ancient proto-language, which seems suspicious. A more detailed analysis helps to reveal several concrete meanings in each case. For instance, Nostratic *baɣV 'much, enough' and Sino-Caucasian *bVHV 'big, many' indicate an underlying meaning 'big/many/much'; the meaning 'big' is present in other Nostratic words: *wolV, *malV, *magV, *bara, *č'ok'V, some of them having Sino-Caucasian parallels. We can conclude that *wolV was rather a collective ('big crowd, many people') because the development in IE shows the meaning 'throng', 'press'; *malV indicated a positive aspect ('big/good', 'big/plenty'); *magV meant 'big/firm'; *bara meant 'big/voluminous'; *čok'V maybe 'big/thick'.

5. Ancient ways of thinking as expressed in Nostratic

There are some clearly archaic features both in grammatical and in semantic development in such well-reconstructed ancient languages as Nostratic and Sino-Caucasian. Respective observations are supported by linguistic data from other ancient languages.

There is an ancient dichotomy of the kind "our: non-our" (oppositions) and "Object A_1: Object A_2" (pair; coupling of twin objects). It is expressed throughout the language (this includes the grammar). Here we find the opposition of personal pronouns ("we-inclusive" vs. "we-exclusive"), of dual vs. both singular and collective (> plural, in some cases), etc. Even Nostratic, a rather developed language (with an emerging category of plurality at different levels, including personal pronouns), holds these oppositions. They are also reflected in Nostratic words, including the system of kinship terms.

Ancient ways of thinking are expressed in many ways in Nostratic:

[D1] There are only two autonomous color terms, namely 'dark/black' (originally just 'dark') and 'bright'. Other words which we translate as color terms, are not autonomous: 'brown', or rather 'grey-brown', is a color of certain animals (see above); 'yellow' is the color of inner organs (and also means 'pale'); 'red' means, firstly, 'blood' (at least, in Dene-Caucasian; in Nostratic, the corresponding word [of a different origin] meant 'raw meat').

[D2] We find an ancient opposition of the type 'vital force' (>'life, alive, live') vs. 'spirit of the dead' in Nostratic.

[D3] There are several words for 'male/man' (in all cases, the designation of an animal is primary) and one of the words is used to designate 'young males/men'. Accordingly, there was a word for both a female animal and a woman.

[D4] Another opposition, connected to the designation of parts of plants (broader: natural objects), is in accordance with the body structure: 'branch' < 'arm', or 'leg', etc. In the language which preceded both Nostratic and Dene-Caucasian, there was a word for both 'blood' and for 'breath, soul', also 'live'. It is possible that both Nostratic and Dene-Caucasian had such an archaic word. The kind of semantic change in 'fat' > 'healthy'; 'thirst, be thirsty' > 'desire, love'; 'sick' > 'upset' is also typical of Nostratic languages, though it can find analogies elsewhere. (Cf. also 'think' > 'be(come) sad'; 'pain' > 'grief'.)

There was a clear development, in many cases, of the underlying "complex"/concrete meaning into several more "simple" meanings:

[E] Nostratic *ʔesA 'settle [for a relatively short time] during a roaming/migration' becomes simplified in daughter languages (> 'be in a certain place' > 'dwell', 'live', 'sit', 'be').

[F] The word with the meaning 'to open one's mouth' partly retained this meaning, and partly changed it in languages (> 'opening, open').

[G] Nostratic *wot'V seems to have a concrete, "complex" meaning 'to overtake an animal (game) during a hunt' (still preserved in Uralic). Other meanings ('obtain', 'get', etc.) might be further developments.

[H] Nostratic, *c'äjha originally meant 'glimmer/twinkle/shimmer/dim light'. This meaning developed into 'bright, clear' (AfAs; Ural) and 'pale, shade' (in IE). Still, these meanings do not have the component 'fire'. This component is present in Sino-Caucasian *c'AjV 'shine, light, fire'. As in several other cases, the moot ancient meaning does not show any difference between 'shine/light' and 'fire/burn/smoulder'.

[I] Nostratic *ńowda had an originally complex meaning which is not easily definable. Illich-Svitych simply gives 'move quickly'. It was certainly more specific. Cf. the developments in IE ('do battle'), Uralic ('pursue, chase, follow'), and also 'be agitated', 'carry', 'deliver' etc. in individual languages.

[J] Nostratic *pelHi 'shake, fear' apparently shows a nondifferentiated meaning which arose from 'shake'. Daughter languages still preserve traces of such development (cf. 'shake' > 'start [because of something unexpected]' > 'get frightened' in Altaic).

[K] There was a Nostratic word, *ńaʔV, which apparently designated both the time when young animals were born, and the these young animals themselves. In a group of languages, this word acquired the meaning 'be born in spring, young animal'; in another group, the meaning is 'season, year'.

[L] Nostratic. *mu(n)k'V had a complex meaning: 'heavy, difficult, requiring much strength'. Further developments in different languages were: 'strain oneself', 'torment', 'strenuous work', etc.

[M] Nostratic. *qot'i (or *Xot'i, after Starostin) meant 'burn', 'fire'. Its cognate was also present in Sino-Caucasian (cf. Proto-Yeniseian *Xot- id.). Since this Nostratic word also meant 'set fire to' it was, apparently, a designation of a fire made by men, and thus different from other words for 'fire/burn'.

[N] Nostratic *gara originally meant 'a thorny branch of a certain tree [apparently, a pine-like tree]'. It acquired, in daughter languages, "simpler" meanings: 'thorn', '[sharp] point', etc.

[O] Nostratic *dalqu seems to mean 'wave/wavy, become wavy/ rough'. Its developments in daughter languages show the meanings 'get rough/choppy'; 'wave'; 'be agitated, upset, stir up'. Altaic shows just 'sea'. Apparently it was a new word for Altaic people. The inner form is clear: 'wavy/rough'.

[P] Illich-Svitych (1990: 145) reconstructed a Nostratic word *Haja (reflected in IE, Uralic, and Altaic) with the meaning 'pursue, chase', but he himself saw that, apparently, there was a more archaic, more "concrete" [= more complex from the point of view of our semantics] meaning which, actually, should be reconstructed for Proto-Nostratic: namely, 'to pursue an animal'. Indeed, in Altaic *aja- 'to hunt, to procure a [wild] animal', this ancient meaning is still reflected. "Individual" Altaic languages contain indications of the whole complexity of the primordial meaning. The word in question means 'a long march' in Buryat, 'travel' in Written Mongolian, 'hunt, campaign' in Middle Mongolian, 'kill a [wild] animal' in Evenki, etc. It is easy to assume that *Haja 'pursue an animal' was an onomatopoetic word.

[Q] Another East Nostratic word, *HütV, is cited in Illich-Svitych's Nostratic dictionary as having a meaning 'time of rest/relaxation', which is rather complex. Illich-Svitych remarks that the Nostratic word might have had an even more complex meaning, compare: 'time of rest/relaxation' in Altaic, but 'evening, night' in Uralic.

[R] A Nostratic word *ʔaĺa 'food' means 'fat (food)' in Afro-Asiatic, 'feed, rear' in IE, 'fat, strength' in Dravidian, and just 'food' in Altaic. Illich-Svitych comments that the underlying meaning might actually have been 'meat food, meat fat'. It developed to 'fatten' in IE and to fattened [= well-fed]' and further to 'strong' in Afro-Asiatic and Dravidian.

The specificity of the thinking of our remote ancestors seems to be reflected in the ancient identity of the words 'year' and 'go/pass' as well as 'year' and 'old/worn out'. The word for 'young (animal)' was also a designation of 'spring'. There were no special words for 'summer', but there was a word which meant '(summer) heat'. There were different words for 'body hair' and 'facial hair', and a special word for 'thick hair' (sing.) (as on some animals) which also designated a 'thin stalk'.

There was a word which meant 'bring [home] a wife'; it also meant 'lead', 'drag'. There was a rather elaborate system of kinship terms, though much less elaborate than in some modern "primitive" languages. There were different words for some concrete actions which we could classify as having the general meaning 'to move': 'run/walk quickly'; 'start moving, jump'; move violently, fight, chase'; 'move rhythmically, shake, dance, be agitated'. Some ancient rituals might be reflected here.

6. Word stability in language evolution

There is one more way to conduct research in the field of the most ancient languages. Beside studying the reconstructed lexics and semantics of Nostratic, Dene-Caucasian, their (and other languages') parent-language, "Proto-World", one may study Dolgopolsky's (1986) hierarchy of stability of the most stable words, as well as shorter lists of stable words (35, 100, 110 units, etc.). Dolgopolsky's list is made up of the material of some 200 languages from different parts of the world. It includes fifteen most stable words. Ten were added from his own data. These words are never, or very seldom, replaced by other words of the given language (not by borrowing, though). They are not subject to borrowing, at least, in the vast majority of cases. Yakhontov lists 35 words. It is not a hierarchy, but simply a list of 35 stable words from all over the world. I will first cite Dolgopolsky's then Yakhontov's list:
Dogopolsky: (1) I/me, (2) two/pair, (3) thou/thee, (4) who/what, (5) tongue, (6) name, (7) eye, (8) heart, (9) tooth/teeth, (10) no(t), (11) fingernail/toenail, (12) louse, (13) tear, (14) water, (15) dead, (16) nit, (17) moon, (18) hand, (19) night, (20) blood, (21) horn, (22) full, (23) sun, (24) year, (25) salt.
Yakhontov: (1) blood, (2) bone, (3) die, (4) dog, (5) year, (6) egg, (7) eye, (8) fire, (9) fish, (10) full, (11) give, (12) hand, (13) I, (14) know, (15) louse, (16) moon, (17) mouth, (18) name, (19) new, (20) nose, (21) one, (22) salt, (23) stone/rock, (24) sun, (25) tail, (26) this, (27) thou, (28) tongue, (29) tooth, (30) two, (31) water, (32) what, (33) who, (34) wind, (35) year.

Yakhontov's list adds the following ones to Dolgopolsky's list: nose, mouth, bone, tale, egg, fish, dog, one, wind, year (this latter might also have the meaning 'old', 'worn out'), stone/rock, fire, new, this, give, know, die. The following words from Yakhontov's list are absent in Dolgopolsky's list: no(t), heart, fingernail/toenail, horn, tear (noun), night. If we exclude 'tear' as a word which frequently represents a compound ('eye-water') and consider 'dead' and 'die' as one word, we would wind up with 40 most stable words, mostly organized as a hierarchy of stability. Note that each unit can be represented in any given language by one, two or more "concrete" words.

Among these words, the most stable ones are personal pronouns and some particles, as well as the word for 'two/pair' (also important for grammar). Most words represent designations of body parts. There are many nouns, but only few adjectives/adverbs and verbs. (Actually, Dolgopolsky's list contains no verbs. It is interesting that about two dozen grammatical elements [= pronouns, particles] common to both Nostratic and Sino-Caucasian, contain almost no verbal elements, though there are many such elements in both Nostratic and Sino-Caucasian grammar.) The "nominal" nature of the oldest lexicon might indicate that Proto-World was, initially, a tool of self-expression and not of communication.

It seems that we do indeed deal with the oldest, most archaic part of human lexicon. Note that in Blažek's Proto-World reconstructions most words designate body parts (hence pairs like 'eye/see', 'nose/smell', etc.). There are many words among this oldest stock which used to have alternative (including metathesized) forms. We discussed words for eye, tooth, tongue above. Cf. also *Ham/maH- 'hand' (also with the suffix *-r/n- in quite different languages), also 'take'. The word for 'nose' (e.g., Nostratic *ŋEqla 'nose' [AfAs: Semit. *nxs, IE *nexs-, Altaic *ŋiaksV 'nose']; Austric: Austronesian *ŋusuq, Amer. *ŋuɬiq, etc.). Many such words, as it was noted above, are of descriptive origin.

References

Blažek, Vaclav
1989 "Materials for global etymologies", in: Vitaly Shevoroshkin (ed.), 1989b, 37-40.

Dolgopolsky, Aaron B.
1986 "A probabilistic hypothesis concerning the oldest relationships among the language families of Northern Eurasia", in: Vitaly Shevoroshkin – T.L. Markey (eds.), 27-50.

Illich-Svitych, V.M.
1990 "The Nostratic reconstructions", in: Vitaly Shevoroshkin (ed.), 138-167.

Nikolaev, Sergej – Sergej A. Starostin
1987 *North Caucasian roots*. Unpublished manuscript.

Shevoroshkin, Vitaly (ed.)
1989a *Explorations in language macrofamilies*. Bochum: Brockmeyer.
1989b *Reconstructing languages and cultures*. Bochum: Brockmeyer.
1990 *Proto-languages and proto-cultures*. Bochum: Brockmeyer.
1992 *Nostratic, Dene-Caucasian, Austric, and Amerind*. Bochum: Brockmeyer.

Shevoroshkin, Vitaly – T.L. Markey (eds.)
1986 *Typology, relationship, and time*. Ann Arbor: Karoma.

Starostin, Sergej A.
1984 "Gipoteza o geneticheskikh sv'az'akh sinotibetskikh iazykov s eniseiskimi i severnokavkazskimi iazykami" [= Hypothesis about the genetic ties of Sino-Tibetan languages with Yeniseian and North-Caucasian languages], *Lingvisticheskaia rekonstruktsiia i drevneishaia istoriia vostoka* (Moskow: Nauka), 4: 19-38.

Material time and formal time: Genetically and metagenetically

Roland Harweg

Für Walter Koch
in alter Freundschaft
und Verbundenheit

1. Material time and formal time

When, nowadays, one thinks of the phenomenon of time, one almost automatically thinks of clocks and calendars. If one wants to make an appointment, one normally will do this in the form of clock time and/or calendar time expressions. In this, one principally seems to have two possibilities, a deictic one and one which is proper-name-like. More precisely, however, there are even three possibilities. The third lies between the two others and manifests, as it were, a mixture of them. I designate this possibility as 'semideictic' (cf. Harweg 1985).

Deictic clock time and calendar time expressions are, e.g., expressions such as *in this minute* or *today*, respectively. Semideictic clock time and calendar time expressions are expressions such as *at ten o'clock* or *on the 20th of August*, respectively. On the one hand, these expressions are proper-name-like. On the other hand, they are part of a domain, namely that of the day and that of the year, which is deictic. Proper-name-like temporal expressions are expressions such as *in the year 54 after Christ*.

Clock and calendar times are, however, not the only kinds of time we know. There are also times of states of affairs. Temporal expressions designating such times of states of affairs are, among other things, temporal clauses and their nominalizations and pronominalizations, i.e., expressions such as *when Charles came back from America* or *after Charles's return from America* or *after that*, respectively. I designate clock time and calendar time as formal time and the time of states of affairs as material time.

In much the same way as formal time, material time can be expressed through deictic, semideictic, or proper-name-like expressions. But if it is expressed by deictic or semideictic expressions, its deictic or semideictic nature is an extremely weak one, or as I have called it elsewhere (Harweg 1990: 250ff.), a "superweak" one. In this case, it owes its deictic nature to deictic components of other deictic dimensions, namely to deictic words of the local and/or the personal dimensions, such as the deictics *here* and/or *I* and the semideictic expressions *the President* and/or *the university*. An example of a superweak deictic material time is the time expressed by the temporal clauses *after Charles had come here* or *when I was in Paris*. A superweak semideictic material time is the time expressed by the temporal clauses *when the President was in Poland* or *when the university was founded*. Material times resembling the denotata of proper names are material times which are expressed by temporal clauses which contain proper names instead of deictic or semideictic expressions. Such temporal clauses are clauses like *when Goethe was born* or *when Rome was founded*.

Whilst the units of material time are states of affairs, the units of formal time are individual entities, metrical units of clock time and calendar time. But these entities are units of formal time not only when they are expressed by temporal phrases, such as *at ten o'clock*, *in 1980*, *for ten minutes* or *for two years*. They are also units of formal time when they are expressed by subject or object phrases, i.e., by expressions such as *1980* in the sentence *1980 was a leapyear* or *this day* in the sentence *This day I shall never forget*. The same thing applies to states of affairs. They are units of material time not only when they are expressed by temporal clauses or their nominalizations and pronominalizations, but also when they are expressed by complete sentences such as *It is raining*, *The sun is rising*, *Charles has written a letter* or *Peter will visit his father*.

The fact that the states of affairs expressed by such sentences are units of material time does not depend on the fact that these sentences contain tenses. Even states of affairs represented by sentences of tenseless languages are units of material time, or what is more, states of affairs need not be expressed at all in order to gain the status of units of material time. If one makes a distinction between the time of expressed and unexpressed states of affairs, one must even say that the time of states of affairs in its strictest sense is only the time of the unexpressed states of affairs. In this article, however, I will not make such a distinction. The time of states of affairs and, for that matter, material time is the time of states of affairs irrespective of whether or not they are expressed. The question is only whether or not all states of affairs are units of (material) time or whether, like Plato in his *Timaios* (37c ff.), one should rather say that only those states of affairs are such units which are in motion. An argument for contending that *all* states of affairs are to be subsumed under the concept of material time, even those which are not in motion, could be derived from the fact that these states of affairs can be *compared* with states of affairs in motion and thus be interpreted as a boundary case of being in motion and, for that matter, as a boundary case of material time.

2. The genetic relationship of material time and formal time

Genetically, material time is the primary form of time. Being the time of states of affairs, it is given together with the states of affairs themselves. At least it is given together with those states of affairs which are in motion, but these constitute by far the greatest part of all states of affairs.

The fact that material time is the time of states of affairs and that formal time is the time of individual entities is in itself, however, not sufficient evidence for the genetic primacy of material time as opposed to formal time. It will be sufficient evidence only if the states of affairs themselves can be regarded as primary with regard to the individual entities.

The view that states of affairs in the world are primary in relation to individual entities or "things" is the view Ludwig Wittgenstein put forward in his *Tractatus Logico-Philosophicus*. Wittgenstein (*Tractatus* § 1.1) states: "The world is the totality of facts" (i.e., states of affairs), "not things" (i.e., entities). Such a view implies, although Wittgenstein does not say so, that entities are only the result of the analysis of states of affairs, rather than that states of affairs are only the result of the combination of entities.

At first sight, Wittgenstein's thesis seems to gain support from language. Expressions of states of affairs are, among other things, sentences, whereas expressions of individual entities are words, more precisely, a certain subcategory of words. It is also obvious that the use of language consists in the use of sentences, or more precisely, texts (Hartmann 1971: 10), rather than in the use of isolated words. On closer inspection, however, we see that there are also sentences whose denotata are not exclusively states of affairs. If, for example, we compare a sentence of the type *There is an eagle* with a sentence of the type *This is an eagle*, it is in fact undeniable that both sentences designate states of affairs. However, it is also obvious that the state of affairs designated by the latter sentence is different from the state of affairs designated by the former sentence. The state of affairs expressed by the sentence *This is an eagle* subsumes an unnamed entity under a named one. Thus, if projected onto the plane of the state of affairs designated by the sentence *There is an eagle*, it is nothing but a mere entity. At the ontological plane, we have, in the case of the denotatum of the sentence *There is an eagle*, really a state of affairs, i.e., the state of affairs of there being an eagle, whereas, in the case of the denotatum of the sentence *This is an eagle*, we have solely the entity of the eagle at that plane. The subsumption of the denotatum of the deictic *this* under the denotatum of the class name *eagle* belongs to another, to a higher, to a logical rather than to an ontological plane.

This argument shows that language does not support Wittgenstein's thesis of the primacy of facts (states of affairs) as opposed to things (entities) all along the line. Searching for further evidence, one will see that the analysis of language can supply still more arguments against this thesis. Such arguments may, for example, be drawn from the analysis of the use of temporal and local "relation words", i.e., prepositions and subordinating conjunctions. It can be shown that, at least at a certain level, temporal "relation words" can only be combined with expressions of states of affairs and that local "relation words" can only be combined with expressions of entities. Thus, we can say *while I was in Italy* or *behind the house*, but not **during the house* or **behind I was in Italy*.

As to the distinction between states of affairs and entities we may conclude from these observations that it will be wise not to consider the world as a whole, but to look at it as compartmentalized into time on the one hand and space on the other. For, as is shown by the above examples, states of affairs are not what constitutes the world as such, but only what constitutes its temporal dimension. The local dimension, accordingly, is constituted by entities.

As we have seen, the distribution of temporal and local "relation words" on the one hand and expressions of states of affairs and entities on the other supplies, in fact, no evidence as to the general primacy of states of affairs in relation to entities. However, it does supply evidence as to their primacy in relation to entities within the dimension of time, and it is only this dimension with which we are concerned in this article. But wherein lies the primacy of states of affairs in this dimension? Obviously, it lies in the fact that states of affairs, at least those that are in motion, i.e., by far the majority of all states of affairs, are the original components of this dimension, and the reason for this lies in the fact that they are states of affairs.

Whereas states of affairs are the original elements of the dimension of time, entities are the original elements of the dimension of space, and they are so by virtue of their being entities. But there are also entities which function as the elements of a certain kind of time. This kind of time is the kind which, in contradistinction to material time, I have called formal time, and the entities which function as the elements of this kind of time are days, months, or years. Obviously, these entities form a very small subclass among the totality of entities and their "entityhood" alone can scarcely be made responsible for their function as elements of formal time. For this function they must have a specific additional property, a property which, for instance, is not found in entities such as human beings, animals, plants, mountains, rivers, towns, houses, tables, or chairs. From this fact we can deduce that a kind of time based on entities rather than on states of affairs, i.e., the formal kind of time, must be a secondary form of time, secondary with regard to material time, the time based on states of affairs.

There is, however, a more direct kind of evidence as to the genetically secondary status of formal time. This may be discovered when we examine the differential semantic quality by which the few entities constituting formal time are distinguished from the many entities constituting space. We will then discover that this quality is nothing but a *reification* of certain *states of affairs* or, more precisely, of movements. We will see that this reification is the factor because of which the units of formal time are genetically secondary to the states of affairs underlying them and, indirectly, secondary to states of affairs, the elements of material time, in general.

The reifications leading to the creation of units of formal time are of three kinds, direct, semidirect, and indirect. As direct, I designate those reifications which go back to simple states of affairs. As semidirect, I designate those reifications which go back to natural pairs or natural parts, i.e., phases of states of affairs. As indirect, I designate those reifications which go back to artificial multiples or artificial parts of states of affairs.

Hence, direct, though partly only approximative, reifications of states of affairs include the formal units of the day, the lunar month, and the solar year. The day can be regarded as a reification of the rotation of the earth around its own axis. The lunar months can be regarded as (approximative) reifications of the motion of the moon around the earth. The solar years are (equally approximative) reifications of the motion of the earth around the sun.

These reifications, however, are reifications which can be formulated only on the basis of scientific findings of astronomy. Thus, they cannot be referred to in the attempt at describing the original, the primitive genesis of units of formal time. For the description of the primitive genesis of these units - and their primitive genesis is what we are concerned with - we can only take into account states of affairs which are apparent even to primitive human beings. Such states of affairs are those which we nowadays call the apparent motions of the celestial bodies. Henceforward, I refer to them as the visible motions of these bodies.

Apart from the visible motions of the celestial bodies themselves, there may, however, be still other visible "motions" which have possibly contributed to the creation of the units of formal time. These are visible changes of shape and visible changes in the duration and intensity of the light shed by the relevant bodies, i.e., the sun and the moon.

The visible motions of the celestial bodies themselves may be subdivided into primary and secondary ones. Secondary motions are motions which result from the combination of primary ones. It is on the basis of this distinction that one can say that the day is the reification of the visible primary motion of the sun, that the lunar month is the reification of the visible secondary motion of the moon, and that the solar year is the reification of the visible secondary motion of the sun. However, the primary motion of the sun and the secondary motion of the moon are, in our latitudes between the polar circles, only partially visible. The motion of the sun between sunset and sunrise and the motion of the moon during the phase of the new moon must be inferred. The only motion that is completely visible in our latitudes is the secondary motion of the sun, i.e., the sun's pendulous motion between the solstitial points.

The disadvantage of the necessity of inferring the invisible phases of the primary motion of the sun and of the secondary motion of the moon is compensated by the advantage that these two motions can be interpreted as simple states of affairs, namely as a kind of circular motion. The advantage of the complete visibility of the secondary motion of the sun is reduced by the disadvantage that this motion has to be interpreted as a pair of states of affairs consisting of the motions of ascension and descension. The day and the lunar month, interpreted as reifications of the primary motion of the sun and the secondary motion of the moon, respectively, may therefore be regarded as direct reifications, whereas the solar year, interpreted as a reification of the secondary motion of the sun, may be regarded as a semidirect reification.

Instead of, or besides, the visible motions of the sun and the moon themselves, there are still other "motions" which may have played a role in constituting the units of formal time, namely "motions" which have to do with the light shed by these two celestial bodies. These motions can be parametrized by means of two oppositions, the opposition between the "motions" of the intensity and the duration of the light and the opposition between a large and a small motion. The combination of these two oppositions results in four types of "light motions", namely

- the large light intensity motion
- the small light intensity motion
- the large light duration motion
- and the small light duration motion.

The reification of these four types of light motion, however, only leads to three units of formal time, since the reification of the two types of small light motion together only leads to a single unit of formal time, namely that of the lunar month. The reification of the large light intensity motion, the motion between intense brightness and deep darkness, may have led to the creation of the formal time unit of the day. The reification of the small light intensity motion, i.e., the motion between low brightness and deep darkness, as well as the reification of the small light duration motion accompanying it, i.e., of the motion between variably long moonlit nights (including the nights of the new moon), may have led to the creation of the formal time unit of the lunar month. The reification of the large light duration motion, the motion between variably long days and nights, may have led to the creation of the solar year. More precisely, these reifications may have *contributed* to the creation of these units of formal time.

All of these reifications of light motions are reifications of pairs of motions. Thus, the reification leading to the unit of the day is a reification of the combination of increasing and decreasing intensity of brightness. The reification leading to the units of the lunar months is a reification of the combination of increasing and decreasing intensity of brightness and, at the same time, of the simultaneous combination of increasing and decreasing duration of brightness. The reification leading to the unit of the solar year is a reification of the combination of increasing and decreasing duration of brightness. Therefore, all of these reifications are semidirect reifications.

Finally, the formal unit of the lunar month can also be interpreted as a primitive reification of the "motion" of moon shapes, the so-called moon phases. This reification is a semidirect reification as well, since it is also a reification of a combination of two motions, the motion of the waxing moon and the motion of the waning moon. It is this reification which may have played the main role in creating the formal time unit of the lunar month. It is obviously far more spectacular than the other three reifications which may also have contributed to the creation of this unit, namely the one which is based on the secondary motion of the moon itself, the one based on the motion of the intensity and the one based on the motion of the duration of its brightness.

In addition to the formal time units which are reifications of pairs of certain motional states of affairs, there are also formal time units which are reifications of certain parts of such motional states of affairs. As mentioned above, these also belong to the category of semidirect reifications. Such units of formal time are the seasons of the solar year and possibly the weeks. Thus, the seasons of the solar year are reifications of either of the two halves of the secondary upward motion and the secondary downward motion of the sun between the solstitial points, reifications of those parts of these two motions which are created by the equinoxes. The weeks, according to one of several hypotheses, go back to the four moon phases and thus to the dichotomies of the motions of the waxing and the waning of the moon which are marked by the half moons.

The times of the day are also semidirect reifications based on dichotomies of certain states of affairs, i.e., motions of a celestial body. But unlike the seasons and the weeks, the times of the day are not reifications of dichotomies of simple motions, but reifications of dichotomies of a pair of motions. This applies, in the first place, to daytimes in the wider sense, i.e., to the day in the narrower sense of the word and to the night. Day and night are reifications of a dichotomy of the pair which consists of the upward motion of the sun from the midnight point to the midday point and of the downward motion of the sun from the midday point to the midnight point. However, it also applies to certain times of the day in the narrower sense, i.e., to the forenoon and to the afternoon, since these are reifications of the dichotomy of the bright phase of the first dichotomy.

Whereas the seasons and the weeks are first degree semidirect reifications, the times of the day are semidirect reifications of a higher degree. The daytimes in the wider sense are semidirect reifications of the second degree, and the daytimes in the narrower sense are semidirect reifications of the third degree. The two degrees of reification which underlie the daytimes in the wider sense, the forming of a pair of motions and its first dichotomy, as well as the three degrees of reification which underlie the two daytimes in the narrower sense, the forming of the pair and its twofold dichotomy, are based on natural phenomena. The formation of the pair of motions is based on the complementarity of the upward and the downward motions of the sun, the first dichotomy is based on the rising and the setting of the sun, and the second dichotomy is based on the midday point of the sun.

The formal time units which are semidirect reifications of a higher degree are more distant from their genetic basis than those formal time units which are semidirect reifications of the first degree. The latter, in turn, are more distant from their genetic basis than those formal time units which can be regarded as direct reifications of their genetic basis, i.e., than the day and the lunar month if interpreted as reifications of the corresponding visible motions of the sun and the moon, respectively. Most distant from their genetic base, however, are those formal time units which I have designated as indirect reifications of their genetic basis. Such reifications are those which are based neither on simple motional states of affairs nor on natural pairs or natural divisions of them, but on artificial multiplications or artificial divisions of natural reifications of motional states of affairs. Indirect reifications based on an artificial multiplication of a natural reification are, e.g., the formal time units of the decennium, the century, or the millennium. Indirect reifications based on an artificial division of a natural reification are formal time units such as hours, minutes, or seconds. The former are artificial derivations from the natural basis of the year, and the latter are artificial derivations from the natural basis of the day.

To be precise, however, there is only one natural reification of a natural motional state of affairs which is precise, and that is the day. All others are only approximative. Thus, with regard to their length, weeks, lunar months, and solar years deviate from their natural basis by fractions of a day. The months of our solar year deviate by more than one day. Lunar years deviate by an even greater number of days. Moreover, two of the units of formal time, the weeks and the months of our solar year, perpetually shift with regard to their natural basis.

As to the motives which may have led the primitive human beings to a reification of certain motional states of affairs and thus to the creation of units of formal time, one may think of the possibility of the arithmetization of time, which is provided by the reifications and which Aristotle (*Physica* 219b) has made a definiens of time. Indeed, in connection with the units of formal time, arithmetization plays the dominant role. True, even material time is not wholly impervious to arithmetizability, but a certain form of reification is a necessary precondition even in this case. The products of such reification are units of time designated by such expressions as *three times* or *for the first time*. These units differ from formal time units like the day, the month, or the year by their indeterminacy and irregularity. These characteristics are based on their immediate connection with indeterminate and irregular states of affairs. They are, as it were, empty and variable forms of reification, forms which have to be filled out by concrete states of affairs. They are a reification of material time not so remote from the states of affairs as units of the type of day, month, and year. They may therefore be classified as semiformal. Presumably, they also precede these units genetically.

Arithmetization, however, was surely not the only motive for the creation of units of formal time. Another motive must have been the endeavor to render time an intersubjective phenomenon. With regard to this aim, the true gap lies between the states of affairs of material time and the semiformal units of time in the sense of the German notion of *Mal* on the one hand and formal time units like the day, the month, or the year on the other. Nevertheless, formal time cannot completely emancipate itself from its natural base. Thus, both the proper-name-like use of expressions of formal time, as in datings, and the deictic use of such expressions must ultimately refer to states of affairs, the former in determining the beginning of the respective era, the latter, though only implicitly, in the determination of the respective deictic origo. A state of affairs determining the beginning of an era is, for example, the birth of Christ, and the state of affairs which determines the deictic origo is the utterance of the speaker.

3. The metagenetic relationship between material time and formal time

In various writings, Walter A. Koch (e.g., 1974a, b) has contrasted the phenomenon of genesis with a phenomenon which he has called metagenesis. Probably in analogy to the notion of metalanguage, Koch uses the term metagenesis to refer to the process of speaking about genesis or, more generally, the process of analyzing genesis, or even of merely becoming conscious of this process. Metagenesis is a process which, as its designation suggests, takes place on another plane, a plane above that of genesis. But metagenesis does not only take place, as is obvious, at a different plane. It also takes place, and this is by no means obvious, in a different direction. Metagenesis, this is Koch's thesis, is the reversal of genesis.

Koch describes this phenomenon in most detail with respect to modern linguistics on the one hand and the phylogenesis of language on the other. The course of modern linguistics and hence the course of linguistic metagenesis, he says, has led from the phoneme, via the morpheme, the word, the sentence and the text, to more complex textual phenomena, such as the dialogue (cf. Koch 1974a: 276ff.). The path of the phylogenesis of language, by contrast, has led from the dialogue via the monologue down to the word and finally to the phoneme. Word and phoneme, according to Koch's account of language phylogeny, seem to have become isolated through writing, i.e., through ideograms on the one hand and through letters on the other (cf. Koch 1974a: 284f.). In the course of genesis, Koch sees a tendency from structural openness to structural closure, whereas in the course taken by metagenesis he detects a tendency leading from closure to openness of structures. Koch admits that his exposition of these theses is extremely simplified, but he rightly stresses their heuristic value (Koch 1974a: 285). Appealing to this value, I shall now try to apply Koch's thesis that metagenesis is the reversal of genesis to the relationship between material and formal time.

From the genetical point of view, material time is, as we argued in detail, primary, whereas formal time is secondary. Formal time, the time of entities, has developed from material time, the time of states of affairs. From the metagenetical point of view, however, it is, in accordance with Koch's thesis, the other way round. In Koch's account, metagenetical processes characterize scholarly activities, especially the development of modern linguistics. However, such an interpretation of the notion of metagenesis is too narrow. Undoubtedly, the development of parts of the consciousness of the man in the street is also characterized by the phenomenon of metagenesis. In the following, I will first consider such processes. However, I will not consider the ontogenetic acquisition of time concepts by the child, which may possibly recapitulate the phylogenetic development from material to formal time. Instead, I will only investigate the conception of time which the man or woman in the street has as an adult.

This conception of time seems to be guided, especially in civilized countries and nowadays more than ever before, by the units of formal time. As a rule, whoever thinks of time, thinks of dates, and dates are fixed in terms of units of clock and calendar time. Units of clock and calendar time are, however, units of the time of entities and, for that matter, of formal time. In this respect, it does not matter whether the units of clock and calendar time are determined through the use of proper-name-like expressions or the use of deictics. Deictic expressions, such as *in two weeks* or *tomorrow*, although they do not function as dating expressions, are units of formal time as well. Even such deictic expressions as *in the past* or *in the future*, being expressions of the time of entities, are expressions of formal time.

By contrast, units of material time play a subordinate and secondary role in time awareness and time conceptions of modern man. In this respect, however, we have to distinguish between two kinds of material time units (states of affairs), those which underlie the formal time units of the day, the week, the month, and the year, and those which have not been used to create such formal time units. From a conceptual point of view, material time units underlying formal time units are relatively prominent, but they are only rarely expressed. Nobody, except in situations of scientific discourse, will say that the earth rotates around its own axis or that it travels around the sun. Material time units which have not been used to create formal time units, however, are frequently expressed, but they are time units without conceptual prominence. Thus, we often utter sentences such as *I have been downtown* or *Then it began to rain*, but we are not aware of representing time units in uttering such sentences.

With regard to the conceptual representation of states of affairs as time units in the consciousness of normal adults in modern societies, one could distinguish three degrees of remoteness from the level of awareness which characterizes modern man with regard to formal time units.

The states of affairs whose conceptual representation as time units is highest and thus nearest to that of the formal time units are the states of affairs, namely motions, underlying the formal time units. In the case of day and year, these are nowadays probably true motions, the rotation of the earth around its own axis and the motion of the earth around the sun. However, in the case of the weeks and the months, the states of affairs are still apparent ones, namely those which have to do with the changes in the moon's shape.

The states of affairs whose conceptual representation as time units is second highest and thus less near to that of the formal time units are certain occurrences of states of affairs not used for creating formal time units, namely those which are used for determining the time of other states of affairs. These are those occurrences of such states of affairs which are expressed by temporal clauses, such as *when I was in France* or *after I had come home*. Such temporal expressions are obviously less determinate than the usual expressions of formal time. Perhaps they are also less frequent, but this is not the main reason why they play a subordinate role in the awareness of time in modern man. The main reason seems to be that their denotata fail to be integrated in a time system.

The states of affairs whose conceptual representation as time units is lowest and thus most remote from that of formal time units are probably those which are used neither for creating formal time units nor in determining the time of other states of affairs. These are states of affairs expressed by normal sentences rather than by temporal clauses. They not only fail to be integrated in a time system, but also draw our attention to themselves rather than to the constitution of (material) time, which they also provide.

Formal time turns out to be the primary form of time not only in the time awareness and conceptions of the man in the street, but also, as far as I can see, in the time awareness and conceptions of the scholars concerned with temporal phenomena. The awareness of material time plays a clearly subordinate role even in the case of the investigators of time. It is presumably highest among linguists and grammarians dealing with phenomena of time. These scholars not only deal with the units of formal time expressed by adverbial expressions. They also engage in the sphere of material time. They deal not only with temporal adverbials, but also with temporal conjunctions and tenses. But despite this, they do not seem to have developed a real awareness of this sphere. This may be due to the fact that the denotata of temporal conjunctions and tenses are only certain parts of the units of material time, not these units themselves. The units themselves are whole states of affairs, and for these, even linguists and grammarians dealing with temporal conjunctions and tenses seem to have shown little interest. And when they have shown interest in such states of affairs, then, as a rule, only within single sentences, not within the context of longer sequences of sentences. But the phenomenon of material time is seen most clearly in sequences of sentences within texts, especially narrative texts. Whereas formal time is the realm of systematic, or more precisely, systemic time, material time is the realm of textemic time, both of potential and of actual textemic time. As potential textemic time, material time is the time of the sequences of states of affairs underlying actual and potential texts. It is the time of those sequences of states of affairs of which only very few, often discontinuously and sometimes in inverse direction, are narrated at all. As actual textemic time, it is the time of those sequences of states of affairs which are actually narrated. It is true that the narration of sequences of states of affairs also refers to units of formal time, since it also uses expressions of calendar time. However, in most texts, especially in narrative ones, these only play a relatively unimportant role.

The dominant role of formal time even in the consciousness of the linguist and grammarian dealing with phenomena of time manifests itself perhaps most clearly in the fact that the conception of time which they have is a unilinear one, for the unilinear conception of time is the conception of formal time. It is an outcome of the fact that the different units of formal time can all be metrically compared with, and integrated into, each other, so that they can be, and are, projected onto one and the same line. This is not the case with normal states of affairs, the units of material time. Normal states of affairs have the most diverse lengths, and what is still more important is the fact that their lengths, as long as they are not subjected to formal measurement, are temporally wholly indeterminate and not integratable into each other. What can be determined is merely their simultaneity, their anteriority or their posteriority with regard to other states of affairs, and these relations are not purely temporal. They are interwoven and impregnated with a number of other kinds of relation, including local, personal, causal and others. The effect of all this is that the phenomenon of material time is not unilinear, but multilinear (cf. Harweg 1991: 47ff.). It is, as it were, a multilinear phenomenon in which, for example, every lifetime, i.e., the lifetime of every individual, forms a special line of material time. The awareness of this multilinearity, however, is all but common. Normally, the life-spans of all individuals are projected onto one and the same line, the line of formal time. The metagenetic conception of time has not yet detected the multilinearity of material time, and the reason is obvious: material time is too diffuse a phenomenon. The diffuse, Koch would say, the open, is genetically primary, but metagenetically it is secondary.

References

Aristotle

[1950] *Physica.* (Ed. by W.D. Ross.) Oxford: Clarendon.

Hartmann, Peter

1971 "Texte als linguistisches Objekt", in: W.-D. Stempel (ed.), 9-29.

Harweg, Roland

1985 "Semi-Deiktika", *Zeitschrift für Phonetik, Sprachwissenschaft und Kommunikationsforschung* 38: 376-386.

1990 *Studien zur Deixis.* Bochum: Brockmeyer.

1991 "Erzählte Zeit und Sachverhaltsfolgenzeit", *Folia linguistica* 25: 41-73.

Koch, Walter A.

1974a "Tendenzen der Linguistik", in: W.A. Koch (ed.), 190-311.

1974b "Semiotik und Sprachgenese", in: W.A. Koch (ed.), 312-346.

Koch, Walter A. (ed.)
1974 *Perspektiven der Linguistik.* Vol. 2. Stuttgart: Kröner.

Plato
1902 "Timaeus", in: *Platonis opera.* (Ed. by J. Burnet.) Oxford: Clarendon.

Stempel, Wolf-Dieter (ed.)
1971 *Beiträge zur Textlinguistik.* München: Fink.

Wittgenstein, Ludwig
1921 *Tractatus logico-philosophicus.*
[1955] [Reprinted London: Kegan Paul.]

The agent and the sentient: A dissymmetry in linguistic and cultural encoding

Yoshihiko Ikegami

1. "Agent" vs. "sentient" and "source" vs. "goal"

"Agent" and "sentient", the pair of notions I am going to discuss in the present paper, are not evenly treated in linguistic literature. There is a substantial amount of discussion centering around the notion of "agent" or "agentivity", while the term "sentient" is hardly ever heard of, and the term "sentiency" (corresponding to "agentivity") is virtually not used at all. The reasons for this unequal treatment are not very far to seek. The agent is typically "human" and is "dynamically" characterized, i.e., characterized as being engaged in causing a change in the situation (rather than remaining inactive simply as part of the situation, thus being "statically" characterized). Thus, combining two of the most important attributes of a cognitively salient entity, namely "being human" and "being dynamic", the agent is deservedly one of the most prominently elaborated notions in language and consequently attracts the linguists' full attention. Another reason for the popularity of the agent in academic discourse is that the agent occupies a privileged position in the syntax of Indo-European languages. Here are two quotations, one from a linguist, the other from psychologists, which testify to the point:

> Quite a few of the present-day Indo-European languages agree with English in using an actor-action form as a favorite sentence-type. (Bloomfield 1933: 172)

> In Indo-European languages, the model used for connoting states of affairs and articulating them linguistically is the *human action model*. A total event is basically articulated into agent, action, and object; the relationships between these are portrayed in sentences in which the vehicles for the referents are related to each other through a "syntax of action". (Werner and Kaplan 1963: 57)

What I propose to call "sentient" here, on the other hand, principally concerns those uses which are generally discussed in such diverse terms as "experiencer" (as in *John's speech pleased MARY*), "beneficiary" (as in *John bought MARY a new hat*), and "recipient" (as in *John sent MARY a letter*), but covers more. The role of "sentient" is also relevant in those cases which are sometimes discussed under the heading of "affected agent" (Saksena 1980) (as in *John got MARY to type the letter*).[1] I will assume that there is a conceptual unity behind all these apparently diverse uses. The sentient, thus defined as an integrated notion, contrasts with the notion of the agent. It will be shown, however, that the ways the agent and the sentient are linguistically encoded are highly unequal, suggesting that there is an underlying cognitive dissymmetry between the person as agent and the person as sentient.

Underlying the contrast between the agent and the sentient are both their conceptual similarity and contiguity. They are similar in that they both primarily refer to a person. They are conceptually contiguous in the same way that the source and the goal are related to each other.

The idea that the agent and the sentient are related to each other as the source and the goal are needs no detailed discussion here (cf. Ikegami 1987). The close conceptual relationship between the two pairs of notions is amply supported by linguistic evidence: the agent is known to be often accompanied by a source marker when it is demoted (as in *Hans wurde VON Marie geschlagen*, as contrasted with the unmarked version, *Marie schlug Hans*). Similarly, the sentient is known to be generally characterized as the goal (as in *Mir gefällt das Buch*). Thus, underlying the notions of the agent and the sentient is the source-goal schema: the agent is conceived of as the source *from* which an action starts and the sentient is the goal *to* which a stimulus comes. Thus, the two notions contrast with each other in terms of the person as source and the person as goal.

From a purely logical point of view, the source and the goal constitute a pair of notions standing on equal footing. As constituents of a locomotion of limited duration, the source and the goal represent the two extreme points marking the beginning and the end of the motion, with the notion of the path mediating the two points.

The agent and the sentient, conceived of as human source and human goal respectively, do not constitute a pair of equally valued terms. Linguistic evidence indicates (as will be discussed in the ensuing sections) that there is a dissymmetry between the notion of the human source and that of the human goal, with the former predominating over the other.

There is something remarkable about this dissymmetry. At the level at which the notions of the source and the goal are not limited to persons and applied more generally in locational senses, the dissymmetry is manifested in exactly the opposite way: the goal predominating over the source. Again, there is a sizable amount of linguistic evidence testifying to the point (Ikegami 1987). Thus, the goal is often found unmarked, while the source generally remains marked (as in the contrast between *go FROM THERE* and *go THERE*). The goal expression may replace the source expression which would logically be expected (as in *DEM KIND das Spielzeug wegnehmen* as contrasted with *take the toy FROM THE CHILD* or *be different TO IT* said side by side with *be different FROM IT*). With a number of verbs of motion which are semantically neutral between the source and the goal, the goal expression can complete a sentence semantically but not the source expression (as in *walk TO THE STATION* as contrasted with the semantically incomplete collocation *walk FROM THE STATION*). Thus, linguistic evidence seems to suggest that the goal is a cognitively more available notion than the source.

2. The bodily basis for semiosis in action and in sensation

In view of the general predominance of the notion of the goal over that of the source, it seems to be highly curious that the notion of the human source (agent) manifests itself in a more dominant way (suggesting its higher cognitive salience) than the human goal (sentient). Whence is this reversal of predominance?

The reason for this different pattern of dissymmetry must be sought in the different semantic domains in relation to which the notions of the human source and the human goal are prototypically conceived. Thus, the human agent is prototypically associated with the semantic domain of action, while the human sentient is prototypically associated with that of sensation.

In the prototypical case of action, the person acts on something other than him- or herself, affects the latter physically, and may possibly cause a change in the latter. Here the person is conceived of as the locus *from* which the affecting force issues. In the prototypical case of sensation, on the other hand, the person receives some stimulus from something other than himself and is affected by it psychologically. The person is thus the locus *to* which the affecting force gets.

In terms of "force dynamics" (Talmy 1985), either case involves two entities and the force issuing from one entity reaches the other. Only the direction of the force is just the reverse: the agent is the starting point of the force, while the sentient marks the reaching point of the force. At this maximally abstract level of schematic representation, the agent involved in action and the sentient involved in sensation can still be conceived of as symmetrically related to each other:

SOURCE --------→ GOAL	GOAL ←-------- SOURCE
(human)	(human)
‖	‖
Agent	Sentient

As soon as we start off from this point, however, the different ways in which the human source and the human goal behave become apparent.

The first point to be discussed concerns the way in which the entities involved behave differently in action and sensation. In either domain, two entities are involved, at least one of which is a person, and the force issuing from one of the entities contacts and affects the other. But the nature of the force involved is not the same: it is physical in action and psychological in sensation. This contrast in the nature of the force involved has a number of interesting consequences. Thus, in action, the force issuing from the agent physically contacts and affects the patient, and this is something that can be observed and confirmed by a third party as well as by the agent himself. Action typically is an objective event involving two optimally discrete entities. The same account, however, does not seem to apply to sensation. In sensation, the whole event is abstract in the sense that the force and its contact with the sentient can neither be observed nor confirmed by a third party. It is a private event internal to, and therefore, available only to, the sentient himself. Hence, it is subjective. Moreover, although the stimulus may very well be located outside the sentient, there is a good sense in which one can say that its locus is in the sentient, since it is only in the sentient (and by the sentient, for that matter) that the stimulus is confirmed to exist. Consequently, the two entities involved in sensation, the sentient and the stimulus, are not as discrete as the two entities involved in action; one tends to merge with the other.

Considered in this way, action and sensation represent the bodily bases for human semiosis. The former serves as the basis for the type of semiosis in which the subject is clearly separated from the object - which again will serve as the basis for the type of semiosis with a focus on clearer articulation. The latter underlies the type of semiosis in which the subject is not clearly separated from, but is merged with, the object - which in turn underlies the type of semiosis in which the orientation is toward blurred articulation (Ikegami 1989a, b, 1991).

Another pair of contrast associated with action and sensation is "actual" and "potential". Consider the semantic contrast between *I strike him* and *I can strike him*, on the one hand, and *I see him* and *I can see him*, on the other. The contrast in meaning between the two sentences in each pair will seem to be much less marked in the latter pair involving a verb of sensation than in the former involving a verb of action. In action, the human agent is typically presumed to be in control of himself in relation to what he does; hence, there is a difference between what the agent actually does and what he or she potentially can do (but actually does not do). In sensation, on the other hand, the human sentient is not in control of him- or herself in relation to what happens to the sentient; the distinction between "can" and "cannot" here thus simply reduces itself to whether or not the given stimulus actually reaches the sentient, hence the virtual equivalence of *I see* and *I can see*.

The point discussed above accounts for the fact that sensory verbs are essentially stative, while actional verbs are dynamic. An unmistakable mark of actuality is a change taking place in the situation, hence the encoding verb can most appropriately be dynamic. Potentiality can best be characterized by no change taking place in the situation, hence the encoding verb is appropriately stative.

The above discussion was focused on the contrast between action and sensation. According to this analysis, action and sensation may appear to be two conceptual domains with no intersection between them. Actually, however, there is one section of either domain where the two domains meet and intersect.

Thus, consider the sensation of touch, which is the most primitive type of sensation in the sense that living organisms not provided with other sensations such as sight and hearing are nevertheless at least sensitive as to touch. Typically, sensation of touch arises when some physical entity comes into contact with a sentient organism. This situation is superficially similar to the one in which an organism in action comes in contact with some physical entity.

A situation involving an organism and a physical entity coming into contact is thus ambiguous in exactly the way discussed above: it can either be a situation involving action or one involving sensation. Which way the ambiguity is to be resolved is dependent on two factors: (1) whether the organism or the physical entity has the initiative in making contact and (2) whether the contact is physical or psychological. Judging from the way such an ambiguous situation is linguistically encoded, however, the resolution seems to be in favor of action rather than sensation. To illustrate from English, the two sentences, *I touched Mary's hand* and *Mary's hand touched me*, will primarily be associated with an actional and a sensory process respectively. The former sentence, however, can also refer to a situation in which the touching was accidental and in which, therefore, the subject did not qualify as agent. What happens here is thus the use of the agent-involving schema for encoding a sensory process. Even in the subdomain of touch, where action and sensation intersect superficially, the linguistic encoding is in favor of the agent rather than the sentient.

The reason for this dissymmetry is again quite clear. For one thing, the speaker, being a person, tends to encode a situation in terms of the one having the initiative - a perfectly understandable anthropocentric orientation characterizing human beings. For another thing, the physical process is concrete and externally manifested and as such much more readily available to linguistic encoding than the sensory process, which is internal and therefore abstract.

3. Extensions of the agent and the sentient

A further dissymmetry will be noticed when we check the ways in which the notions of the agent and the sentient are extended. As an initial stage in category extension, we consider those cases in which the contact is made not with a first-order entity (i.e., a concrete physical entity), but with a second-order entity, namely an event.[2]

The counterpart of action at this stage will be causation. In action, we have the agent as source and a first-order entity as goal and the agent affects the physical entity. If we substitute a second-order entity as goal, the agent's role in relation to it will have to be rephrased as causing some sort of event - an event involving the first-order entity affected by the action. A typical situation will be one in which the affected first-order entity undergoes a change. This is what is conceptualized as causation. The agent is now reinterpreted as causer.

The close relationship between action and causation is amply manifested in the way these notions are linguistically encoded. On the one hand, there are a number of cases of polysemy involving these two notions (cf. French *faire*, Japanese *suru*) as well as semantic shift from one to the other (cf. English *do*, German *machen*). There is, on the other hand, a continuity involving a series of syntactic constructions encoding action and causation (cf. *strike him - strike him down - strike him dead*).

The counterpart of sensation at the stage we are now considering will be best illustrated by what is sometimes characterized as "adversative passive" in certain languages. Here the sentient is affected by some event, and the important point here is that the event in question can just be of any kind (no matter whether it is a transitive or an intransitive type, for example). Thus this passive is different, for example, from the English passive, which is normally limited to the transitive type of event and is thus to a large extent syntactically rather than semantically conditioned, or from the intransitive passive in German, which is not associated with an implication of being adversely affected.

It is interesting to note that the sentient-involving passive described above is apparently limited in distribution among the world's languages. It is known to be characteristically attested in languages of Southeast Asia (including Japanese, where it has traditionally been termed *ukemi*). What we find in European languages instead is the passive focusing on the affected first-order entity and thus being syntactically conditioned (at least on the face of it) as converse of the active construction encoding action. Thus, in these languages, we have the agent-involving schema for action underlying the passive construction - only the point of the passive construction is to focus on the affected entity and defocus the agent.

The situation of the passive in European languages can be seen as one of the notion of action encroaching on the domain of sensation. In fact, the same encroachment is found in other respects of encoding as well. Take English, where, as mentioned above, the sentient-involving passive is not readily encoded. But consider sentences like *John bought a book FOR ME* and *It rained ON ME*. The prepositional phrases with *for* and *on* here encode a sentient in terms of how (i.e., advantageously or adversely) the speaker was affected by the events described (i.e., *John's buying a book* and *raining*). Thus, the sentences we have here certainly represent the sentient-involving construction analogous to the adversative passive discussed above. But note that the construction is in the active, rather than passive, voice and the sentient involved is encoded syntactically only as an optional element of the construction. A situation of a sentient affected by an event is processed here in terms of the prototypical schema for action.

For another manifestation of the predominance of the agent-involving schema over the sentient-involving one, consider the use of the English causative verb *have*. Considering the original meaning of the verb, the construction involving the verb *have* now in question may very well have referred to a situation in which the subject of the verb is affected by an event and thus functions as sentient (as is still noticed in a sentence like *I had so many people coming to me for advice*). But the meaning of *have* has clearly shifted to "causative" (especially in the construction in which it is followed by the object and an infinitive, as in *I had him come*). Here again, we have the agent-involving schema replacing the sentient-involving one.

Category extension of the prototypical schema for actional and sensory process is also brought about by shifting away from the notions of the agent and the sentient properly defined. Cases for such extension for action are found when the encoding schema is applied to construe an event in which no agent in the proper sense of the word is involved. Thus, consider a potentially ambiguous sentence like *John broke his leg*. A first interpretation associated with the sentence will be that John broke his leg unintentionally, perhaps by mere carelessness. Another possible interpretation will be that John did so intentionally for certain purposes. In the first interpretation, John was not an agent in the proper sense, since he was in control of himself. But the way the situation is linguistically encoded is not different from the way the situation is encoded in which John was fully in control of himself. Thus, an actional process not controlled by an agent is still encoded in terms of the schema for action.

Is an analogous extension found for the schema encoding the sentient? Prototypically, the sentient as such is not in control of him- or herself. What happens if a sensory process is characterized by self-control on the part of the sentient? Does the prototypical schema for sensory processes still apply? The answer seems to be "no". A sensory process controlled by the sentient is encoded in terms of a schema for action. English is typical in this respect in providing separate verbs for some controlled sensory processes: *look at* (cf. *cast a glance at*) a person, *listen to* (cf. *give ear to*) a person. Sensory processes are here construed as instances of action or activity.[3] As we have already seen, sensory processes can be construed in terms of an actional schema even when they are uncontrolled, as in English *I see a dog* and *I heard a dog* (schematically construed in the same way as *I strike a dog*). Languages provided with this mode of encoding can readily extend the same schema for controlled sensory processes, as in the case of Japanese and in English sentences like *I smell the meat* and *I taste the meat*. The result is one and the same schema serving both for uncontrolled and controlled sensory processes. In either case, the picture we find is a dissymmetry strongly in favor of the agent rather than the sentient.

The agent and the sentient behave differently vis-à-vis metonymic extension. Consider the following alternation: *He hit my nose* and *He hit me on the nose*. The two sentences may apply to one and the same situation, but they encode the same situation in different ways: while the first sentence focuses on the particular locus of my body which was physically affected by being hit, the second implies that my whole person was affected, psychologically as well as physically, by being hit on the nose. The extension from a particular locus on the body to the whole person here is of course mediated by a sensory process, a characteristic attribute of an organism (prototypically, a person). When a non-organism is in question, the same alternation is normally not possible: *He hit the edge of the table* and **He hit the table on the edge*. Barring sensory transmission, the metonymic extension from a part to the whole is not envisaged in the case of a table. Notice, on the other hand, how natural this extension is for a person. Consider the passive versions of the pair of sentences above: *My nose was hit* and *I was hit on the nose*. The second sentence will seem to qualify better as the unmarked choice than the former. This is so because a person normally has a higher topical value than a body part and hence more readily and naturally qualifies as topic of the sentence. Thus given a person as sentient, metonymic extension from body part to the whole person is quite natural.

What is the counterpart of this extension for the agent? Consider the following pair of sentences: *I hit my stick on the door* and *I hit my hand on the door*. The parallelism in the encoding schema here suggests that the way in which the relationship between the agent and the instrument (*my stick*) is conceived is analogous to the way in which the relationship between the sentient and the body part (*my hand*) is conceived. A person's instrument, in other words, is conceived of as analogous to a person's body part. Notice that only the direction of extension here is different as seen from the viewpoint of the person involved. We have noted an extension from the body part to the whole person when the person is a sentient. When the person is an agent, however, the extension is from the whole person to something contiguous to the person. This difference in directionality is fully motivated. Sensation is a process internal to the human body, while action is a process in which the human body interacts with some entity external to it. Notice also in this connection that in the sentence involving an instrument (*my stick*) the unmarked interpretation will be that the subject was fully in control of him- or herself, while the one involving a body part (*my hand*) more readily lends itself to an interpretation that the event happened in a way not fully controlled by the subject - again reflecting the characteristic traits of action and sensation, respectively. But the tilt in favor of action can be noticed here, too. As soon as the two sentences are rephrased as *I hit the door with my stick* and *I hit the door with my hand*, it is the actional interpretation which is likely to be the unmarked choice.

4. Cultural coding of the agent and the sentient

A maximal case of metaphorical extension is the rhetorical figure called "personification". Personification arises when a nonhuman noun is collocated with a phrase or a clause which is normally associated with a human noun. Thus, depending on whether the associated human noun is one functioning as agent or sentient, two types of personification are envisaged. The two types, however, do not seem to function rhetorically in exactly the same way.

Compare *The daffodil danced* and *The daffodil is happy*. (To foreground the difference, the tenses in the two sentences are intentionally made different.) Notice the different ways in which the effects of personification are achieved. In the former sentence involving an agent, some of the characteristic behaviors of a person in dancing are mapped onto the daffodil. In the second sentence involving a sentient, personification does not simply consist in analogous mapping. Because it is a sensory process of typically private nature that is in question here, the reader must empathize with the daffodil. If we talk about mapping at all, it is the mapping of the sensory process as experienced by the reader onto the daffodil to which the same sensory process does not apply. Here are again manifested the contrastive features characterizing action and sensation. The agent-based personification primarily concerns external features, while the sentient-based personification is primarily internal. The latter presupposes empathy on the part of the reader, and to the extent that the empathy is successfully applied, it will have certain emotive effects which may not be associated with the agent-based personification.

It is interesting to note that speakers of different languages may differ in preferring one type of personification over the other. B. Chamberlain, an early English scholar who taught linguistics in Japan in the late nineteenth century, wrote to the effects that the agent-involving personification was avoided in Japanese and that this was a negative trait of the language:

> Another negative quality [of the Japanese language] is the habitual avoidance of personification - a characteristic so deep-seated and all-pervading as to interfere even with the use of neuter nouns in combination with transitive verbs. Thus, this language rejects such expressions as "the *heat makes* me feel languid", "*despair drove* him to commit suicide", . . . etc. One must say, "being hot, I feel languid", "having lost hope, he killed himself", . . . and so on. (Chamberlain 1936: 275)

To be contrasted with this is a remark by Soseki Natsume, a celebrated native novelist, who said in effect that the agent-involving personification was so artificial that he could not simply bear with it:

> Whenever I come across an instance of personification (like "Pity cries"), I am nauseated by its artificiality and snobbishness. I have come to the conclusion that personification is something to be detested . . . It looks as if a monkey is trying in vain to look like a feudal lord by placing a crown on his head. (Natsume 1906: 261-262, my translation)

Finally, the contrast between agent-based and sentient-based schema can be manifested at a yet higher level of discourse organization, namely the folktale. In prototypical Western folktales (i.e., those classified under the traditional name of "eigentliche Märchen"), the main plot of the story evolves around "the hero" who demonstrates what he is worth by performing a series of "heroic" acts. The hero then is a clear exponent of the notion of the agent. It has, however, been noted that the hero as dramatis personae is curiously lacking (with some exceptions, of course) in traditional Japanese folktales. Hayao Kawai (1982), a Jungian psychologist, suggests in his book on the folktale that the Japanese counterpart of the Western hero is a woman who patiently endures. The heroine is beset with all kinds of miseries and persecutions, lives through them without losing her integrity, and finally emerges out of them as one blessed with a happy marriage. (The Western Cinderella might be quite close an analogue.) Common to the Western hero and the Japanese heroine are the facts that they are not happy at the beginning, go through a series of trials and tribulations, and end up with a happy marriage. The trials and tribulations they undergo are not of the same kind: the hero demonstrates his worth by successfully overcoming a series of difficulties, acting as "agent", while the heroine does the same by successfully enduring a series of hardships patiently as "sentient". The two types of coping with problems of life are of contrastive nature, but they are certainly both necessary conditions for success and happiness in life. It is interesting to see that cultures may have different preferences in focusing on one rather than the other of the two fundamental aspects of human nature.[4]

Notes

1. Only a person is typically qualified to function as affected agent. Unless the entity affected is capable of sensing the stimulus which instigates it, there is no possibility of it acting as agent in response to it.
2. For the terms "first-order entity" and "second-order entity", see Lyons (1977: 442ff.).
3. An analogous process is found in those languages which normally assign dative case to the sentient but switch to nominative (the case form which also marks the agent) when the notion of control comes in. Cf. Sridhar (1979).
4. Reference can also be made to Chun and Zubin (1990).

References

Bloomfield, Leonard
1933 *Language*. New York: Allen & Unwin.

Chamberlain, Basil H.
1890 *Things Japanese*.
[1939] (6th edition.) London: Paul, Trench & Trubner.

Chun, Soon A. – David A. Zubin
1990 "Experiential vs. agentive constructions in Korean narrative", *Proceedings of the Berkeley Linguistic Society Annual Meeting* 16: 81-93.

Dirven, René – Günter Radden (eds.)
1987 *Concepts of case*. Tübingen: Narr.

Ikegami, Yoshihiko
1987 "'Source' vs. 'goal': A case of linguistic dissymmetry", in: René Dirven – Günter Radden (eds.), 122-146.
1989a "Culture and semiotics", in: Walter A. Koch (ed.), 1989b, 13-26.
1989b "Homology of language and culture: A case study in Japanese semiotics", in: Walter A. Koch (ed.), 1989a, 388-403.
1991 "'DO-language' and 'BECOME-language': Two contrasting types of linguistic representation", in: Yoshihiko Ikegami (ed.), 285-326.

Ikegami, Yoshihiko (ed.)
1991 *The empire of signs: Semiotic essays on Japanese culture*. Amsterdam: Benjamins.

Kawai, Hayao
1982 *Mukashibanashi to Nihonjin no Kokoro* [Folktales and the Japanese mind]. Tokyo: Iwanami.
Koch, Walter A. (ed.)
1989a *Culture and semiotics*. Bochum: Brockmeyer.
1989b *The nature of culture*. Bochum: Brockmeyer.
Lyons, John
1977 *Semantics*. 2 vols. Cambridge: University Press.
Natsume, Soseki
1906 *Bungakuron* ['A Theory of Literature']. Tokyo: Okura Shoten.
[1975] [Tokyo: Iwanami]
Saksena, Anuradha
1980 "The affected agent", *Language* 56: 812-826.
Sridhar, S.N.
1979 "Dative subjects and the notion of subject", *Lingua* 49: 99-125.
Talmy, Leonard
1985 "Force dynamics in language and thought", *Papers from the Parasession on causatives and agentivity at the 21st regional meeting of the Chicago Linguistic Society. Part 2*: 293-337.
[1988] [*Cognitive Science* 12: 49-100]
Werner, Heinz – Bernard Kaplan
1963 *Symbol Formation*. New York: Wiley.
[1983] [Reprinted New York: Erlbaum]

Language and brain

Helmut Schnelle

1. Revisions in the foundation of linguistics

1.1. Traditionally and conventionally, languages are conceived of as external to human beings. They are sets of sound patterns related to meanings, and they are used as tools of communication. The fact that human beings are able to produce and to receive these sound patterns with meanings does not make them internal.

There have been several attempts to correct this commonsense notion, by Herder, Humboldt, the Neo-Humboldtians, the Neo-Grammarians, Wundt, Jespersen, Bühler, Jakobson, to mention just a few. For all of them, language was part of the internal constitution of a human individual or of a group of humans. However, with the exception of Jakobson, none of these linguists addressed the problem of the relation of language and brain. For Jakobson, it was a matter of special interest but not a fundamental question. It was only recently that the problem was emphasized by Chomsky as part of a larger program of the reevaluation of linguistics (cf. also Schnelle 1991).

For Chomsky (1986a: 10), linguistics becomes "part of psychology, ultimately biology. It will be incorporated within the natural sciences insofar as mechanisms are discovered which are revealed in these more abstract studies; indeed, one would expect that these studies will be a necessary step toward serious investigation of mechanisms" (cf. also various passages in Chomsky 1986b and 1988).

This integration of linguistics and psychology into the framework of the natural sciences is mediated by referring to the brain: "Regard talk of mind as talk of the brain undertaken at a certain level of abstraction at which we believe, rightly or wrongly, that significant properties and explanatory principles can be discovered" (Chomsky 1986a: 7). Thus, linguistics is the psychology of certain human abilities realized in the mind/brain, namely the ability to use language in a regular way and the ability to acquire the ability of its use. Language may be analyzed abstractly in terms of structure representations (Chomsky's view) or it may be studied functionally by describing speakers' and hearers' acts of referring and predicating, speaking and understanding organized by underlying linguistic structure (Bühler, Jakobson, Halliday, Searle, etc.). In both cases, language is studied more abstractly and more globally than in the more detailed descriptions of brain structures which enable a human to speak, understand, and acquire a language, just as the biological structures are more abstract and lie on a more global, viz. higher, level than those of the chemistry or of the quantum theory of cells, proteins, and the like.

1.2. Integrating all of this into a common framework means that our research must contribute to the characterization of the relations between the levels of abstract structure, functional description of behavior, and causal descriptions of physical processes on various levels of detail, whether these relations are called reductions or strict correlations. We should insist on reduction or strict correlation. No structural or functional analysis should be considered satisfactory without an (idealized) analysis of the underlying physical causalities and vice versa (cf. the levels of structure, locutionalization and localization in Schnelle 1991: § 5.1).

However, the citations from Chomsky reveal an underlying attitude which contrasts with our requirements. He gives priority and even independence to abstract studies over more concrete analyses. In fact, his linguistic descriptions are made, and intended to be made, without any consideration of biological facts that might be relevant. He gives only a superficial consideration to the status and to the possibilities of the relation of abstraction.

As to the latter, Chomsky (1986a: 8) writes: "The statements of grammar and the statement that R(P,L) [person P knows language L] are similar to statements of a physical theory that characterizes certain entities and their properties, in abstraction from whatever may turn out to be the mechanisms that account for these properties: say a nineteenth century theory about valence or properties expressed in the periodic table." In other words, linguistics is similar to nineteenth century chemistry, i.e., it is in a stage in which an explanation of its relation to physical mechanisms is still too early. It was only twentieth century quantum physics which provided the reduction, thus justifying the abstract representations of nineteenth century chemistry. But, as Chomsky explains later in the article, the relation will not be quite of this type, since biological explanations of the brain mechanisms will be causal and functionalist (in Kripke's sense) whereas linguistic representations are - according to Chomsky - purely descriptive. This was not the case with nineteenth century physics whose structural description had a causal and functionalist basis in its experiments. Chomsky does not provide any further explanation, not even a discussion, of whether this relation of abstraction is to be conceived of as definitional, translational or of the kind that Davidson has specified as supervenient. Nor does he consider the other possibility, that idealized causal and functionalist systems are introduced as an explanatory framework in the sense that their dynamics is transparently and explicitly constrained by linguistic structure descriptions. My own attempts at reconstructing linguistic processes in terms of quasi-neuronal nets are certainly of the latter type (cf. Schnelle 1991: chap. 4 - 7).

1.3. But let us return to the other problem which Chomsky does not discuss at all. Why should abstract and independent studies of linguistics be a necessary step prior to serious investigation of biological mechanisms? Why not start research with assumptions about a framework for relating linguistic and biological brain research and then develop statements of linguistics and statements of brain structure in mutual interdependence and correspondence? This is what happened in theoretical chemistry after the advent of quantum theory. The time of correlational modeling has come in linguistics like it came for chemistry by the modelings of Lorenz, Bohr et al. eighty years ago. Why not try to find a correlation of linguistic and psychological descriptions of the knowledge of language with brain-architecture and brain-dynamics and then proceed correspondingly? The standard argument against such an undertaking is that "this mode of description is at present of little help in answering the question of how we talk about what we see" (Jackendoff 1983: 4). However, as Jackendoff (1983: 7) concedes, "a little help" can be found in theorizing about some global conditions imposed by the brain: finite resources for storing information and empirical conditions of knowledge acquisition. But the point is not whether the neurosciences are of any help to linguistics or psychology, but rather whether it is possible to come to a common understanding of the phenomena. One is warned that "a premature demand for unification can have the effect of stultifying research" (Jackendoff 1983: 5), and it is said that the neurosciences do not offer sufficient substance for fruitfully correlated research. But this argument no longer holds, as I shall try to show.

1.4. However, there are some uncomfortable surprises to be expected on the route to interdisciplinary cooperation. The linguist may have to change his or her familiar types of theorizing, the methods derived from logic, fundamental mathematics, abstract algebra, as well as from formal and symbolic linguistics. Instead, dynamic and genetically oriented biological methods of theorizing - based on the mathematics of dynamic systems - will have to be learned and applied.

Let us briefly contrast the methods. Since Saussure, a language has been seen as a "monostructure" system. Whether simple or complex, all phenomena in a language exemplify a unique underlying system of structuring principles. In this respect languages are like algebras or like games determined by their constitutive rules or laws. Significant properties and explanatory principles are defined by stating the corresponding systems of rules or laws - or at least the principles from which they can be derived. Linguists have learnt to theorize in this framework. It has become their second nature. Since the system is conceived of as holistic and monosystemic in determining the simplest and the most complex phenomena by the same set of principles, it tends to mark the differences. Something which is not structured by the system of rules or principles is essentially and basically different. Thus, structures underlying other kinds of human behavior or even the behavior of animals are seen as completely different and incomparable to language.

In contrast to this, a biologist tends to think of structures as emerging from evolution. Higher complexity evolves from previously available simpler systems, and components developed in the organism for specific applications are put to use in other contexts. In principle, there are no large steps in the evolution of complication. Natura non facit saltus! As everywhere, this also holds for the brain. "Neurobiological analyses serve to remind us that the brain is evolving piecemeal, each piece serving some need in the animal's natural world, but in ways compatible with already existing structures" (Nadel et al. 1986: 246). Priority for genesis and kinesis (or dynamics) rather than for monosystemics and mere syn-taxis!

Applied to language faculty, we must be prepared to find biologically determined structural levels of development. The language of simple word patterns used in the context of, and with reference to, the actual situation may well be one of the earlier levels of the earliest presyntactical and prereferential linguistic structure. This structure may in turn be closely related to perceptual and motoric structure, e.g., organizational patterns of vision and gesture acquired in the infant's first year. Starting from a level of simple language which is mainly determined by the simultaneously perceived situation, more complicated uses are made possible later on by exploiting memory capacity and special memory circuits which organize interconnections of sentential patterns. Finally, the organization of complicated, interlaced constructions provides for the full efficiency of syntactically highly organized and situation independent language. A layered description of such an evolution has already been given by Bühler (1965) with reference to the development starting from the uses of deictic particles (in demonstratio ad oculos, deixis ad phantasma, anaphora) and leading to those of the article and the relative pronoun in German and some other languages (cf. the detailed study and organismic analysis by Cheang 1990: §§ II.3 and III.2). The linguistic framework is very similar to the psychogenetic explanatory framework of the logician Quine (cf. esp. 1974, 1990: end of § 35, cf. Schnelle forthcoming b).

This layered developmental account is closely related to the different parts of the brain that may come to be used by the faculty of language. According to Nadel et al. (1986: 243) we must distinguish, within a perceptual system, subsystems of concept formation and prototype extraction. These are mainly represented in the neocortical projection areas. Beyond these, there are parts of the limbic system, in particular the hippocampal formation, that establish a temporary "relation" between percept and concept representations in that "it helps bind them together (literally and figuratively ...), and it imposes on them a [structural] interpretation" (Nadel et al. 1986: 246). In the visual process, these systems are "central to the construction of cognitive maps - internal representations which capture the spatial layout of scenes and events" (Nadel et al. 1986: 245). The same organization receives a more elaborate presentation in Edelman (1989, pt. III: "Memory, Ordering, and Concepts"). Moreover, the frontal brain areas subserving the working memory (cf. Goldman-Rakic 1992) will play an important role in the development of syntax transcending mere pattern combinations.

1.5. In the case of linguistic analysis, thus, we may first consider neocortical representations of word class information related to percepts and concepts as well as to compositional features, as provided in argument structures (or lexical sentences, cf. Schnelle 1991: 503ff.). Our first problem will be the neocortically motivated classification of word semantics. The account must be genetic, i.e., start with prelinguistic perceptual structures available in the individual brain (of animals and infants) and show how these structures are related to the first words learned egocentrically by the child and then gradually transformed through the requirements of the organization of intersubjective communication, first in actual situations and then in situation-free literary communication of growing complexity.

The basic idea is that language learning is not a process of the setting of parameters in a fixed framework but a process of gradual reorganization of levels of acquired structures (cf. also Quine 1974: §§35, 36). Chomsky's concept of learning as the setting of the parameters of a single uniquely determined framework of principles is at best plausible for the latest stage of literary language. His arguments derive their plausibility only from restricting his consideration of language to the developmental level of more complex sentences.

When the procedures postulated are applied to simple sentences, they are sometimes rather implausible since they then require the same complication of analysis as when applied to complex sentences. Take Chomsky's discussion of wh-movement. It is not explained at all why wh-expressions are in fact obligatorily fronted. The explanation would have to relate to the communicative role of these sentences. But the communicative role is central to the structural determination of the simpler sentences directly. Their organization should be direct and not via movement. In other words, the conditions of communicative adequacy should directly determine syntactic structures, as for instance proposed by the Prague syntacticians (Sgall et al. 1986).

Let us come back to the problem of the classification of word classes and their argument structures. There are reasons to believe that the classification depends on the position of the word representations in the cortex. Since sentences are composed from words of different classes, these words must be appropriately correlated. We shall not assume that cortical correlation is represented by actual composition of activity patterns, but rather by binding the arguments of the argument structures (or lexical sentences) of the words in a sentence. The binding is provided by a certain interaction of the cortex with cortical appendages organizing time and space (cf. chap. 7 in Edelman 1989; on the special function of binding, cf. Nadel et al. 1986: 246), and it is controlled by signals in working memory (Goldman-Rakic 1992). The actual arrangement or composition of the words in an utterance is a matter of the input parsing or of output generation, which, as we shall assume, is again not represented in the neocortex alone but by another interaction with parts of cortical appendages (cf. in particular Edelman 1989: 127ff. on succession and sense: the hippocampus).

1.6. These indications illustrate the kind of theorizing that we would obtain by developing theoretical linguistics in correlation with evolutionary brain-architecture instead of isolating it in terms of monosystemic systems of principles or rules. We would have simple lexical entries represented in the neocortex being locally correlated with patterns of perception and action - perhaps through mediating structures (cf. Damasio – Damasio 1992). There are two processes. The words of a sentence are kept correlated as long as the speakers or hearers remain focussed on the sentence in their production or understanding. This correlation is a binding procedure realized by the interaction of the neocortex with cortical appendages. At the same time there is the syntactic organization of parsing or producing the sentence. It also must be realized by an interaction of the lexical units activated in the neocortex with the control procedures of the cortical appendages and the control signals in the working memory of the forebrain. It seems to me that the difference between (a) the (semantical) binding of the word components of a sentence in an internal representational pattern (basic for internal inference and "thought") and (b) of the (syntactical) constitution and communicatively efficient integration in the temporal input and output organization of the words of utterances are not sufficiently distinguished by the neurobiologists (such as Nadel, Edelman and others). In order to allow correlated biologico-linguistic theorizing, a minimum of knowledge in both fields is necessary. I shall try to give some of it in an outline. (For more details see Schnelle, forthcoming a, forthcoming b.)

2. Computational neuroanatomy

2.1. Computational neuroanatomy is the study of the relation between neurobiological structure and computational function. It is acknowledged that details of the field are still speculative, but it may indicate ways in which structure and function might ultimately be linked.

By the end of the nineteenth century, it was accepted that brain functions are brought about by the activities of neurons, determined by their interaction signals transmitted over axons and synapses. Not each of the ten billion neurons of the brain can be connected to all other neurons. Instead, they are primarily connected with neurons in their neighborhood in variable and adaptive patterns of connectivity. To more distant neurons they are only connected through myelinated "cables" genetically fixed and generated during maturation. This is the "neuron doctrine".

After its acceptance, the psychobiological explanation of brain function was based on mechanistic associationism. All psychological phenomena were viewed in terms of sensory inputs, motor outputs, and

the neural associations or connectivities needed to bring about a complex "mapping" of sense data on actions. According to the underlying neurology, the most prominent feature of the human brain was the cerebral cortex, and in particular the neocortex, which was assumed to have evolved most recently in vertebrate phylogeny; within the neocortex, there were areas "dedicated" to particular functions such as vision, audition, movement of the hands, etc. and "undedicated" areas reserved for the formation of associations. These associations were the hallmark of higher mental capacity, and the more undedicated cortex a species had, the more intelligent it was assumed to be. Humans had the most of this kind of brain substrate, as befits our status in phylogeny.
Classical neuroanatomy provided the foundation for these neurological speculations by viewing the functional organization of the brain in terms of several central postulates:

(1) Sensory inputs are relayed over a few subcortical regions to a sole neurocortical receiving area for each modality - primary sensory cortex.
(2) Primary cortex directs its outputs to adjacent secondary or psychic cortex for further perceptual analyses.
(3) Psychic cortex sends its outputs to multimodal association areas where integration across sensory modalities occurs.
(4) These association areas are the highest level of analysis and abstraction. They are the source of outputs to the motor system controlling complex behaviors such as language or object manipulation.

2.2. This view of the structure and function of the brain has been modified in several respects. It has been established that there are practically no strictly undedicated areas in the brain. Most are specifically related to one or the other modality. The view that any stimuli, responses, and consequences could be interchangeably associated has become problematic. As a consequence the laws of associative learning have given way to analyses of the properties of particular knowledge systems and their specific modes of acquiring information. In the alternative theory developed in the seventies and eighties, the special connectivities determining a specific behavior are achieved through selective processes acting on what are essentially sequences of combinational networks of more or less dedicated neurons. Very roughly, we may say that integration occurs first within modalities (unimodal associations), then between modalities (polymodal association), and finally between polymodal inputs (supramodal association). Each of these three levels has in turn a number of layers which subserve different pathways. We will come back to this. In any case, the establishment of connections between neural activities reflects the selective strengthening of preexisting sequences of dedicated circuits (pathways of processing) rather than instructive specification of previously undedicated circuits whose connections depend fully and uniquely on stimuli. It thus turns out that the brain is not a general purpose system adapting in arbitrary ways to the patterns of stimulation but rather a system whose flexibility in learning and behavior is constrained in species-specific ways and realized through selective strengthening of appropriate interconnections within preexisting subcircuits.

2.3. Very detailed neuroanatomic analyses of the visual system (cf. Felleman – van Essen 1991, cf. also the review in Schnelle forthcoming a) have shown that we must distinguish a complicated network of subcircuits which are related to the more global representation of pathways as discussed by Mishkin et al. (1987). For the present purposes of relating linguistic analyses to neurobiological ones, it is sufficient to consider three pathways: first, the pathway of the perception of form and color features as well as of figure, second, the pathway for the perception of direction and degree of visually perceived movement, and third, the pathway of the perception and representation of the locations of the constituents of a visual scene (see Figure 1). All three start from the primary visual area at the back of the cortex (centered around the black dot). All three of these pathways feed into two output pathways, one for the eye (and head), the other for the hand. They control the orientation and movement of the eye when focussing on certain objects in the scene, and of the hand when pointing to or manipulating objects of the scene. Finally, there is a pathway of tactile perception that leads from the sensory representations of contact in the hand to finger movement controlling the grasp of an object.

These neurobiological pathways provide the framework for the semantics of visual and manipulatory situation-determined language. Semantics proper consists of the relation of pathways for language understanding and speech with the visual pathways just mentioned. Thus, we will have a pathway organizing and controlling the articulatory movements (starting at the black dot in the middle) and one or several pathways analyzing auditory input to produce the analysis of word-forms. They become correlated via mediating systems (Damasio – Damasio 1992) comprising the network of lemmata and general concept terms with simultaneous patterns in the pathways for visual and manipulatory input and output (cf. the branching connections in Figure 1).

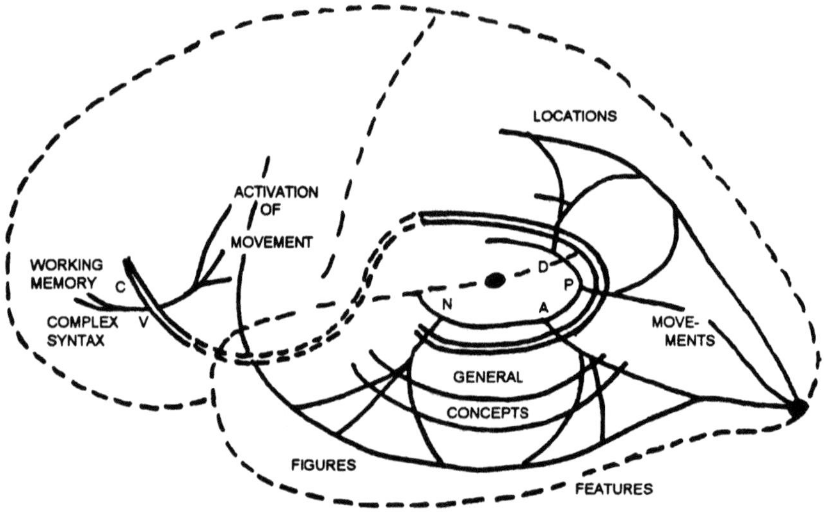

Figure 1. Neurological pathways and positions

Figure 1 indicates these pathways and their positions in the cortex. The figure is a slightly revised version of figure 6 in Schnelle (forthcoming a) where its relation to neurobiological research is discussed. As has been indicated above, these pathways provide only for the analysis, organization and representation of simple patterns (such as correlated with words with or without their argument structures, i.e., the lexicon). The binding of the patterns forming a sentential structure and the syntactic organization of the input and output of utterances require two types of the interactions of these pathways with cortical appendages and the working memory in the forebrain.

3. A logico-bio-linguistic framework

3.1. Let me now turn the tables and discuss logical formats for linguistic representations appropriate to be put in relation to the neurobiological architecture. I shall tentatively assume that the lexicon is represented in the neocortex, whereas syntactical organization of efficient input and output of utterances and the fixation of internal semantic patterns as a basis of thought require two types of interaction between the neocortex and its cortical appendages.

It seems to be appropriate to start with the discussion of those symbolic linguistico-psychological approaches which, though being purely mentalistic, have already made steps in the right direction - for instance, Jackendoff (1983). I fully agree with his basic thesis: "To study semantics of natural language is to study cognitive psychology" (Jackendoff 1983: 3) if and insofar as cognitive psychology is related to neurobiology. Percepts and concepts must be understood as instantiated by activation patterns of neurons! I do not agree, however, with some of his specific theses, namely (a) his conception of argument structure, which is derived from a critique of representations of standard predicate logic and is an unjustified transfer of syntactic structure conceptions to semantic representations and (b) his conceptual structure hypothesis, which insists on a single level of mental representation, at which linguistic, sensory, and motor information are compatible. Assumption (a) is an inappropriate mixture derived from the conditions of efficient internal representation (i.e., the semantics of reference and inference) with those derived from the conditions of efficient organization of the input and output of utterances (the problem which defines syntax). Assumption (b) is in conflict with our basically genetic position, in which several layers of gradually reorganized mental structure emerge during language development. In fact, the earliest and simplest layers are still used after more complicated layers have developed. They are activated in every simple situation that permits simple word use. By contrast, the later forms of syntax are only used in literary or educated communication, typically in written form.

My critique follows from relating mental (psychological) considerations to neurobiological ones. Before discussing the differences, let me briefly mention a methodological point in which I agree with Jackendoff and others: the relation between language and visual perception should be considered a paradigmatic case for analysis.

3.2. I shall start with the discussion of assumption (b), the assumption that there is a single level of mental representation, conceptual structure, at which linguistic, sensory, and motor information are compatible. With Fodor, it is argued that conceptual structure must be rich enough in expressive power to deal with all things expressible by language. Typically, Jackendoff (1983: 17) assumes "that the possible conceptual structures attainable (my emphasis, H.S.) by a human being are characterized by a finite set of conceptual well-formedness rules." It is here that we find the core of the problem. Are we only concerned with the structure of the final system attained? If not, the presupposition seems to be that any system developed earlier than the final system is merely a fraction or a partially still unspecified section of the final system. If the system is determined by a single and unique set of rules (or principles) the acquisition process is an approximation process to the final stage. But this is not the case as we know. In the case of irregular verb forms, the child passes through three phases, the acquisition of forms as mere patterns, the generalization of the regular generation to all forms (with the result of incorrect overgeneralization of the irregular forms), and the final stage in which the irregular verbs block the regular inflection process. Thus, in the third stage, two forms of generation coexist: the pattern memorization and the regular generation. If this were generally the case, several principles of organization would coexist. We should assume that there is not a single conceptual structure, but several structures governed by different types of principles. A corresponding situation is illustrated by Bühler's classical genetic analysis of the deictic pronoun (1965: §6.7), the article (1965: §20), anaphora (1965: §§8, 26), and the relative pronoun (1965: §27). It is obvious that different principles of organization coexist and that an analysis which concentrates exclusively on combinational and compositional organization (i.e., its constitution and transformation) is misguided. The plausibility of a single syntactic and conceptual structure is an artifact of the limitation of theoretical organization principles to constitution and transformation.

The alternative is not the narrow conception imputed to Piaget (Jackendoff 1983: 17) that certain rules must be learned by generalized observational procedures. The alternative is that there are stages of systems of increasing complexity in which varying functional principles, memory access and memory organization come to be used, each of these stages having consequently its own system of principles. This does not seem to be very economic for the organism in terms of memory space but it seems to be very efficient in terms of attentive energy and in terms of a smooth acquisition process since it allows early communication with very limited resources that only gradually increase in the process of acquisition. In the context of the brain, memory space is not a decisive factor. There are plenty of neurons! What counts is the organization of temporal efficiency of communication. I conclude that we need a system of developmentally layered subsystems which are each acquired at an appropriate time of maturation, involving in each case "phase transition" adaptations of the levels to make them compatible.

3.3. Let me turn to the problem of the format of word classes and argument structures. Jackendoff derives his conception from a critique of standard predicate logical representations. His own alternative is, like many of the recent proposals for argument structures, determined by representational principles of syntactic constitution and transformation. Whereas Jackendoff argues that standard predicate argument structure as introduced by standard logic, is utterly inappropriate, I hold, by contrast, that unorthodox interpretations and extensions of the orthodox standard format of predicate logic are the best framework for a representational structure that can be correlated with neurobiological architecture. In view of my program of a layered perceptual and conceptual genesis of language understanding, I shall make use of representations on several layers of perceptual and conceptual genesis. They all have predicate logical form, but layer specific interpretations of their constants and variables. I shall distinguish the following layers:

(1) perception of features and feature patterns and the primitive semantic correlation of visual patterns and sound patterns (associations in cortex?)
(2) perception of mutually interdependent feature configurations (primitive "argument structure" type, involving basal ganglia?),
(3) conception of things and situations (implemented as in 2.),
(4) use of more complex perceptions and conceptions of things and configurations which transcend mere patterns (requiring use of working memory), and
(5) situation independent ("literary") language (requiring extended use of working memory and of stored patterns).

It is only on the fifth level that our discussion can be directly related to current linguistic analysis. But already on the first level we can distinguish perceptually motivated major word classes and from the third level onward, we need a system of types for variables to be introduced in critical correspondence with the requirements of recent theorizing on argument structure. In the present article, I cannot discuss all details (I refer to Schnelle forthcoming b). To illustrate the spirit, it should be sufficient to discuss some aspects of the first two levels, even if this discussion does not yet bridge the gap between the proposed logico-bio-linguistic kind of theorizing and the classical linguistic approaches.

3.4. At the first level, we merely assume that the brain is able to register features of perception, on the one hand simple features, such as those of color, form, and figures in vision (and also propriocepts of own movements of eye, head, limbs), and on the other, phonetic features of speech as well as patterns in both areas, such as complete visual figures or

morpheme and word patterns. Furthermore, certain patterns of speech become correlated with certain features of vision and perceptual and reactive movement. The result is the semantic ability that is typical for the one-word phase of language acquisition in which the child stays for about three quarters of a year (cf. inter alia Bühler 1965: 72). The acquisition of word-perception associations obviously presupposes the structured acquisition of pattern perception during the first year of life.
If we require a formal symbolic representation as an idealization of the achievements of this level, it is a simultaneous use of several simple terms, which, corresponding to Quine's first level of language acquisition (Quine 1960: §19; 1974: §14), should be conceived of as similar to single mass-terms used as complete sentences. Quine insists in almost the same words as Bühler that at this level, a distinction between words (terms) and sentences is impossible. If we insisted in representing this stage symbolically, we would have to take sets or conjuncts of sententially used terms.

Let us assume that the child sees a red square and a blue triangle. The registered perceptual features of the scene are RED, BLUE, TRIANGLE, SQUARE - understood as the conceptual, i.e., mental information in the sense of Jackendoff (1983: 32), without presupposing, as he does, that this mental information mediates projection into awareness. In my view, the mental terms are only shorthand for patterns of neural processing (cf. Quine 1974: §9, cf. also Cheang 1990: esp. §§2.2/3). They serve their purpose in communicating with traditional manners of speech, such as Jackendoff's, relating them nevertheless to connectivities in the brain. Let us thus assume, in particular, that the applicability of the mental terms in a situation corresponds to activations of the distinct cell assemblies F,G,H,I in the brain. Registration is thus represented by the truth of the statements that cell assembly F registers the presence of the mental feature BLUE or BLUE(F) in short. Correspondingly, G, H, and I represent the presence of their corresponding mental features RED(G), TRIANGLE(H), SQUARE(I).

It has often been asked what the biological status of the mental(!) predicates in these statements is. Primarily they seem to be labels without direct functional relevance to the brain: The cell assemblies cannot "see" their own labels nor the labels of their neighbors (cf. Fodor 1986: 86). Along with Fodor, I take it that there are information states of the whole network connected with the cell assemblies mentioned. These information states are represented by the connectivities, i.e., "patterns of relations among distinguished states of the objects that constitute the network" (Fodor 1986: 83) that specify the activation of the respective cell assemblies causally, given the appropriate sensory stimuli and reactivities. (In following the chain of causal reaction further, the states of the cell assemblies are even related to the appropriate states of the entities themselves in the external world.) Let us denote the corresponding connectivities of the neural network, which detect the presence of the features, respectively by BLUE', RED', TRIANGLE', SQUARE' to indicate their close correspondence to the mental features given above. Note that the connectivities may have identical parts and form a subnetwork of the complete network of perception. At this primitive level of the system, there is a one-to-one correspondence between mental features and connectivities of the neural network. Formally, however, we now have a correspondence between an activated cell assembly and the network whose operation regularly causes that same activation rather than an abstract predication representing the intended representational role of the cell assembly. Instead of the statements, we should write the correspondences between specific connectivities or networks on the one hand and cell assemblies specifically activated by the operations of the networks on the other:

{BLUE'≈F, RED'≈G, TRIANGLE'≈H, SQUARE'≈I}.

A similar structure applies to the perception of words. When a child has learned the words *blue*, *red*, *triangle*, and *square*, it must be able to register their presence in an utterance heard by cell assemblies, say *F*, *G*, *H*, *I*, and corresponding connectivities leading from nerve activations in the inner ear to these cell assemblies, namely, *BLUE*, *RED*, *TRIANGLE*, *SQUARE*. If seeing red and hearing the word *red* are correlated the registration of color perception G and the registration of word perception *G* should be connected. A color stimulus of red is processed by RED' to yield an activation of G which matches someone else's utterance of *red* via hearing in the ear, sensory-cortical pathway and brain processing by *RED'* and registration in *G*.

3.5. In the previous paragraphs, we only used percepts which are typically expressed by nominals (nouns or adjectives). Is there any neuronal criterion distinguishing nouns, adjectives, and verbs that apply to seen objects at the primitive level of acquisition? The answer seems to be that it is the location relative to the pathways of visual analysis.

As indicated above (2.3), detailed neuroanatomic analyses of the visual system have shown that we must distinguish three pathways for visual perception: first, the pathway of the perception of form, color and figure, second, the pathway for the perception of direction and degree of visually perceived movement, and third, the pathway of the perception and representation of the locations of the constituents of a visual scene. All three of these perceptual pathways provide input to two output pathways: one for the eye (and head), the other for the hand. They control the orientation and movement of the eye when focussing on certain objects in the scene, and of the hand, when pointing to or manipulating objects of the scene. Finally, there is a pathway of tactile perception that leads from the sensory representations of contact in the hand to thumb movement controlling the grasping of an object.

These neurobiological pathways provide the framework for the semantics of visual and manipulatory situation-determined language. Let us presuppose the pathways for word perception with outputs in distinct cell assemblies (such as F, G, H, I illustrated above). Semantics consists of relating these cell assemblies to the outputs at the pathways just mentioned.

Figure 1 above indicates these pathways and their positions in the cortex. Obviously, the pathways for linguistic understanding are encircled by these visual and motor pathways. It is a possible suggestion that during the acquisition of words, the cell assemblies whose activation represents recognized words become localized near their semantic correlate (as already indicated in figure 1). Thus, cell assemblies for nouns related to recognized figures will be placed near the middle parts of the inferior temporal lobe and proper nouns towards the temporal pole, adjectives denoting features of form or color, nearer to the posterior parts of the temporal lobe. By contrast, verbs, being more concerned with activities, are instead represented in relation to the lower part of the frontal lobe. These speculations about nouns, adjectives, and verbs have been fully confirmed by the empirical findings of Damasio and Damasio (1992: 70). It is made clear by Damasio (1992: 535) that the words of these word classes are not necessarily represented near these locations, but the relations to these locations must in any case be intact. Now, there is the third pathway serving the analysis of location and locational direction passing over inferior and posterior parts of the parietal lobe. Which words should be considered as being placed near this pathway? Speculations obviously point to demonstrative deictics and prepositions. The latter should be in between this region and the area MT (marked "MOVEMENTS") since they are representing location and direction of movement (the latter being analyzed by area MT).

Let us emphasize, however, that these locations are primitive locations, which do not determine fully the later use of these words. In later developments, these words will also be related to representations of a conceptual scheme and its uses in thinking and planning instead of merely being determined by the patterns of perception. Their status will then be more and more determined by the complex requirements of the conceptual framework on the one hand and by the requirements of complex syntax for efficient thought and communication on the other. Working memory should essentially come into play. It may be that the conceptual determinations for thought will then require a reorganization or replacement of the early placements realized in primitive processes of language acquisition. Damasio's remarks (1992) mentioned above are thus very pertinent.

3.6. Primitive classification of words into major word classes is not enough. We are confronted with an important additional problem related to cross-references and interdependencies between words in a sentence. Since the problem is treated in linguistics by means of argument-structures, we are now concerned with the primitive basis of this notion.

As presented so far, the same pattern would have been registered by the cell assemblies if the situation presented contained a blue square and a red triangle instead of a red square and a blue triangle. After all, the cell assemblies registered only the simultaneous presence of all four features without indicating which color is paired with which form. But in fact, we know that children can make this distinction before starting to speak. They must have acquired a distinct representation of the two situations during the first year of their lives. It is interesting, however, that aphasics have difficulties exactly with these distinctions, as typically shown in the standard token test. This points to the fact that additional processing capacity comes into play which is partially disturbed or lost in patients with language deficits. It may be that the basal ganglia are involved (more especially the head of the left caudate nucleus, cf. Damasio, 1992: 536. Cf. also Edelman 1989: 133ff. for the importance of the basal ganglia).

In any case, what the child must register neurally in addition to the features themselves is which of the features are related to each other. This registration could be represented neurally by some relation between the cell assemblies which register the perception of properties. Seen abstractly, there are two solutions to this problem, grouping or coindexing. To the linguist, the normal or basic solution appears to be grouping, just as language keeps sentential conjunction and inner-sentential predication or attribution separate. The solution would thus be:

{[BLUE(F),TRIANGLE(H)],[RED(G),SQUARE(I)]} vs.

{[BLUE(F),SQUARE(I)],[RED(G),TRIANGLE(H)]}

This grouping might, in addition, be explained as a result of a completion of components, here TRIANGLE and SQUARE, characterized as incomplete (cf. Frege, Tesnière, X-bar theorists like Harris, Jackendoff etc.).

The problem is, how could grouping - or incompleteness and completion - be represented in the brain? Due to the correspondences indicated above, we would have to designate groupings of cell assemblies or networks as follows:

{[F,H],[G,I]} vs. {[F,I],[G,H]} or perhaps {[BLUE', TRIANGLE'], [RED',SQUARE']} vs. {[BLUE',SQUARE'], [RED', TRIANGLE']}

But how could the grouping of cell assemblies (or networks) indicated by square brackets be represented by cell assemblies and connectivities between cell assemblies?

The other solution is coindexing. The solution would then be (indexing this time the correspondences of networks and cell assemblies simultaneously):

$\{[\text{BLUE}'\approx F]_1,[\text{RED}'\approx G]_2,[\text{TRIANGLE}'\approx H]_1,[\text{SQUARE}'\approx I]_2\}$ vs.
$\{[\text{BLUE}'\approx F]_1(p),[\text{RED}'\approx G]_2,[\text{TRIANGLE}'\approx H]_2,[\text{SQUARE}'\approx I]_1\}$

Here also we would have to investigate the possible representations in the brain.

Let us note a formal feature. Whether to present coindexing the way we did or whether to write

{[BLUE'≈F](x), [RED'≈G](y), TRIANGLE'≈H](x), [SQUARE'≈I](y)} vs. {[BLUE'≈F](x), [RED'≈G](y), [TRIANGLE'≈H](y), [SQUARE'≈I](x)}

or also

F(x)&G(y)&H(x)&I(y) vs. F(x)&G(y)&H(y)&I(x)

is formally equivalent. The latter are obviously formulae of first order predicate logic. They differ from the standard predicate logic by their interpretation. The variables x and y take values in certain distinct correlational properties or activities of the brain, still to be identified - perhaps in cortical appendages like the hippocampus or by Malsburg's dynamic adjustments of synaptic weights - whereas the constants designate cell assemblies.

This cross-referential interpretation of the role of variables in formulae of predicate calculus is very fruitful, and it agrees with basic considerations by logicians (cf. e.g., Quine 1966: 103 or the typical introductory remarks to Lambda-abstraction). Interpreting the variables merely as taking individual objects as values - the orthodox interpretation also assumed by Jackendoff - leads to confusion about the implications of using the predicate calculus as a representational scheme, as is shown by Jackendoff's critique mentioned above (3.1). Indeed, Jackendoff's critique is not well-founded.

Let us first discuss the status of these representations since there has been much discussion on argument structures during recent years, usually connected with attacks on proposals of standard predicate logic. However, much confusion resulted from the linguistic motivation to use argument structures as interfaces between syntax and semantics whose task is not only to represent semantic aspects but also certain aspects of syntax. This conception is ill-founded if the cortical considerations given above should be correct. We should separate semantic from syntactic organization.

In connection with introducing his own version of argument structures, Jackendoff (1983: §4.1) advanced arguments against deriving predicate-argument structures from standard predicate logic:

(a) The use of individual constants and variables merely refers to individual objects;
(b) In logical representations, resemblances to natural language syntax have vanished;
(c) Words of categories other than verbs are translated into predicates by combining them with the verb *to be*;
(d) Allegedly unmotivated connectives are introduced (an argument having also implications for (b));
(e) Prepositions are not analyzed properly;
(f) Prepositional phrases cannot be used referentially.

As can be shown (cf. Schnelle, forthcoming b), (a) and (f) are only justified for the orthodox interpretation of the predicate calculus, not for the primarily cross-referential interpretation of variables indicated above.

Arguments (b), (c), and (d) depend on the presupposition that structure representations should resemble representations of syntactic structures. In view of current neurobiological research, we should distinguish between syntactic organization which is concerned with the efficient compression (or encoding) of information for communication in a linear channel and the semantic interdependencies between the words in a sentence controlling their reference, coreference, and inference. The former most certainly are controlled by interaction with cortical appendages, whereas the latter need not be established by such an interaction; it could also be implemented by dynamic synaptic adjustments à la Malsburg (1986).

Argument (e) is considered as a central one by Jackendoff. Taking into account what has been said above (3.5), we see that prepositions are located at positions in the brain where location and direction of movement is represented. Thus they are determinations of the same type as the nominal determinations just discussed, i.e., they are predicates over variables which are correlators of predicates. The phrase *on the table* will be represented in activations of the cortical lexicon by a cell assembly for *table*, located near cell assemblies representing typical percepts of tables and a cell assembly for *on* represented near detectors for location or gradients of movement. Both are cross-referenced - this is essential - by the same cross-referential variable.

4. Arrangements of argument structures and their binding in the cortex

Let us illustrate what has been discussed so far by another simple example. The German boy Christian Morgenstern sees a gray weasel on a round pebble amidst a floating brook. He addresses Christa Palmstroem in pointing to them and says, "Schau, das graue Wiesel auf einem runden Kiesel inmitten Bachgeriesel!" whereupon Christa turns her head and eye in the direction indicated by Christian. This is our global and external characterization of what Christian and Christa are doing.

A possible internal semantics of what is going on would have to refer to their internal states. We may well agree with Quine that without detailed empirical analysis of the neurophysiology of the brains, it would be impossible to characterize the internal state uniquely (Quine 1990: 62). Due to the indeterminacy of translation hypothesis, we may even doubt whether there are any internal correspondences between Christa and Christian (cf. Quine 1960: 8, 74, 94). "Yet, each perception is a single occurrence in a particular brain and is fully specifiable in neurological terms . . . Perceptions are neural realities, and so are the individual instances of beliefs . . . Physicalistic explanation of neural events and states goes blithely forward with no intrusion of mental laws" (Quine, 1990: 70-1). Since this is so, we are allowed to analyze Christa's and Christian's neurological brains provisionally as identical, at least in those parts that realize language understanding without any conflict with Quine's philosophical assumption that in fact their brains may have very different connectivities. Neurobiological results seem to indicate that important variation can be found, but only variations of functionally determined architectural topologies (Edelman 1989).

But how do we come to build hypotheses about these structures? We could start from linguistic conceptualizations which are symbolic and thus surely mental. Here also, we stay in accord with Quine: "Mental entities are unobjectionable if conceived as hypothetical physical mechanisms and posited with a view strictly to the systematizing of physical phenomena" (Quine 1974: §9, p. 33-34). In this perspective, we shall first introduce symbolic means of representation describing the symbolic organization of cross-referenced predicate-argument structures. Our symbolic patterns are globally constructive schemata for hypothesizing neuronal networks.

Let us try an analysis of the example. We have the following notions: *weasel*, *pebble*, *gray*, *round*, *brook*, *floating*, *on*, *amidst*, *Look*. Let us take their corresponding percepts and specify their binding (E=there is):

> E at-p E at_q E at_r : WEASEL (at-p) & GRAY (at-p) & PEBBLE (at-q) & ROUND (at-q) & BROOK (at-r) & FLOATING (at-r) & dominant_DIRECTION_OF_my [Christian's]_REGARD [= "my_Look"] (at-p) & slightly_BELOW_the_dominant_direction_of_my_REGARD (at-q) & PERCEPTUAL_ENVIRONMENT_of (at-r, at-p and at-q).

Do we have any idea where these percepts could be represented in the neocortex? As already indicated above (2.3 and 3.5), the process splits up into three pathways, an inferotemporal pathway for perception of form, color and figure, an occipital pathway of movement perception, and the parietal pathway of location and orientation. On the other hand, the frontal cortex organizes the controlled and directed movement of the eye. Omitting the details, we may nevertheless postulate that GRAY and ROUND are registered along the posterior parts of the inferotemporal pathway, whereas the figure of WEASEL and PEBBLE as well as the banks of BROOK are registered at the more intermediate or frontal parts of the temporal pathway. FLOATING, being mainly a perception of movement, is to be found in MT (the pre-occipital pathway), and the locational marks DOMINANT_DIRECTION_OF_REGARD, BELOW, ENVIRONMENT are registered at the parietal pathway. Finally, the action control of LOOK! is frontal near the prerolandic area assigned to the muscle activations of eye and head. In order to simplify representation, we shall in the following denote the cell assemblies by the same terms as the mental percepts, and not by uncommitted terms, such as P,Q,R,S above. For the arrangement of the cell assemblies see again Figure 1 above.

What remains to be specified is cross-referencing. It will be indicated by the variables x,y,z and corresponding quantifiers. Obviously, they are essential. Without them, we could not distinguish a scene of a gray pebble and a round weasel from our scene where we have a gray weasel and a round pebble. How could the cross-referencing be represented in the brain? Let us recall the indications given by Nadel et al. (1986). They insist that the hippocampal formation "helps bind them [the percept and concept representations in the neocortex] together (literally and figuratively ...), and it imposes on them a [structural] interpretation" (Nadel et al. 1986: 246). Damasio's remarks (1992) rather point to parts of the basal ganglia. Whichever, our cross-referencing should consist in the activity of hippocampus or basal ganglia, organizing three different positions x,y,z. These are primarily connected to the particular orientations of parts of the scene as registered in the parietal pathway and secondarily to the cross-referenced characteristics of form, color and figure in the temporal pathway, as well as to the movement perception of floating in area MT. The representation above should be transformed in a representation marking the cross-referential variables:

> E x E y E z : WEASEL(x)&(at-p)(x) & GRAY(x)&(at-p)(x) & PEBBLE(y)&(at-q)(y) & ROUND(y)&(at-q)(y) & BROOK(z) & (at-r)(z) & FLOATING(z)&(at-r)(z) & dominant_DIRECTION_OF_my(Christian's)_REGARD [= "my_Look"](x)&(at-p)(x) & slightly_BELOW_the_dominant_direction_of_my_REGARD(y) & (at-q)(y) & PERCEPTUAL_ENVIRONMENT_of(z)&(at-r, at-p+at-q)(z)

Only those registrations that, in our symbolic schema, are marked as cross-referenced are appropriately correlated through their connections to the binding organ. This organ could be the hippocampus or parts of the basal ganglia. The alternative could be a binding procedure built up via dynamic build-up and phasic activation of connectivities according to Malsburg (1986) (which might well be controlled by the subcortical structures mentioned). All of this requires certainly further elaboration that cannot be given here.

As indicated above, we presuppose that the word percepts, such as weasel, pebble, gray, round, are represented near their corresponding percepts. This is due to the learning process, which, we speculate, will go as follows: if, according to Quine's design, a parent has verified that a child's view is directed towards, say, the pebble, he or she may say "Look! A pebble." Let us assume that the child identifies the phonetic pattern of "pebble" and that, since its view is directed towards the pebble, it registers in its temporal pathway the figure of the pebble by the activation of PEBBLE. It may be that the child has first a rather diffuse representation of its word patterns. Given the correlation with the PEBBLE-activation the registration of the identification of the word is concentrated near PEBBLE, i.e., near the visual registration in the more frontal part of the temporal pathway.

5. Summary

The topic of language and brain requires new ways of theorizing. Symbolic representations and rule systems must be related to networks of interactive units: nerve cells, cell assemblies, cortical areas in the brain. The time has come to work on the details of relations between abstract and concrete representations. New insights from the neurosciences about cortical pathways, their function and organization as well as about correlations or "bindings" between features occurring on different pathways are encouraging. A concrete interpretation of the perceptual and conceptual differences of word classes, of morphology, and of syntactic principles of organization will be possible. Linguistic and extra-linguistic perceptual processes are systematically related in the genesis of language, which is an organic multisystemic process rather than merely the setting of parameters in one and only one homogeneous system governed by one set of principles only.

References

Brand, M. – R.M. Harnish (eds.)
1986 *The representation of knowledge and belief.* Tucson: The University of Arizona Press.

Bühler, Karl
1965 *Sprachtheorie.* (2nd edition.) Stuttgart: Fischer.

Cheang, Kiseang
1990 *Semantik der Deixis.* Opladen: Westdeutscher Verlag.

Chomsky, Noam
1986a "Changing perspective on knowledge and use of language", in: M. Brand – R. M. Harnish (eds.), 1-58.
1986b *Knowledge of language*, New York: Praeger.
1988 *Language and problems of knowledge*. Cambridge, Mass.: MIT Press.

Damasio, A.R.
1992 "Aphasia", *The New England Journal of Medicine* 326,8: 531-539.

Damasio, A.R. – Hanna Damasio
1992 "Brain and language", *Scientific American* 267,3: 63-71.

Edelman, Gerald M.
1989 *The remembered present: A biological theory of consciousness*. New York: Basic Books.

Felleman, D.J. – D.C. Van Essen
1991 "Distributed hierarchical processing in the primate cerebral cortex", *Cerebral Cortex* 1: 1-47.

Goldman-Rakic, P.S.
1992 "Working memory and the mind", *Scientific American* 267,3: 72-79.

Innis, Robert E.
1982 *Karl Bühler*. New York: Plenum Press.

Jackendoff, Ray S.
1983 *Semantics and cognition*. Cambridge, Mass.: MIT Press.

Malsburg, Ch. von der
1986 "Am I thinking assemblies?", in: G. Palm – A. Aertsen (eds.), 161-176.

Mishkin, M. – T. Appenzeller
1987 [1990] "The anatomy of memory", *Scientific American* [German translation in Wolf Singer (ed.) *Gehirn und Kognition*, Heidelberg: Spektrum der Wissenschaften]

Nadel, L. – J. Willner – E. Kurz
1986 "The neurobiology of mental representation", in: M. Brand – R.M. Harnish (eds.), 219-257.

Palm, Günther – Ad Aertsen (eds.)
1986 *Brain theory*. Berlin: Springer.

Quine, Willard Van Orman
1960 *Word and object*. Cambridge, Mass: MIT Press.
1961 *From a logical point of view*. (2nd edition.) New York: Harper and Row.
1974 *Roots of reference*. La Salle, Ill.: Open Court.
1990 *The pursuit of truth*. Cambridge, Mass.: Harvard University Press.

Schnelle, Helmut
1991 *Die Natur der Sprache*. Berlin: de Gruyter.
forthcoming a "Sprache und Gehirn - Sprachfähigkeit als neuronales Netz". to appear in: *Kognitionswissenschaft*
forthcoming b "Words in the brain".
Sgall, Petr – Eva Hajicova – Jarmela Panevova
1986 *The meaning of the sentence in its semantic and pragmatic aspects*. Prague: Academia.

INTERCHAPTER

Structure as idyll: The genesis of meaning in nature

Roland Posner

Structures of nature are presented in black-and-white photographs. The meanings that these structures acquire in human vision are made apparent in color photographs below. Looking at these pairs of pictures, the beholder can experience in himself or herself the way human vision makes nature meaningful.

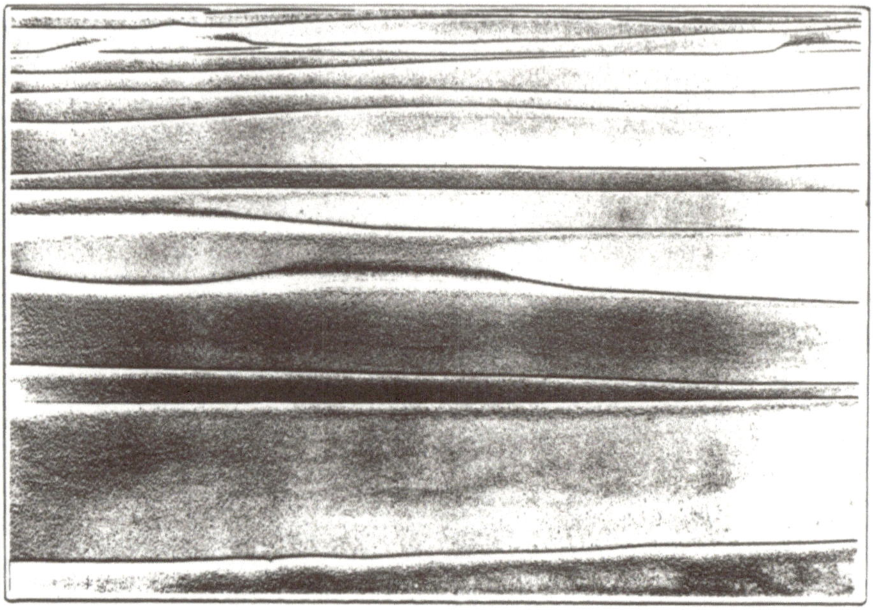

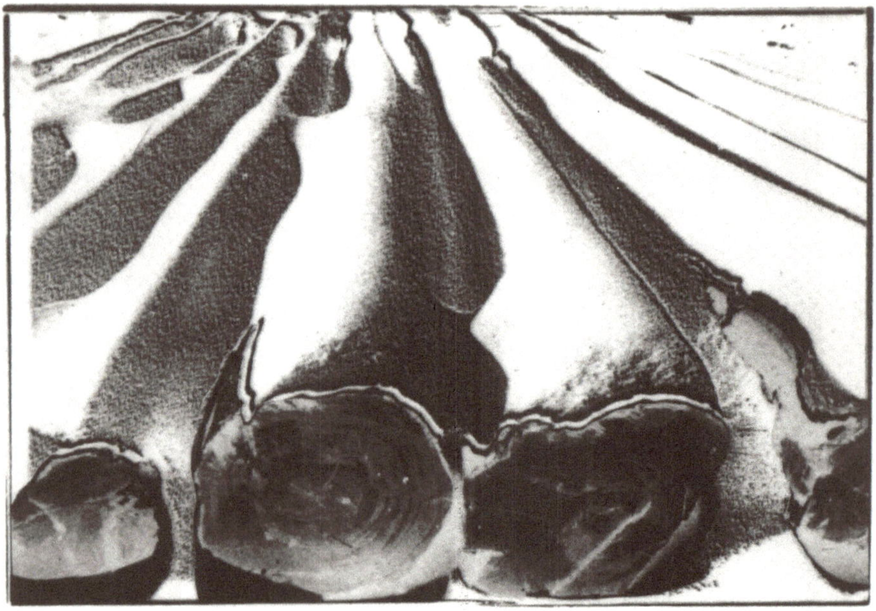

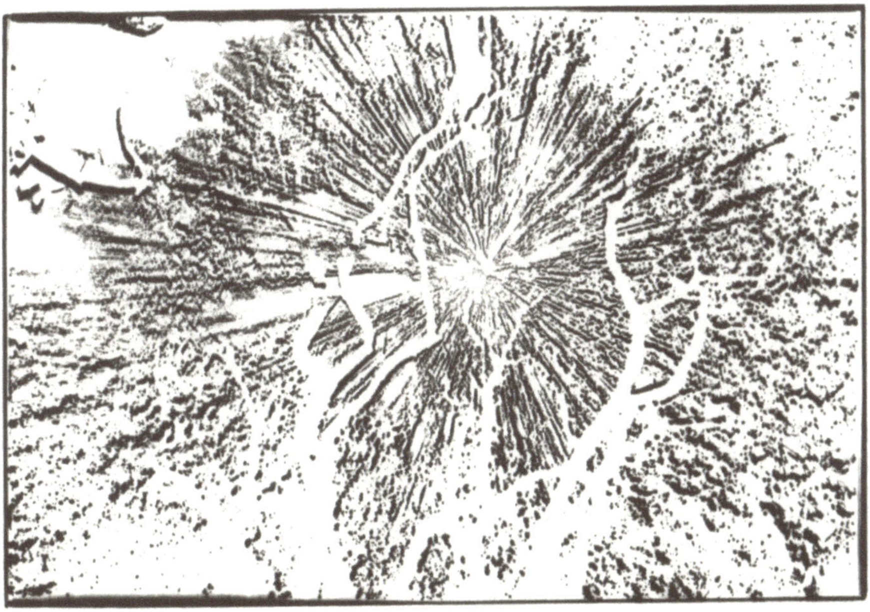

PART IV

Eikonogenesis and graphogenesis

Constants in 40,000 years of art

Emmanuel Anati

The analysis of rock art on a world scale is a new field of research. The present study is based on data collected from over twenty major rock art areas located in all continents in all the areas reached by humankind. Some general trends and a number of working hypotheses have emerged which are summarized in seventeen postulates. Rock art is subdivided into four categories which show different elements of style and content and refer to four different socio-economic backgrounds. Universal characters are identified by the comparative analysis of large concentrations of rock art around the world. The presence of a basic grammar and a syntax in the visual art of nonliterate societies indicates widespread patterns and illustrates the presence of a human "basic logic" and of universal conceptual processes of sequences and associations. Some paradigms and archetypes in the visual arts are sorted and illustrated, and a preliminary world vision of the phenomenon of rock art is attempted. Some of these universal patterns still survive in contemporary art.

1. The earliest art

The hunting way of life has characterized humanity for over two million years and left profound marks on the intellectual nature of the species. The impact of this heritage on human evolution is expressed in Postulate P1:

> *P1.* The fundamental human processes of association and "logic" have developed throughout the ages in which the human species has acquired its basic behavioral patterns.

These millennia are characterized by life in hunting bands, who shared activities and refined communication. This way of life reached a high level of perfection and efficiency in the last 100,000 years with the diffusion of our direct ancestor, *Homo sapiens sapiens*. This ancestor had a well developed technological level and quite refined mental abilities. Capable of producing precise and efficient implements, *Homo sapiens sapiens* was responsible for creating an ideology whose matrix is still present at the core of modern conceptuality. This includes a capacity for synthesis and abstraction which, among other things, led humans to produce art and to develop an articulate and complex language.

Art of the Sapiens Fossil population is the mirror of mind and spirit and constitutes a precious record of the conceptual and psychological matrix of humans. As further elaborated below, we come across archetypes and paradigms in this population which are essential to our being and which are still deep within us.

Populations whose livelihood is based on hunting and gathering are today almost extinct and are confined to the most inhospitable parts of the world. These include the deserts of Australia and Southern Africa, the tropical forests of the Amazon basin, the Congo basin, Southeast Asia, the Arctic tundra of Lapland, Chukatka, Alaska, and the great Canadian North. They occupy territories which constitute more than 20% of the Earth's land surface and yet are less than 1% of the world's population. Only 500 years ago, at the time when America was discovered, 70% of the Earth's land surface was populated by hunting people, who then accounted for more than 20% of the world population. 12,000 years ago, at the end of the Pleistocene, the entire population of the Earth pursued this means of subsistence. This was gradually replaced by economies based on food production. It is only in the last 12,000 years that other ways of life have developed.

Tribal societies around the world share the characteristic of producing art, especially rock art. Their visual output is recorded in thousands of zones distributed in 120 countries on all continents with the only exclusion, so far, of Antarctica. A preliminary evaluation has shown that over 35 million prehistoric figures from about 140,000 sites are known in the world today. Over 70% of all known rock art was produced by hunting and gathering societies, while less than 30% is the work of Pastoralists and Agriculturalists. The growing interest in this art is caused by the light it projects on the collective memory and on universal conceptual processes.

Through these ancient expressions of the human mind, a wide range of submerged memories come back to consciousness, reviving stored chapters of our intellectual heritage. However, its historical relevance is equally important. Rock art is a sort of pictographic writing which constitutes humanity's largest and most significant archive of its history for 40,000 years until the advent of conventional, modern ideographic, and then alphabetic writing. New discoveries increase the record almost daily. Even in Valcamonica, where we have been working for 30 years, the 250,000 figures discovered are only a small part of the cultural patrimony that has yet to be surveyed. Other examples are from Lesotho where over the past 20 years Smits has catalogued thousands of pictures from around 500 caves and rock shelters, but there are still about 5,000 sites that have yet to be catalogued (Smits 1988). Analogous situations exist in most countries (Anati 1981, 1989).

Recent surveys reveal the size of this heritage that is being reappropriated by human culture in such disparate regions as the Americas and the Far East (Grant 1982; Schobinger - Gradin 1985; Chen 1988). Countries like Sweden, Spain, and France that claim to have complete inventories are rare and even in these countries, there are new discoveries every year, demonstrating that exploration is not yet finished (Beltran, 1988). The recording, study, and comprehension of the rock art patrimony is still in its infancy.

So far, cataloguing and making inventories of rock art have been mainly the private work of students and amateurs, sometimes supported by universities and museums. It is only in the last few years that some governments have become aware of the importance of this documentation as an immense cultural and historical heritage belonging to all humanity and are now embarking on systematic inventories. This came about largely as a result of an information campaign conducted by the CCSP (Centro Camuno di studi preistorici, Italy) and CAR (ICOMOS International Committee on Rock Art) with the support of UNESCO and ICOMOS (International Council on Monuments and Sites).

At the same time, it was deemed necessary to stimulate individual scholars to make the results of their research available to the scientific world. For this purpose, we promoted a monographic series of books on rock art, "The Footsteps of Man", published by Jaca Book in Italy and by Cambridge University Press in England. Following the recommendation of a UNESCO consultancy held in 1981, the periodical *Bollettino del Centro Camuno di studi preistorici (BCSP)* has become the "World Journal of Prehistoric and Primitive Art" and devotes much space to current studies and discoveries of rock art.

Mobiliary art, figurines, plaquettes, and decorated objects have received different treatment. Once they are found, they usually end up in museums and state collections, which then draw up the inventory more for the institution than for humanity. A single world inventory is also necessary in the sector of mobiliary art. However, both in terms of the immensity of existing documentation and world distribution, rock art appears to be a global phenomenon which constitutes more than 90% of the known prehistoric art. It has remained *in situ* and is found in all inhabited continents.

In order to reach a valid global assessment of rock art, there is great need of a data bank. Data should be classified using universal standards so that comparisons can be made. This would require collecting the same type of data, having the same cataloguing procedures in different countries and creating a fluid and open exchange of information. With this aim in mind, CAR has produced, since 1984, standard forms which can enter a world data base in different languages and can be read and printed in any language. This data bank is already being compiled in various countries such as Canada, Mexico, Italy, Ireland, Tanzania, Malawi, Israel, and Saudi Arabia (Anati 1984).

The rock engravings in Alto Adige, Italy, and those only a few kilometers away in Austria are analyzed and catalogued in different ways, since scholars are unaware of each other's methods and criteria. This sort of situation should no longer be allowed for it signifies that systematic research is impossible. Coordination should be promoted by international organizations; regional and local institutions as well as individual scholars should be encouraged and given concrete support and guidelines.

2. The social and economic context

Certain elements of style and content have been shown to be a constant on a world level. We have been able to distinguish four categories of rock art with universal characteristics (Anati 1981):

A. Archaic Hunters: the art of hunting populations which do not know the use of the bow and arrow. Figures and signs are connected to one another, but true descriptive scenes are practically absent.
B. Evolved Hunters: the art of hunting populations which use the bow and arrow. Descriptive and anecdotal scenes are present.
C. Pastoralists/Animal Breeders: the art of populations whose main economic activity is the rearing of livestock. The main interest of the artist focuses on domestic animals.
D. Complex Economy: the art of populations with diversified economies, including farming activities. Mythical scenes and schematic groupings of signs are repetitive.

Only a small percentage of documented rock art cannot be put into any of the four categories, and this in itself indicates that almost all of the world's rock art belongs to one of these four categories. The art of hunting populations constitutes about 70% of the total known documentation.

These subdivisions are, of course, schematic. There are transitional phases and groups with mixed characteristics, and within each category there are considerable variations which should be studied further. Nevertheless, at the present state of research and given the immensity of the material already noted, it is becoming necessary to develop guidelines of analysis on themes and styles which go beyond the limits of regional definitions. The methodology requires further sophistication, but even now it is producing results. For example, a preliminary order has been established for the larger groups, the rich concentrations in Central Tanzania, Madhya Pradesh in India, Central Arabia, the Negev and Sinai deserts, and Seminole Canyon in Texas.

Criteria based on themes and typology, the nature of the associations and the presence of scenes, figurative, schematic and abstract styles, the importance given to certain wild or domestic animals, the presence or absence of certain symbols which act as "fossil guides", have all led to the discovery of significant, recurrent elements. This has brought us to the working hypothesis of Postulate P2:

> *P2.* Universal reflections conditioned by way of life influence behavior, thought, ideology, associative processes, and consequently artistic manifestations.

Work in this field is in progress, and we presently have the first results. However, the main work of synthesis requires more time. A thorough analysis of the essential elements of rock art groups at a world level is necessary to determine universal factors and to identify and distinguish them from local factors. Five elements have been taken into consideration: (1) subject matter, (2) types of associations, compositions and scenes, (3) stylistic trends, (4) technical patterns, and (5) type of location on rock surface.

In Valcamonica, for example, it was found that general patterns do exist. The Proto-Camunian period (ca. 8,000-6,000 B.C.) is type A (Early Hunters), while all the successive periods belong to group D (Complex Economy) (ca. 6,000-16 B.C.). Groups B (Late Hunters) and C (Pastoralists) do not appear to be represented in Valcamonica (Anati 1976). For some groups of rock art in Spain, we may make quite clear attributions: Group A for Cantabrian cave art (ca. 26,000-8,000 B.C.); Group B for a large part of Levantine rock art (tentatively dates ca. 8,000-2,000 B.C.); Group C for some complexes of schematic art, and for others Group D (both present since ca. 3,500 B.C.). The rock art of

Mount Bego (France), until recently, had been considered by many to be the work of an agricultural population. In fact the Mount Bego rock art seems to be Group C, that is, the work of a principally animal rearing population (ca. 3,000-1,000 B.C.). It is only in a later phase that the marks and images, including figures of plows, would appear to reflect a partly agricultural mentality (after 1,500 B.C.). The rock art of the Negev and Sinai deserts shows successions of types A, B and C, while no rock art of type D has been detected apart from an isolated case, in a sequence of rock art which displays seven main stylistic horizons from 6,000 B.C. to the present (Anati 1979).

It is likely that more specific definitions have yet to be made within Groups A and B. In various parts of the world, there are extremely schematic groups of paintings and engravings which at first appearance seem abstract. Nevertheless, marks with numerical values, repetitive associations of dots and lines, and network motifs actually follow conceptual rules. In Europe, various groups of Mesolithic art, as the La Cocina type in Spain, Azilian art in France, Magelmoisian art in Germany and Scandinavia, and the Romanellian art in Italy are very close to styles encountered in Turkey and the Near East, North Africa, Malawi and Zambia in Southern Africa, along the Murray river in Southern Australia and in various Polynesian Islands. Several of these groups produced either mobiliary art, rock art, or both.

Many of these groups have two main recurrent features. A certain amount of specific associations of lines and dots which are repeated more or less identically in various parts of the world, and constant proximity to escargotières, or shell mounds. These are sites where human groups camped and where heaps of shells and other mollusc remains are to be found. They reveal a particular diet and economy based on the collection of small river or marsh animals. Thus, it appears that a specific graphic style reflects a way of life with more or less uniform consistency at different ages and in various parts of the world. We tend to find this type of art in the proximity of rivers, lakes, marshes, or the sea-shore. It would appear to reflect a gathering mentality rather than a hunting mentality of people who survived on gathering mollusks. Such activity, for reasons that are not entirely clear, seems to have aroused a particular interest in arithmetic and a tendency for schematism rather than the spirit of observing large animals in great detail as it appears in most type A Groups (those of the Archaic Hunting populations).

3. Major sites

At UNESCO's request, we have attempted to single out the rock art zones of "primary importance". It was necessary to establish objective criteria, and even if these criteria are modified for future evaluation, the attempts have been useful. Many of the initial problems are still unsolved, but working hypotheses support the continuation of research.

The list, which was published in *BCSP* 21 (1985), did not include all of the 120 countries where rock art is known to exist and recognized others as having more than one zone of primary importance. Each of these zones may include numerous sites, such as the Franco-Cantabrian zone or the areas rich in cave painting of Central Tanzania or Lesotho. Based on a series of criteria already pointed out in the UNESCO report (1982), we selected 144 sites which, from the available information, are to be considered of primary importance. Upon studying their distribution, we came to realize that they are more or less homogeneously diffused on all inhabited continents. Indeed, rock art appears to be a truly universal phenomenon. For almost all cases recorded so far, the following postulates are valid:

> *P3*. Rock art is a phenomenon of nonurban, nonliterate societies.

> *P4*. In all zones examined there are areas of great concentration of rock art which do not reflect an analogous concentration of people.

On the contrary, it seems to be a universal pattern that rock art is scarce where the population is more concentrated. There appear to be many zones all over the world where people engaged in artistic or other ritual and social activities of which little or no trace has yet been found besides rock art.

4. Pictograms, psychograms, and ideograms

The following postulates specify the types of signs which are characteristic of rock art:

> *P5*. Each of the four categories of rock art (representing respectively the production of Archaic Hunters, Evolved Hunters, Pastoralists and people with a Complex Economy) display a limited typological range of subjects which recur in the rock art all over the world.

> *P6*. Three types of signs are to be found, differing grammatically from each other, pictograms, ideograms, and psychograms. (These types of signs recur not only in rock art, but also in mobiliary art)

Pictograms (and mythograms) are figures in which we may recognize identifiable forms of real or imaginary objects, anthropomorphic, and zoomorphic figures. *Ideograms* are repetitive and synthetic signs that are sometimes interpreted as zoomorphic and anthropomorphic schemes, sticks, tree-shapes, phallic and vulvar signs, discs, groups of dots and lines.

> *P7*. Repetitiveness and constants of association seem to indicate the presence of conventional concepts in a number of ideograms.

Psychograms are signs which are not recognizable as, and do not seem to represent, either objects or symbols. They are strokes, violent outputs of energy that perhaps express sensations or even more subtle perceptions.

Pictograms are images whose shapes we may identify. They may have metaphorical meanings. Only very rarely do other themes occur, such as vegetation, landscapes or realistic portraits, and only in very specific contexts.

> *P8*. In the rock art of nonliterate people, pictograms contain four main types of themes: (1) anthropomorphic, (2) zoomorphic, (3) topographic and tectiforms, (4) implements and weapons.

Next to pictograms, we often find associated ideograms. The meaning of the figures themselves and of these associations has been studied and discussed by many scholars, and various theories have been suggested (Breuil 1952; Leroi-Gourhan 1982; Lommel 1970; Luquet 1926; Marshack 1972). Nowadays, no one is satisfied with the mere identification of the animal species represented or other technical details, since this is only a prelude to further study.

For Renaissance painters, the dove was undoubtedly a dove, but when Fra Angelico painted it in his Annunciation scene, in order to understand its meaning it is not enough to state that in a particular part of the painting there is a dove. When we know the Christian theme that inspired the artist and when we are familiar with his methodological and conceptual equipment, the dove takes on a metaphoric content and meaning as the Holy Ghost. Similarly, Picasso uses the dove not merely as a dove. He combines this pictogram with the ideogram of the olive branch that we recognize as such only because we are initiated. Twenty thousand years hence, perhaps someone will ask what could be meant by the ideogram near the beak of that strange pictogram that vaguely resembles a bird. It may not be easy then to read it as a metaphor of 'peace'.

A host of other examples could be cited such as the eagle in Roman times. "Yes, the bird represented does seem to be an eagle." Tests will prove this from the shape of the beak or from the wingspan. But what does this eagle mean? Sometimes, when the pictogram is accompanied by four ideograms that we are nowadays able to read as SPQR, some inkling of the conceptual content of the figure will perhaps emerge as a sort of "totemic animal" of the Senate and the People of Rome.

Thus, the splendid pictograms of the Hunters frequently accompanied by their ideograms, must have been clearly legible to anyone who was familiar with their conceptual content. Today, there has been a breach in the direct tradition. It is the work of the archaeologist to assemble the components and observations that may enable us to approach an understanding of the content.

Ideograms, in ancient writings and in many groups of rock art, are signs that convey ideas from the writer to the reader, from the painter to the real or imaginary beings to whom the message is addressed. We have recognized three main types of ideograms which tentatively have been named: (1) anatomic (such as vulvar signs or hand prints), (2) conceptual (such as the cross or the disk), (3) numeric (such as groups of dots or of lines). Most of the ideograms used for millennia by prehistoric groups are later found as characters in the first writings in various parts of the world, in China, the Near East, and in Central America.

> *P9*. In the cases in which the meanings of the ideogram can be traced back to a single common core, a universal pattern is established.

The rectangle or the square means land, site, or territory in most early scripts and in most areas of rock art. Horizontal wavy lines mean water or liquid. The rayed disk means sun, light, or day. The nonrayed disk means sky or air. The triangle with a point turned downward means the pubic triangle, sex, fertility, or birth in some early ideographic scripts from different continents. The problem is whether they can be traced back to a common archetype.

Psychograms are signs that convey sensations from the one who represents to the one who observes. For us, this level is more abstract than the symbolic level which still has its own well-defined meaning. The psychogram works at a subconscious level, as do certain archetypal signs that our conscious memory is no longer able to define. However, deep within the self, these signs release associative and sensory processes on wavelengths that escape the bands of ordinary transmission and are remarkable for their immediacy. Psychograms do not seem to be intellectualizations, but signs that have the power to stir the blood and the mind without conjuring up any specific associations. They are the quintessence of something that is hard to define consciously, but which resonates deep within us.

In the rock art of all continents, landscapes are extremely rare, as are images of plants. They are either completely absent or indicative of some particular characteristic when they do exist. In Tanzania, there is a rock art sequence spanning over 40,000 years which is considered to be the longest known. Yet, all the images of the vegetal world are concentrated in one phase, which is tentatively dated to the end of the Pleistocene and the beginning of the Holocene (ca. 12,000-8,000 B.C.). From the data at our disposal, it appears to be a phase of gathering populations rather than hunting, even though it belongs to a Hunting and Gathering period (Anati 1986). Although these exceptions are rare, they are not to be ignored.

5. Paradigms

In each of the aforementioned categories, elements exist which we call "paradigms" and which we see repeated all over the world. There are preferences for the type of place: cave, shelter, or open air rock surface, and there are preferences for the type of surface, be it horizontal, oblique, vertical, floor, or ceiling. For all four categories, there seem to be precise choices made in the surface on which the painting or engraving is done: the chosen surfaces have preferential characteristics of color and form that seem to mean that the "support" (or background) has been taken into account by the creator of rock art. From the preliminary analysis of a random sample of 1,000 surfaces, it appears that a high percentage (over 85% of paintings and engravings) corresponds to recurring criteria. However, exceptions exist and these require explanation.

In many cases, precise technical methods appear to have been used, whether for painting, engraving, peckings, or graffiti, or for the various types of relief work. In some cases, it seems to be simply the result of a given technological level way of thinking, or both. This leads us to further postulates of rock art research:

P10. Certain rock art techniques are broadly repeated throughout the world which do not seem to reflect processes of acculturation or diffusion.

P11. A few basic colors are used in rock paintings all over the world, red being by far the most common in all continents.

P12: Even in the absence of direct communication between different populations, similar outcomes have been found in places far apart.

Thus, it cannot be a question of external influence in all cases. We may therefore hypothesize that certain convergences exist. These paradigms bring us into contact with our own submerged memory and may open the way to a fresh understanding of the human mind. In Table 1 (see overleaf), the main universal factors are summarized.

In Europe, Asia, Africa, the Americas, and Australia, the same range of figures may be found. Certain aspects of the environment, economy, and social life simply do not enter into the artist's thematic range. The artists, hunters or otherwise, made precise choices in terms of the themes they chose to represent. They did not paint or engrave just anything.

Sometimes we may come upon themes of preference. The large representations of animals that may be life-size or more, are almost exclusive to the hunting populations. With Pastoralists, they are only to be found in certain zones that today are deserts, such as Arabia and the Sahara. They are extremely rare in the rock art of Complex Economy populations. On the other hand, there are other phases in both Archaic and Evolved Hunters' art in which medium sized or even miniature animal representations are found. Another interesting phenomenon has been noted: when large animals are prevalent, human figures are rare. When the animals are smaller, the percentage of anthropomorphic figures is usually higher. In all complexes, primary and secondary subjects are to be found. There is an initial choice and then there are minor complementary themes.

P13: There are repetitive elements, ideograms and pictograms, which appear to accompany the dominant figures.

It is frequently stated that the animal is the most commonly represented figure in European Palaeolithic art and that throughout the world, animals are the main feature of Archaic Hunters' art. However, contrary to what is normally believed, isolated animal figures are practically

Table 1. Dominant characteristics of rock art

Category	Thematic	Associative	Stylistic	Technical	Location
ARCHAIC HUNTERS	Ideograms psychograms, pictograms. Large wild animal figures. Anthropomorphic and zoomorphic images. Human figures are often masked.	Simple associations. Complex compositions exist but are rare. Hermetic and synthetic syntax. Scenes are rare.	Generalized naturalistic engravings. figures and ideographic sequences.	Prevalence of paintings over vertical Monochromatic ensembles. Polychromy is restricted to specific regional phenomena.	Normally on surfaces, and on roof or plafonds, with exceptions.
EVOLVED HUNTERS	Pictograms, animal and human figures. Few ideograms. Psychograms are rare. Constant presence of imaginary beings.	Descriptive and anecdotal scenes are widespread. Simple syntax defined by the associations in the scenes and compositions.	Naturalistic and realistic ensembles.	Prevalently painting, also a lot of engravings. Most paintings in Eurasia are monochromatic. Polychromy widespread in Africa and South America.	Normally on vertical surfaces.
PASTORALISTS BREEDERS	Pictograms, domesticated animals and anthropomorphic figures. Images of huts and other structures. Sporadic ideograms. Absence of psychograms.	Scenes and compositions. Prevalence of descriptive anecdotal scenes.	Realistic groups with frequent tendency to generalization. Idealization of forms.	May be painting or engraving. Varies from zone to zone.	On vertical or oblique surfaces.
COMPLEX ECONOMY	Human, animal, structures, implements and weapons; schemes and ideograms. Pictograms more varied than in previous categories. Psychograms only in some cultures.	Compositions and scenes. Complex syntax with vernacular characteristics. Mythological scenes are widespread.	Schematic and abstract graphemes. Special interest in details. Vast use of metaphoric depictions.	Engravings prevail but painting is preferred in some zones.	On horizontal, oblique or vertical surfaces. Location more varied than in the other categories.

nonexistent, animals are almost always accompanied by ideograms. Frequently, animals occupy the largest space on the painted surface, simply due to their size. But symbols (ideograms) vary in number, and they frequently exceed the number of animal figures with which they are associated.

As was already brought to evidence by Laming (1962) and Leroi-Gourhan (1965), there are frequent cases of animal figures of different species being associated with each other in the Upper Palaeolithic cave art of Western Europe. There are two animal species, the bison and the horse, which are more frequently represented one in front of the other as the main theme; that is to say that they are preferred to other wild animals, repeated more frequently, and would seem to imply something quite specific in the dialectic of associations.

In Tanzania, the Archaic Hunters' art contains representations of the elephant and giraffe which play a similar role, in terms of associations and quantitative proportion to other animals, such as the horse and bison in Europe. They are frequently associated with each other and are by far the most commonly represented animals (Anati 1986). Quite probably, they played the same role within the frame of mind of the Tanzanian Archaic Hunters as the horse and bison did in the Franco-Cantabrian mentality in Europe. These are the premises for yet another paradigm:

> *P14*. The art of hunters is characterized by dominant animal species in a dialectic relationship.

As for the meaning of such associations, Laming and Leroi-Gourhan have proposed a possible connection to the concept of male-female interplay. Other theories have been proposed, but no final conclusions have been reached. It is not unlikely that they reflect metaphors of binary relations. Whatever the case may be, there can be little doubt that the European association of horse and bison shows parallels in the African association of elephant and giraffe.

Such paradigms seem to reveal the existence of "logic constants", the study of which could prove productive. If we were to carry out a detailed analysis on a world scale, a great deal of progress could be made on the study of paradigms. It is particularly important to develop a systematic analysis that defines the associations, compositions, and scenes; that is to say, the type of relationship between one grapheme and another in any given context. This would then provide some indications of the way of thinking and the conceptual dynamics reflected by the associations.

6. Archetypes

In the art of the Evolved Hunters, there are common scenes such as hunting scenes, scenes of dancing, other social events, and warfare. As for the Archaic Hunters, there do not appear to be any real descriptive scenes either in Europe or elsewhere, or if they do exist, we have not yet recognized them as such. But associations in sequence do exist, and they may represent some sort of "metaphorical scenes". Their conceptual structure is in any case the mirror of a mental process which has been submerged in urban societies. This holds true for all continents and is an intriguing psychological phenomenon.

There are certain types of association that resemble one another in different regions. The association between pictograms and ideograms is present in the art of hunters all over the world. An animal that has been added in a certain way in the context of other associated figures does not necessarily reflect a naturalistic reality as our present-day thinking would like to suggest. The animals are to be found repeatedly in similar relationships in accordance with rules that made sense for "them" but which do not correspond to the type of composition and vision of our contemporary culture. For example, there is apparently no concept of a "base" or "ground level" in most of the art of the Archaic Hunters. The large animals are placed on cave walls or on rock surfaces as if they were levitating or suspended in mid-air. It has often been asked if they are not the spirits of the animals rather than the animals themselves. We find this in Europe as well as in Tanzania, Australia, and elsewhere.

The association between animal and symbol recurs with analogies within all Archaic Hunting populations: the animals as the subject and the ideograms (the symbols) as a type of logic and reasoning around them. With the Evolved Hunters, we find scenes which show a completely different mentality. According to our way of thinking, the associations of the Evolved Hunters are more narrative, naturalistic, and less abstract. But is it correct to consider the mentality of archaic hunters as more "abstract" than that of more evolved cultures? Sometimes, the transition between the two is unpredictable. The tradition may be upset at a certain point and something new begins. In other contexts, however, the evolution is gradual and we may therefore detect the transitional phases. Within this process, we may catch a glimpse of the changes in the cognitive mechanism. The following postulate may well apply to all kinds of art, including dance and music, to all places, and to all times (Anati 1983):

P15. Changes in style and associations in the visual arts illustrate changes in the cognitive processes which, in turn, reflect modifications in the choice of priorities.

From an analysis of themes, typology of the figures, marks, and graphemes, patterns emerge which constitute the "grammar" of rock art, as if the images were words in sentences or in composite concepts. Isolated marks are extremely rare, as isolated words are rare in a conversation. In rock art, there are groups which reflect systems of associations that we might refer to as the syntax. They are sentences composed of groupings or sequences of graphemes just like in the spoken and written languages of today. Herein may lie the key to reading the rock art, the key to an ideology with universal characteristics, or at least a sequence of constant associations which go beyond, indeed precede, the limits of linguistics. Thus,

P16. The existence of logical archetypes may be deduced.

P17. Through rock art, we may come to recognize some of the fundamental elements of man's cognitive dynamics.

If, on the other hand, we continue to analyze isolated figures without looking at the context and associations, we will end up with a catalogue in which each figure is separated from the rest. It would be like reading each word in a sentence separately, without seeing the sentence as having grammatical sense and logical syntax. It is not always an easy task because there are superimpositions, composite associations, and cognitive accumulations which come about in the course of time.

Sometimes, the process is extremely complex. Looking at a panel, we may find that there were only three figures to begin with, two animals, one in front of the other, and an accompanying ideogram. After some generations, another figure was added, two thousand years later on, another four marks may have been added. The primordial association should be singled out, the repetitive elements verified, analogies and comparisons should be studied in order to evaluate whether it consists of fortuitous or intentional superimposition.

In many instances, the accumulation of marks appear to be intentional even if they are from different periods. Others appear not to be intentional or are at least unrecognizable as such. But in both cases, we frequently discover that language is expressed by these signs. Sometimes, previous graphemes may have been considered as part of the support, like the natural shapes of the rock. These may have been used or not in the association of graphemes of the superimposed depictions.

Over and above the specific characteristics found in the Spanish Levant, Valcamonica, Tanzania, Madhya Pradesh, or in the decorated Palaeolithic caves of France and Spain, a type of language is found which touches upon prototypes of a universal nature. Provincial characteristics also exist which we refer to as "vernacular". It is, of course, easy to attribute a dialectical nature to all that we do not understand as language. However, we are increasingly aware of the fact that universal paradigms and archetypes occupy an extremely large and important space in all art of all times.

7. Towards a world vision

Sequences with certain constant features have emerged in the Sahara, Tanzania, the Nile Valley, the Middle East, Madhya Pradesh in India, and Russian Central Asia. The four categories already mentioned, the Archaic Hunters, Evolved Hunters, Animal Breeders, and Complex Economy groups, are usually present in the same chronological order. In Australia, however, no Pastoralist-Animal Breeders' rock art has yet been found. The Australian sequence is mainly composed of the Archaic Hunters and in a few areas in the north, of the Evolved Hunters. The Complex Economy category is represented in only a few very recent cases.

In South Africa, the main rock art zones comprise a sequence with little evidence of Archaic Hunters. These appear further north, in Namibia, Zimbabwe, and Tanzania. The greatest artistic activity of this zone, probably the richest in rock art thus far known in the world, emerges as the work of the Evolved Hunters from about 13,000 years ago. They maintained the same visual language, with only minor modifications, up until the present century. Pastoralists-Animal Breeders and Complex Economy groups are also present. Rock art shows the existence of these Pastoralist and early farming populations, while in most of the territory, the Evolved Hunters' way of life persisted (Lewis-Williams 1983).

In Tanzania, during the Archaic Hunters' period, we come upon a short parenthesis of a vegetarian culture whose economy was based on the gathering of fruits and berries. Analogous phenomena occur elsewhere. In the Central Sahara, the same type of rock art seems to reflect a population whose lifestyle depends largely on gathering and who quite probably made extensive usage of drugs. A similar phenomenon is recognized along the Pecos river and in the Seminole Canyon in Texas.

Chronologically, in the areas mentioned, this phase is located between the Archaic and the Evolved Hunters.

In Near Eastern deserts, rock art shows sporadic examples related to a Complex Economy, that is to say of Food Producers and Agriculturalists. In the Negev desert there is even the engraving of a plow. This was a brief episode during the Roman period. The stratigraphy of overlappings and superpositions shows that the population both before and after these artists were Pastoralists. In itself, the existence of a few drawings which indicate a Complex Economy does not always mean the beginning of a new era.

Rock art indeed provides segments of history. However, we have to admit that history is not always as simple and linear as we would like it to be. Food, sex, and defence for survival appear to have been, in all contexts of all times, the main concerns of the human species and the ones which are most frequently expressed by rock art (Anati 1992).

A series of elements in the comparative study of rock art has been set out. It is now necessary to organize our sporadic pieces of knowledge, checking for validity, recognizing which elements are of a universal nature, and which are regional. Local features become more common when a Complex Economy is indicated. Universal paradigms prevail in the art of the hunting populations with variants whose meaning must be considered. Clearly, it would be extremely unlikely to find representations of llamas in Australia or of kangaroos in Argentina, but within the general framework, we come upon animal subjects which are invested with the same kind of values and similar metaphors for different species in different parts of the world.

The clearest evidence for the presence of universal paradigms lies in the ideograms which are found as constants all over the world. Vulvar, phallic, cruciform, stick-like, egg-like ideograms, cup marks, cup and ring marks, hand prints, foot prints, and animal tracks, are to be found in the Palaeolithic art of Europe as well as in the context of Archaic Hunters' art on all continents. The art of the Archaic Hunters appears to have broad universal characteristics in terms of subject matter, types of association, and stylistic trends. The art of the Evolved Hunters has many more local features. The real tower of Babel comes into being when the Hunting and Gathering era approaches its end.

Art, like most other aspects of culture, then becomes increasingly provincial and more conditioned by surrounding influences. However, on the whole, many common denominators persist. The most obvious feature is the fact that rock art is produced at all and that, as we have seen, there are analogous typologies, choices of subjects, types of associations, constant uses of certain colors, and repeated preferences for certain kinds of rock surfaces. Added to these basic elements are details, embellishments, and the different stylistic approaches used by the various categories of people. The most essential differences seem to reside in the seasoning rather than in the substance.

I have attempted to present the general framework of the rock art research that is in progress. It has been opportune to elaborate a methodology requiring information which would have been unthinkable only a few years ago. Rock art has been considered as a phenomenon which should be treated globally and not simply as a series of local phenomena with no connection to each other. Future research will enable us to verify this approach.

References

Anati, Emmanuel

1976 *Evolution and style in Camunian rock art.* Capo di Ponte: Edizioni del Centro.

1979 *L'art rupestre du Neguev et du Sinai.* Paris: L'Equerre.

1981 "The origins of art", *Museum* 33/4: 200-210.

1983 *Gli elementi fondamentali della cultura.* Milano: Jaca.

1984 "The state of research in rock art: A world report presented to UNESCO", *Bollettino del Centro Camuno di studi preistorici* 21: 13-56.

1986 "The rock art of Tanzania and the East African sequence", *Bollettino del Centro Camuno di studi preistorici* 23: 15-68.

1989 *Origini dell'arte e della concettualità.* Milano: Jaca.

1992 *Radici della cultura.* Milano: Jaca.

Beltrán, Antonio

1988 "L'art préhistorique espagnol: Nouveaux horizons et problèmes", *Bollettino del Centro Camuno di studi preistorici* 24: 13-44.

Breuil, Henri

1952 *Quatre cents siècles d'art parietal.* Montignac: Centre d'études et de documentation.

Chen, Zao-Fu

1988 *Arte rupestre preistorica in Cina.* Milano: Jaca.

Grant, Campbell

1983 *The rock art of the North American Indians.* Cambridge: University Press.

Laming-Emperaire, Annette
1962 *La signification de l'art rupestre paléolithique*. Paris: Picard.
Leroi-Gourhan, André
1965 *Préhistoire de l'art occidental*. Paris: Mazenod.
1982 *The dawn of European art*. Cambridge: University Press.
Lewis-Williams, J. David
1983 *The rock art of Southern Africa*. Cambridge: University Press.
Lommel, Andreas
1970 "Shamanism: The beginning of art", *C'A* 11/1: 39-48.
Luquet, Georges Henri
1926 *L'art et la religion des hommes fossiles*. Paris: Masson.
Marshack, Alexander
1972 *The roots of civilization*. New York: McGraw-Hill.
Schobinger, J. – C.J. Gradin
1985 *L'arte rupestre delle Ande e della Patagonia*. Milano: Jaca.
Smits, L.A.
1988 "Recording and deciphering of rock paintings in Lesotho", *Bollettino del Centro Camuno di studi preistorici* 24: 93-99.

Deixis vs. modeling in the phylogeny of artistic behavior

Paul Bouissac

1. Deixis vs. modeling

Semiotic and linguistic traditions have consistently distinguished two categories of verbal as well as nonverbal expressions: indexical ones and referential ones. The former are usually designated by the generic term *deixis* (a transliteration of the Greek δεῖξις, which is the nominal form corresponding to the verb δεικνύναι, 'to show, to point'); the latter can be considered an instance of the general notion of "representation" or "modeling". This distinction arises from the discovery that some linguistic expressions have their meaning strictly determined by the situation in which they are used and that, taken out of any situation, these linguistic forms are so to speak empty. The determination of *I*, *you*, *here*, *now* in the normal use of language varies with each situation, a characteristic that Roman Jakobson (1963: 176-181) captured by categorizing these linguistic forms by the metaphorical term "shifter", thus indicating that their content shifts as situations change. By contrast, referentiality applies to those linguistic expressions whose meaning is determined relatively independently from the situations in which they are produced. Within a linguistic community there is a consensus that, assuming that the speaker is correctly informed and does not lie, a sentence such as *Bochum is West of Berlin* conveys information concerning an actual state of the world, regardless of the situation in which the statement is produced. It is this capacity to convey information about something which is not present, by means of a conventional arrangement of signs, linguistic ones or other ones, which is designated by the term "modeling". The emitter of a referential statement represents a state of the world independently from his/her current situation as well as a possible state of the world or a state of a possible world. Naturally this clear-cut conceptual distinction between indexical and referential expressions can be shown to be actually blurred in concrete instances of linguistic communication as both kinds of expression are blended in pragmatic discourse and as there are several levels of integrated conventions in a given linguistic community. It is well known that even the most explicit and objective statements require some cognitive framework and situational assumptions in order to be fully understood. This distinction does nevertheless stand on theoretical and empirical grounds. There is indeed a great deal of phylogenetic, ontogenetic and neuropsychological evidence that *deixis* and *referentiality* refer to markedly different types of behavior (e.g., Nespoulous 1981, Petitto 1987).

As has been hinted above, this distinction does not apply only to the verbal domain but to the nonverbal as well, and, set in the conceptual framework of evolution it might offer some cues regarding the phylogeny of artistic behavior.

The origin of art has been much debated over the last hundred years or so, principally since prehistoric art was discovered and dated. Ethnographic and ethological data have further fueled a controversy which pitches the philosophers of aesthetics against those who attempt to assess the significance of artistic behavior in socio-biological terms. Sebeok's "Prefigurements of Art" (1981: 210-259) or Alland's "Affect and Aesthetics in Human Evolution" (1989), a sequel to his book, *The Artistic Animal* (1977), illustrate this dialogue, as do Eibl-Eibesfeldt's (1975), Dawkins's (1986), Lumsden's and Wilson's (1981) and Koch's (1984) contributions. The fundamental issue seems to focus on whether there is continuity between some types of animal behavior and human artistic behavior and, in the negative, what is the exact nature of the gap.

2. The biological status of animal "artistry"

Two well documented examples are highlighted in the literature: the decorative display of colorful objects such as flowers and shells by the bowerbirds (Plate 1) and the "paintings" produced by chimpanzees which were provided with canvasses, brushes, and paint of various hues (Eibl-Eibesfeldt 1975: 143-145 and 279-280; Sebeok 1981: 216-217).

The color photographs of both the bowerbirds' and the chimpanzees' works which have been published in the above books are indeed striking and can only evoke, in the mind of the viewer, human artistic productions. On several occasions, the latter even scored very well in the art market when exhibited in art galleries - with, as well as without, indication of their origin. There is little doubt that the former would meet with a similar success if their fragility and ephemeral character could be overcome. But mere formal resemblances constitute a very weak argument, even if one is told that the bird examines the results of its actions from a distance before adding some colorful material at a

Plate 1. Decorative display (flowers, shells) by male bowerbird. (Photo credit George Chaloupka.)

particular spot, suggesting a critical evaluation of its "artistic" work, and that some chimpanzees appear to have a sense of the moment at which their works are completed. The argument suggests variously that animals have aesthetic feelings and are able to make artistic judgments or that humans' artistic activities are a part of their animal heritage rather than a distinctive feature of the species.

However, if these two classical examples are now considered in their ethological contexts, their semiotic status appears to be somewhat different. In the bowerbirds, for instance the satin bird (*Ptilonorhynchus violaceus*) which is found in Eastern Australia, "the male builds beautiful and complicated bowers solely for mating. More than one female may come to the bower and then wander off to her nest, and from then on all the parental responsibilities are hers" (Bonner 1980: 158). Obviously the accumulation of bright and colorful objects is one of the means through which males compete with one another in order to gain reproductive access to the greater number of females. It is an interesting strategic alternative to the development of imposing fans or appendages which can become self-defeating if they are pushed to some cumbersome fitness-reducing limits as it has been shown for some species such as the monogamous swallow (*Hirundo rustica*) in which "females prefer males with elongated tails over those with shortened tails, but [. . .] males with experimentally elongated tail ornaments [capture] smaller, less profitable preys than those with shortened tails" (Møller 1989: 132). Having conducted extensive experiments, Møller concluded that "impaired foraging efficiency of tail-elongated males increased the frequency of fault bars in their tail feathers, probably as a result of food deficiency during moult" (1989: 132). It can be assumed that the relatively short, inconspicuous tail of the bowerbirds optimizes its viability. The evolution of this courtship behavior enables the bigger and stronger males to build more imposing bowers and to accumulate "decorative" elements which make their mating territories more conspicuous and attractive to females. The exact nature of the visual impact of the bowers upon the female may be difficult to determine, but it seems obvious that it is of the same sort as the impact produced by fans and appendages in some other species like the peacock and the widowbird. The more efficient breeders among the male bowerbirds can devote more time and energy to their decorating, i.e., courtship advertising, hence reproductive activities.

The case of the "artful" chimpanzees is more complex, yet it ultimately belongs to the same paradigm. The literature underlines the striking similarities between their pictorial works and modern art, particularly what is diversely labeled "action painting", "tachysm", "abstract expressionism", and the like. But an examination of the context in which these individual apes engage in such activities is illuminating. First there seems to be a relation between the formal qualities of the "works of art" produced and the social status of the animals. Commenting the results of an experiment conducted at the Munich Zoo, Eibl-Eibesfeldt notes that the characteristics of the "pictures" can be correlated to the social rank of the animals within their group. "The higher-ranking male covered thc entire canvass and began turning the brush after each stroke, including curved lines in the picture. The higher ranking female also filled all available space with bold strokes from the center to the periphery . . . The lower-ranking female . . . limited herself to painting a small spot at the lower edge of the canvass. She pressed down so hard that the paper became roughened" (Eibl-Eibesfeldt 1975: 279-280). It would therefore appear that such "artistic" actions belong to the general category of territorial marking, a behavior which, in many species, consists of smearing surfaces with special body excretions or with urine and faeces. But even if this natural propensity is a factor in the above mentioned experiments, it should not be forgotten that the chimpanzees' actions are not entirely spontaneous. The keepers and experimenters prompt the animals and create a motivational framework and, in a manner which has been amply documented in the case of the language-learning apes (Sebeok – Umiker-Sebeok 1980), often unwittingly cue them. Those who have commented on these paintings in particular marvel at the presence of curves and fan-like patterns. But this is exactly what can be expected given the mechanics of muscles, bones and joints if the animal is sitting or squatting while "painting". What would be truly amazing would be straight horizontal or oblique lines because it would demonstrate an adjustment of the whole body dynamic to an intended result. Training, of course, could have an impact on the finished pictorial product, but this is an entirely different matter which has little to do with the phylogeny of artistic behavior.

3. From landmarks to maps: The imagination of space

In the case of the bowerbirds as well as in the case of the chimpanzees, we are witnessing forms of communicative behavior whose visual components (artefacts) have no meaning outside the situations in which they are implemented and displayed, from the point of view of the individuals involved. They can be considered parts of intra- or inter-specific symbolic interactions endowed with adaptive value. The displayed "statement" of the male bowerbird makes sense only if the animal is inside or near the decorated bower; the smearing of surfaces by primates and other mammals has meaning only as long as the territory is inhabited and defended. This is not to claim that these complex organisms do not possess a cognitive representation of space, but it appears that, for all practical purposes (i.e., their survival), this representation strictly pertains to the limits and topographical structure of their own territory and possibly to the territories of their preys which are encompassed by their own. Behavioral and neurophysiological evidence "strongly suggests" that humans and other species, such as rodents, share the capacity "to keep track of their directional heading relative to an inertial and hence allocentric coordinate system" (McNaughton et al. 1991: 190). Natural and artificially enhanced landmarks map space to a certain degree, but these visual or olfactory artefacts ultimately refer to nothing else but the individual organisms which produce them. They closely fit the definition of deixis which has been discussed earlier. Moreover, there does not seem to exist a significant gap between the types of behavior described above and humans' use of beacons, landmarks, and sign use for dead reckoning, as well as warning and decorative signs destined to achieve survival purposes, such as seducing a prospective mate or warding off competitors and, in the context of more complex societies, asserting status, delimiting boundaries, attracting customers and the like. Whatever can be said, metaphorically, to "make a statement" through its visual impact, does so in relation to particular individuals and is an instance of deixis. Some early evidence of human art likely belongs to this paradigm, but it would probably be a mistake to consider it primitive, because modern humans have obviously conserved this competency which developed earlier in evolution and has not ceased to be adaptive. This means that more ancient types of behavior, and the neurophysiological architectures which make them possible, can coexist with more recently evolved ones, as it is the case in biological evolution. For instance, the vestibulo-oculomotor system and circuitry in vertebrates was conserved after the development of an optimal solution. It is considered a largely conserved phylogenetic plan in vertebrates (Baker et al. 1987). As Graf (1988: 166) comments:

> An . . . important characteristic of central nervous system operations is seen in the peripheral mechanisms utilized to simplify central operations. This feature has been demonstrated by the uniformity of the vestibulo-oculomotor system in regard to its peripheral spatial organization, including its central nervous connectivity, and by the general and specific constancy of central visual world representations. Thus, the

> workload for the central nervous system is decreased in favor of animal economy or presumably higher-order operations (e.g., learning). When considering how the brain works, we should look into similarities among apparent differences of expressions or behaviors. Disregarding obvious similarities of sensory-motor operations across species would mean disregarding one significant biological aspect of brain function.

The hypothesis propounded in this paper is that two human competencies must be distinguished with respect to "artistic" productions: one which evolved earlier in vertebrates and was conserved in *Homo sapiens sapiens*; the other which evolved more recently. The former enables an organism to structure its environment with respect to a self-centered perspective and to manipulate signs in relation to this context along the flux of changing interactive situations. The latter makes it possible for the brain to engage in modeling, i.e., constructing and objectifying in a more or less abstract manner spatial representations irrespectively of the position of the self or in which the self does not occupy a privileged position. The main evolutionary advantage of this capacity - which is correlated to a weight increase of the brain and involves more complex neurological computations - is probably the flexibility it affords in planning strategies and in knowledge transmission. Reference to space in this mode is strikingly different from the process involved in deictic behavior. It does not concern space here and now, but a representation of space from which the structuring organism (*Homo sapiens sapiens*) is twice remote, both as ego and as other, and within which various kinds of relations can be expressed. From the point of view of evolution, it seems irrelevant to distinguish between a map, a schema, a blueprint, the topological symbolism of an architectural layout, a classical or a geometrical abstract painting. These examples are all instances of spatial representation which are variously categorized and valued within the framework of socio-historical changes and the institutions they generate. This, of course, does not exclude the possibility of interferences with the deictic function. Individuals or cities, for instance, can compete with one another through ostentatious architecture.

It seems that while some pictorial remnants of prehistoric humans are patent examples of deictic behavior, a great number of them are referential in the sense outlined above. Raphael (1945), Anati (1960), Marshack (1972), Leroi-Gourhan (1958), and Gamble (1980), among others, have convincingly argued that many of the pictorial displays which have survived spatially represent myth structures, social organizations, or other coded knowledge. Such instances of modeling presuppose cognitive capacities that cannot be inferred from the observation of the deictic behavior of the most advanced animals. Data from clinical neuropsychology indicate indeed that the neocortex is critically involved in spatial analysis and planning (e.g., Gazzaniga 1979, De Renzi 1982). As Dimond (1980: 250) puts it:

> The structuring of the brain for interrelationship with external space can be seen as one of the significant steps forward in the development of intelligence . . . Primitive space-mapping is fundamental to the intellect, because any orderly scheme of things demands arrangement and order within some kind of framework . . . Spatial mapping is the raw fabric of the intellect, the means by which the intellect takes off and the starting point of intellectual exercise, whatever other considerations apply. In other words, abilities other than those directly relevant to spatial functions themselves still employ mapping facilities provided by the brain, whether they be flower arranging, following a complex argument or playing a musical instrument. Many aspects of science, for example, depend on the capacity to construct spatial vision.

Dimond concurs with Lorenz's argument that "when organisms were able to register and to map space, some new development had taken place which resulted in the generation of a new potential for creative function" (1980: 250).

An understanding of human artistic behavior in an evolutionary framework requires, it seems, that deictic and referential competencies be carefully distinguished. Evolutionism is both a theory of continuity and a theory of change. Many of those who have tackled the issue have failed to make this distinction and have arrived at conclusions that seem difficult to hold. Sebeok (1981) relates art to categorization and parallelism in an essay cautiously titled "Prefigurements of Art" in which he extensively draws from zoological data. Koch (1984) attempts to explain human artistic behavior by two well documented tenets of animal ethology: territoriality ("The fundamentals of art originated as early as the territorial behavior of animals", Koch 1984: 283) and displacement: "Thus, courting rituals and *delays* as well as displacement activities and *diversion* seem to be the molds of subsequent aesthetic development. While *delay* (tension plus détente) points to subservient forms of art, *diversion* (diversion plus attraction) is mainly characteristic of the evolution of *art for art's sake* . . . Yet, for the time being, we should be content with similarities" (Koch 1984: 293, 298).

Alland (1989: 6) endeavors to identify aesthetic universals and holds that "art is the accidental result of a series of evolutionary unrelated adaptations each of which has adaptive significance in domains other than aesthetics". In his earlier work on this issue, Alland (1977) had pinpointed five such human adaptations which, in his opinion, contributed to the emergence of aesthetic behavior, namely: (1) environmental exploration, (2) play, (3) attention to spatial configuration, (4) ability for fine-grained perceptual discrimination and (5) metaphorical language. But given the fact that all, except the last one, are well represented in many other species and, in most cases, more brilliantly than in *Homo sapiens sapiens*, only the capacity for metaphorical language would be criterial, something which would amount to restating the problem rather than solving it, because if indeed as some persuasively claim (e.g., Sebeok 1985) language is based upon modeling abilities, both representational art and language have the same cognitive prerequisites, and the presence of either cannot explain the emergence of the other.

4. Conclusion: Steps toward an evolutionary theory of art

It seems that an evolutionary theory of art could be formulated under the following conditions:

(1) The issue should be disentangled from the philosophical aesthetic discourse; typical behavior and culturally conditioned interpretations should be sorted out in order to reduce the confusion arising from ill-defined assumptions; the notion of "aesthetic feeling", for instance, seems to cover a great variety of behavioral phenomena which are lumped together in cultural and ideological constructs; this notion indiscriminately encompasses spontaneous preferences for particular morphological configurations (gestalten), specific interest in selective aspects of the environment or their simulations, reactions to information variations, appetitive attitudes towards, and bonding tendencies with, conspecifics, fetishistic valuations of objects, and all the institutional elaborations which have taken place over space and time determining a rich historical and cultural diversity. All these aspects have been abundantly investigated and documented (e.g., Frijda 1986: 355-361). Secondary ideological productions, such as the concepts of "aesthetic value", "genius", "art for art's sake", "beauty" or "the beautiful", which have emerged in the context of technologically advanced complex societies, are elements of the problem rather than fundamental givens than can be used as cornerstones of an explanatory theory.

(2) It should not be assumed that the global expression "artistic behavior" refers to a homogeneous and unique type of behavior, nor that

the boundaries between art and non-art are reliable indicators of aesthetic universals. Ethnocentric definitions of art are nothing but types of artistic behavior among others which have to be explained, notably the fact that the dividing line is subject to so great a variation between the poles of technology and religion. An inquiring into artistic behavior cannot be satisfactorily conducted if it merely skims ethological and ethnological data for anecdotal *ad hoc* examples. Too much seems to be taken for granted in the current literature dealing with evolutionist explanations of cultural productions. Moreover, on the psychosocial level, insufficient attention is paid to the relative part of normative "artistic behavior" and all the social constraints at play in both passive and active definitions of art in the context of schools, museums, secular and religious cults, and more generally social stratifications.

(3) Once the broad category of "artistic behavior" has been replaced by well-defined types of behavior questions concerning their phylogenetic evolution, their ontogenetic development and their evolutionary significance can be formulated with a certain measure of clarity. The hypothesis which has been sketched in this paper - namely that the deictic and modeling competencies are two phylogenetically distinct evolutionary phenomena - could help clarify the issue at stake, mainly if it could be demonstrated that they involve distinct brain architectures and processes as clinical neuropsychological data seem to indicate. However, it is important to keep in mind, as it has been pointed out above, that neurobehavioral structures which have evolved at different points of evolutionary time can perfectly well coexist as long as their coexistence does not impair the fitness of the organisms which carry them, but rather enhances it.

(4) Finally, an evolutionary theory of artistic forms - and the correlate behavior they imply on the individual and societal levels - must address the issue of the competitive mode of their creation, replication and diffusion. Critical aesthetic discourse and institutional tools of selection, conservation and reproduction, as well as mercantile strategies, record and amplify the success of certain forms over others rather than determine their rate of survival in human societies. Only recently the *sui generis* dynamism of "works of art" and the issue of their seemingly irrational attractiveness, akin in many respect to semiotic parasitism (Bouissac 1993) - but has not the artist's status always been tainted by this association? - have been confronted from a theoretical point of view inspired by Darwinism rather than by philosophical idealism. The concept of "meme" (Dawkins 1976: 206) casts indeed the whole issue of cultural evolution - including its "artistic" components - in a novel

theoretical framework, as it was early recognized by Koch (1986) and Dennett (1990), for instance. The general notion that some informational structures can replicate and evolve in the brain environment like genes do in the biosphere may be more than an exciting metaphor. If this were actually the case, the capacity to represent an allocentric space, while affording an invaluable strategic advantage to those organisms which had evolved the brain structures making it possible, could also be considered a liability since it would provide *ipso facto* an opportunistic milieu for the evolution of parasitic structures, including fictional and abstract constructs, but also, ironically, deictic marks, thus distracting or reorienting the biological resources available toward the unforgiving, but pleasurable, spreading of these "memes". Perhaps this is the vision that René Char (1974) intuited when he wrote *Le monde de l'art n'est pas le monde du pardon* ('The world of art is not the world of forgiveness.') The literal translation may sound as cryptic as the original is until we relate it to a popular saying it echoes: *L'art, ça ne pardonne pas.* ('Art is unforgiving.') Unforgiving as indeed the very laws of evolution are.

References

Alland, Alexander, Jr.
1977 *The artistic animal.* New York: Doubleday.
(1989) "Affect and aesthetics in human evolution", *The Journal of Aesthetics and Art Criticism* 47: 1-14.

Anati, Emmanuel
1960 *La civilisation du val camonica.* Paris: Artaud.

Baker, R. – J.M. Delgado-Garcia – W. Graf – R.F. Spencer
1987 "Vestibular ocular organization in vertebrates is a largely conserved phylogenetic plan", *Society for Neuroscience: Abstracts* 13: 49.10.

Bonner, John Tyler
1980 *The evolution of culture in animals.* Princeton: University Press.

Bouissac, Paul
1992 "The construction of ignorance and the evolution of knowledge", *University of Toronto Quarterly* 61.4: 460-472.
1993 "Semiotisches Wettrüsten: Zur Evolution der artübergreifenden Kommunikation", *Zeitschrift für Semiotik* 15: 3-21.

Char, René
1974 *Le monde de l'art n'est pas le monde du pardon.* Paris: Maeght.
Dawkins, Richard
1976 *The selfish gene.* Oxford: University Press.
1986 *The blind watchmaker.* London: Longman.
Dennett, D.C.
1990 "Memes and the exploitation of imagination", *Journal of Aesthetics and Art Criticism* 48.2: 127-135.
De Renzi, Ennio
1982 *Disorders of space exploration and cognition.* Chichester: Wiley.
Dimond, Stuart J.
1980 *Neuropsychology.* London: Butterworths.
Eibl-Eibesfeldt, Irenäus
1975 *Ethology: The biology of behavior.* New York: Holt, Rinehart & Winston.
Frijda, Nico H.
1986 *The emotions.* Cambridge: University Press.
Gamble, C.
1980 "Information exchange in the Palaeolithic", *Nature* 283: 522-523.
Gazzaniga, M.S. (ed.)
1979 *Handbook of behavioral neurobiology.* New York: Plenum.
Graf, W.
1988 "Motion detection in physical space and its peripheral and central representation", *Annals of the New York Academy of Sciences* 545: 154-169.
Jakobson, Roman
1963 *Essais de linguistique générale.* [Transl. by N. Ruwet.] Paris: Minuit.
Koch, Walter A.
1984 "Art: Biogenesis and semiogenesis", *Semiotica* 49: 283-304.
1986 *Genes vs memes: Modes of integration for natural and cultural evolution in a holist model.* Bochum: Brockmeyer.
Lande, R.
1980 "Sexual dimorphism, sexual selection, and adaptation in polygenic characters", *Evolution* 34: 292-305.
Leroi-Gourhan, André
1958 "La fonction des signes dans les sanctuaires paléolithiques", *Bulletin de la Société de Préhistoire française* 55: 307-321.
Lumsden, Charles J. – Edward O. Wilson
1981 *Genes, mind, and culture: The coevolutionary process.* Cambridge, Mass.: Harvard University Press.

McNaughton, B.L. – L.L. Chen – E.J. Markus
1991 "'Dead reckoning', landmark learning, and the sense of direction: A neurophysiological and computational hypothesis", *Journal of Cognitive Neuroscience* 3: 190-202.

Marshack, Alexander
1972 *The roots of civilization: The cognitive beginnings of man's first art, symbol, and notation*. New York: McGraw Hill.

Møller, A.P.
1989 "Viability costs of male tail ornaments in a swallow", *Nature* 339: 132-135.

Nespoulous, J.-L.
1981 "Two basic types of semiotic behavior: Their dissociation in aphasia", in: P. Perron (ed.), 101-125.

Perron, P. (ed.)
1981 *The neurological basis of signs in communication processes*. (Toronto Semiotic Circle Monographs.) Toronto: Victoria University.

Petitto, L.A.
1987 *From gesture to symbol: The relationship between form and meaning in the acquisition of personal pronouns in American sign language*. Bloomington: Indiana University Linguistics Club Dissertations.

Raphael, Max
1945 *Prehistoric cave painting*. Princeton, N.J.: University Press.

Sebeok, Thomas A.
1981 *The play of musement*. Bloomington: Indiana University Press.
1985 "On the phylogenesis of communication, language, and speech", *Recherches sémiotiques / Semiotic inquiry* 5.4: 361-367.

Sebeok, Thomas A. – Jean Umiker-Sebeok (eds.)
1980 *Speaking to apes: A critical anthology of two-way communication with man*. New York: Plenum.

Children's drawings: Ontogenetic aspects and phylogenetic roots

Martin Krampen

1. Children's drawings and semiotics

Children's drawings are semiotic objects like verbal utterances. While the developmental aspects of language have been studied in depth within the framework of semiotics (cf. Jakobson 1942), only few studies on the development of iconic performance have been published (e.g., Krampen 1989, 1991). This seems to be due to the predominantly linguistic interest of semioticians.

Semiotics should, however, as Sebeok (1977) postulated, not only assume the task of the synchronic or diachronic study of different kinds of current sign systems (semiotics of the "normal"), but also of their developmental aspects (semiotics of the "adnormal" in children and of the "denormal" in the ageing).

2. Children's drawings and Piaget's developmental stages

As for children, it was Piaget who made a decisive contribution to the study of iconic development in children. For Piaget, the development of the mental image is an important phase in the ontogeny of intelligence comprising the four stages of sensorimotor intelligence, preoperatory representations, concrete mental operations and formal operations.

For the developmental semiotics of iconic signs, Piaget's classification of "symbols" (i.e., icons in Peircean terminology) is of special interest because it, too, follows a developmental sequence. This classification is proposed in *The psychology of the child* (Piaget – Inhelder 1969) in a chapter explicitly titled "Semiotic Function and Imitation". There are five classes of semiotic functions of imitation that develop successively, starting during the stage of preoperatory intelligence:

(1) Delayed imitation (e.g., a child imitates the behavior of absent persons, such as the temper tantrum of a playmate hours after it has happened).

(2) Symbolic play (e.g., a child mimics her or his own behavior; thus, when tired, closes the eyes, sucks the thumb, and then puts the teddy bear in a toy bed, making the latter a symbol of self).
(3) Drawing (in children's drawings, acquired graphic symbols - e.g., circles for a head, a ball, or a wheel - are used at first in the same way as toys in symbolic play, to assimilate, and later to accommodate reality).
(4) Imagination proper (an object or a process is imitated internally without the appearance of outside gestures; the child accommodates to reality rather than assimilating it).
(5) Mentioning (e.g., a child mews without a cat being present; the absent cat is verbally represented by "arbitrary" signs).

From this classification it can be seen that children's drawings are considered a type of symbol appearing in the transition between symbolic play and imagination.

3. Children's drawings and primitive art

Interestingly enough, it has been argued that children's drawings, as they emerge in the course of developing imagination in the individual, are akin to "primitive" art (e.g., Luquet 1922). This would mean that art has a development, fitting in some way the hypothesis that the stages of ontogeny follow a sequence similar to that of phylogeny. As in the ontogeny of the child, the individual at some point must "invent" imitative drawing, in phylogeny at a given state of development primitive man must have "invented" the representation of relevant environmental aspects by drawing. For both types of invention, it is argued, first meaningless traces are left on surfaces (scribbles). Some of them fortuitously resembling certain aspects of the environment are endowed with meaning until finally meaningful traces are produced at will. Obviously, this simple a hypothesis can hardly be maintained. The connection between the ontogeny of drawing and phylogeny must be looked for elsewhere.

4. Phases of development in children's drawings

The ontogenesis of children's drawings follows certain stages which are described differently by different authors. Luquet (1927) outlined a sequence of five phases. Although these phases have met with criticism in some of their detail, they still constitute an acceptable approximation to what can be observed in the development of children's drawings. Luquet's phases are:

- scribbling (ages 2-3)
- fortuitous realism (i.e., the discovery of similarities between certain features of scribbles and objects in reality) (ages 3-4)
- failed realism (i.e., synthetic incapacity) (ages 4-5)
- intellectual realism (i.e., the child draws what she or he "knows" about reality, e.g., what is inside the house as well as its exterior) (ages 5-8)
- visual realism (i.e., the child draws what is visible only from a certain point of view in reality, e.g., with a certain perspective) (ages 8-12)

Typical and most important in Luquet's scheme of phases (as opposed to proposals by other authors) is his basic hypothesis that behind the development of children's drawings there is a tendency towards realism. He devoted an entire chapter (titled "Le realisme") to this idea in his 1927 book *Le dessin enfantin*. In the end, the goal of drawing would be a realistic translation of the visual properties of objects into graphics. But according to Luquet (1913, 1927), children's drawings do not directly transmit the properties of objects. Children do not sit in front of objects copying them but, rather, put onto paper the properties of "internal models" of external objects.

Considering his own interest in children's (nominal) realism and his concept of increasing accommodation to reality during the development of intelligence, Piaget could not but sympathize with Luquet's ideas. In fact, Piaget and Inhelder (1948) incorporated Luquet's phases into their own schemes of the development of children's drawing. Piaget and Inhelder describe three stages in the development of free drawing.

The first stage (up to the age of 4) is that of "synthetic incapacity". As soon as the child abandons her or his interest in observing the traces of her or his own motor activity left by rhythmic scribbling, (s)he begins to take account of simple topological relationships in drawings, placing shapes close together that (s)he associates in his or her mind (proximity) and separating them, if this is how (s)he imagines them. Order is achieved to the degree of sequences of pairs. Drawing surrounding or enclosing forms is mastered as long as shapes are simple, but it still presents some difficulties when more complex forms are dealt with. Thus, eyes may be placed outside the head, buttons outside the body. The roof may project into the house instead of skyward. At this stage, the principles of continuity are dealt with only in a summary fashion.

One of the principal indicators of synthetic incapacity is, in fact, that the parts of a figure are simply juxtaposed at their borderlines instead of being continuously linked together (e.g., an outline). This stage in free drawing is thus characterized by an unsystematic application of primitive topological relationships to complex shapes, such as human figures, animals, and houses. The predominance of topological features that are not completely mastered is not due to a lack of drawing technique, but constitutes a law of development.

Following the stage of synthetic incapacity is that of "intellectual realism" (4-8 years). Intellectual realism refers to the fact that children draw everything they know even if it is not actually visible from a certain vantage point. Topological relationships are completely acquired during this stage, whereas projective and metrical relationships (i.e., perspective and proportion) just begin to emerge without being coordinated. Proximities are correctly rendered: thus, arms are attached to the trunk, and eyes are placed side by side in the head (even if this is shown in profile). Order is present in the drawings of landscapes or houses although not yet in accordance with the coordinates (e.g., the chimney is attached perpendicularly to the roof). Surrounding and enclosure become so important that in many instances the inside of an object (e.g., a house) is shown together with its surrounding outline in so-called transparencies. Some projective and metrical relationships of space are shown, but in an incoherent fashion. Since there is no coordination of viewpoints or perspectives, a single drawing in this phase presents many different viewpoints at once. This is also the reason for the so-called pseudo-rotations which show, for example, a cart head-on, the wheels attached in plan (circles), and the little man pushing it in profile - three different viewpoints at once! Straight lines and geometric shapes are used without regard to measurement and proportion. In a group of buildings, each may be shown from a different view. In intellectual realism, the elements of space just begin to become projective and metrical, but their relationships still obey the topological laws of proximity, separation, order, and continuity. Between the objects in space and the drawing there can be only an element-to-element correspondence of the features. But this correspondence is qualitative, and no coordination of the emerging projective or metric relationship can be seen. There is no understanding of perspective or proportion. No external horizontal or vertical coordinate systems combine perspectives and proportions into coherent wholes; hence, the result is a confused (though charming) medley of viewpoints.

In stage three (average ages 8-12), free drawings start to unite perspective, proportion, and distance. This stage is called "visual realism". Visual realism appears relatively late, since it requires the concepts of projective and metric space (perspective, proportion, metrical distance), which in turn presuppose advanced concrete mental operations. The projective and metrical concepts of space develop interdependently, and not one after the other. The projective relationships preserve the true relative position of shapes and figures as against the pseudo-perspectives and pseudo-rotations of intellectual realism. True proportions and relative or coordinate distances between shapes and figures are guaranteed by metrical relationships. Systems of mental operations merely replace empirical constructions.

By combining Piaget's and Luquet's phases, the following developmental sequence emerges:

- Scribbling (ages 2-3) is a late manifestation of sensorimotor intelligence.
- Fortuitous and failed realism is the same as Piaget's synthetic incapacity (ages 3-5) and is connected with the preoperational stage of mental development.
- Intellectual realism (ages 5-8) parallels the transition from the preoperational stage to the stage of concrete mental operations.
- Visual realism (ages 8-12) presupposes the developmental stage of concrete mental operations.

5. Children's drawings in nursery school

To discover the phylogenetic roots in children's drawings we must look, however, more closely at its components. This can be done by studying in detail the formal inventories of the various stages as proposed by Lurçat (1979) for nursery school children.

In the first phase of nursery school (3-4 years), children begin to abandon rapid motoric and unidirectional scribbling in favor of first curves, lines, and forms. In the course of time, these take on the character of iconic signs, beginning with "ideograms" (take for instance the first "tadpole men"). Prerequisites for this development are two types of control, both consequences of muscular maturation (especially that of the flexor in the wrist): motor control and perceptual control.

Motor control leads to the fragmentation of continuous lines by "breaking" rapid scribbling movements and to the rotative isolation of single lines out of tight scribbles. Perceptual control can graft itself onto the slowing down and fragmentation of motor activity. Formerly, the eyes just followed the traces made by the pencil. Now perception of the beginning and the end of a trace allows the eye to control a "deliberate" return of the pencil to the beginning of the line just executed. From here the pencil may continue that line in another direction (simple control, Figure 1) or "close" an angle or a curve by a well-aimed trace concerning both ends of a line (double control, Figure 2).

It is interesting to note that the chimpanzee Congo, one of a number of "painting apes" that have been studied, was capable of double control (Figure 3; Morris 1962).

Figure 1. Simple control

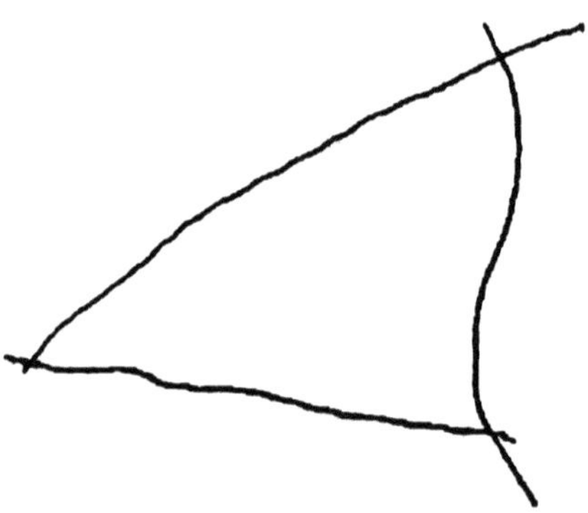

Figure 2. Double control

Figure 3. An example of double control produced by Congo, a chimpanzee.

The formal inventory, between 3 and 4 years of age, consists, at the motoric level, of curved open and closed lines (round forms) and continuous loops. At the level of perceptual control, circles, rectangles, spirals, hybrid slopes, arabesques, and crosses become possible. Hybrid slopes exhibiting reversal of wrist rotation become the fundament of writing (Figures 4 and 5).

Figure 4. Continuous hybrid slope

Figure 5. Reversal of line direction produced by Congo

In the second phase of graphic activity in nursery school (4-5 years), the repertory of graphic forms expands considerably. This expansion is a consequence of further development of three kinds of abilities: motor abilities (e.g., the controlled combination of distal and proximal arm-wrist movements), perceptual abilities (e.g., visual identification of forms and identical reproduction of specific "geometric" forms), and representational abilities (figurative drawings).

Whereas the first phase was characterized by an oscillation between a priori and a posteriori verbal interpretation of impulsive graphic gestures and the repetition of simple schemata, graphic activity now becomes more organized.

The drawing surface is explored systematically in all directions. This leads to orthogonal realizations such as the following: vertical or horizontal lines and their combinations; parallels; rectangular grids and ladders (Figure 6); oblique realizations such as rays in different directions (Figure 7); curvilinear realizations such as elaborated arabesques (Figure 8).

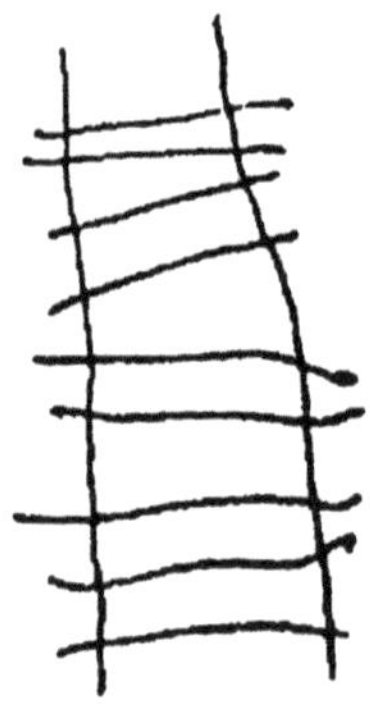

Figure 6. Orthogonal realization - a ladder

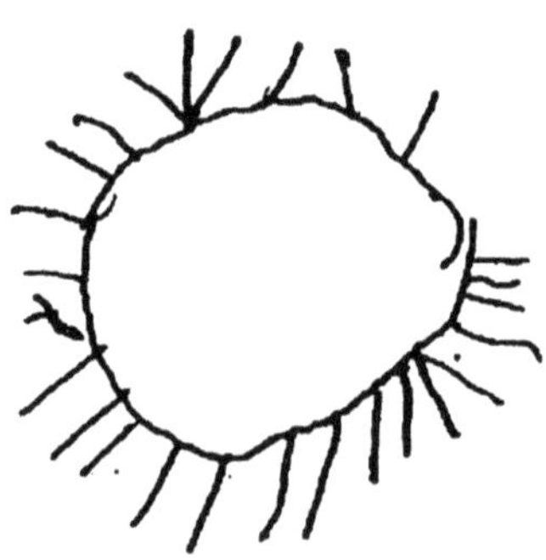

Figure 7. Oblique realization - rays in different directions

Figure 8. Curvilinear realization - an arabesque

Furthermore, orthogonal and oblique lines may be simply combined, or more decorative patterns are generated with oblique lines and arabesques. Finally, the drawing surface is organized in an even more complex and "intellectual" way. Symmetrical operations are tried outside the form by juxtaposing it with the same graphic elements (Figure 9) and inside the form by subdivision.

Figure 9. Symmetrical juxtaposition

Parts of the subdivided forms are filled in with pencil or color. Forms are combined or laid out in a row and connected by a single dividing line. Parts of these subdivisions may again be filled in. At the end of this phase of development, we find large compositions that combine all the operations mentioned above: spatial factors such as directional lines; spatiokinesthetic factors such as arabesques; and visual factors, such as symmetry, subdivision, and filling in. All these operations work together in a highly decorative and aesthetic fashion.

In the last phase of nursery school (5-6 years), graphic activity is shaped by different competing influences. Thus, nursery school exercises become visible in spontaneous drawings. The interests of the developing child's personality appear as drawings of individual fantasies and images. Consequently, there is a parallel tendency toward increasing graphic control and free graphic play, while the antagonism between abstract form and realism diminishes. The period between the ages of 5 and 6 is the "age of the copy". Direct copying from examples of letters, shapes, and figures in nursery school has its repercussions in spontaneous graphic activity. External models are internalized and produced at will. Thus the stock of forms increases.

The previous conflict between abstract forms and realism in spontaneous drawings comes to an end because a certain inhibition in form production takes place. The forms produced are often miniaturized or present the usual geometric shapes - circles, squares, triangles. Subdivision of these geometric shapes is further spontaneously exercised. The production of arabesques and cycloids is continued. The abstract tendency attached to the exploration of the drawing surface is now integrated with figurative drawing and is expressed, for example, in the form of a highly decorated treatment of the human figure (Figure 10).

Figure 10. Highly decorative treatment of human figure

6. The development of graphemes

For Lurçat, graphic activity is, at the outset, a unitary behavior that begins to split into two functions between the ages of 4 and 5: symbolic (iconic) representation and writing with conventional signs. Another French author, Olivier (1974), goes one step further by asking the challenging question "Le dessin enfantin est-il une écriture?" The general hypothesis under which his research was conducted states that all children between the ages of 4 and 6 acquire the same quantitatively limited stock of graphemes. These graphemes can easily be produced and combined. Like phonemes in language, they are meaningless in themselves; but during the phase of "intellectual realism", they are put together according to certain rules to form an unlimited number of meaningful drawings. Between the ages of 7 and 9, however, new rules begin to substitute for the preceding ones and the grapheme system dissolves to make room for other forms of graphic production.

In order to test his hypothesis, Olivier conducted two separate studies regarding the inventory of forms used in children's drawings. The first consisted of a longitudinal study of his own three children. The second was a quantitative test with a sample of 180 nursery school children. The results of the longitudinal study may be summarized as follows:

First Tracings

Quite in accordance with Piaget and Inhelder (1948) and Lurçat (1979), Olivier (1974) found a first differentiation of children's graphic activities during the second year of life: rapid and rhythmic rotative and hatched translative movements. By slowing down, miniaturizing, and fragmenting these two types of movements, two types of graphic traces are generated, continuous modulations of lines and discontinuous stereotyped graphemes.

Modulations (M)

Resulting from increasingly slow execution and better control of the first differentiation in rapid and continuous rotative or translative traces, three kinds of modulation become possible:

- arabesque-like, predominantly rounding, continuous curves (Figure 11)
- angular continuous traces (Figures 12 and 13)
- modulations exhibiting both round and angular continuous traces (Figure 14)

Figure 11. Predominantly rounding continuous curves

Figure 12. Angular continuous traces

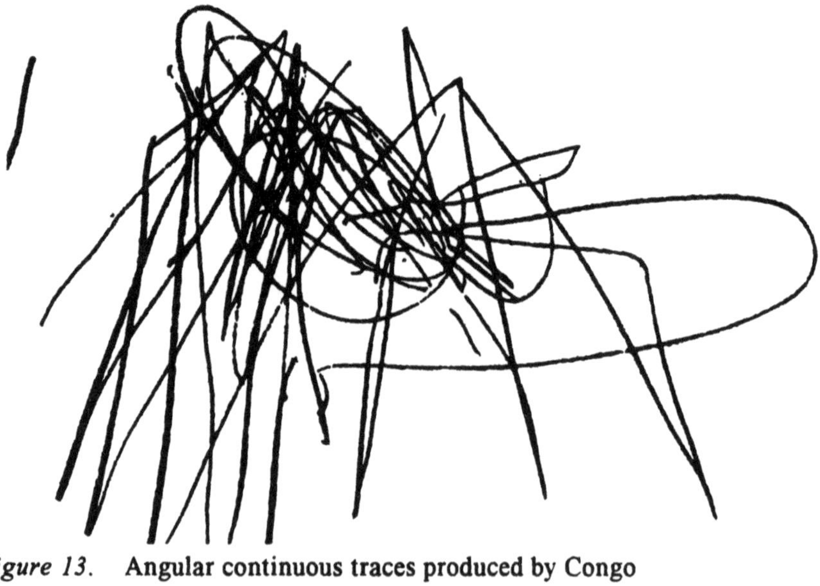

Figure 13. Angular continuous traces produced by Congo

Figure 14. Modulations combining round and angular continuous traces

First Grapheme System (G1)

Out of the modulations there emerges the first set of graphemes based on the slow execution and control of discontinuous traces. This is the result of the opposition of three pairs of distinctive features:

- rotative (R) versus translative (T) traces
- closed (C) versus open (O) traces
- repeated alternation of line direction (rA) versus nonrepeated alternation (A)

The eight graphemes resulting from the combination of these distinctive features are given in Figures 15 and 16 (overleaf). These first graphemes may be combined in three modes:

- intersection
- inclusion
- repetition

Second Grapheme System (G2)

The second set of graphemes is added to the first one by a new and distinctive feature that is a result of reducing the intersection of two graphemes to their controlled junction in one or more points. Thus the presence of junction (J) and the increased inhibition of repetition in alternation (A) differentiates between the eight additional graphemes and those of the first system.

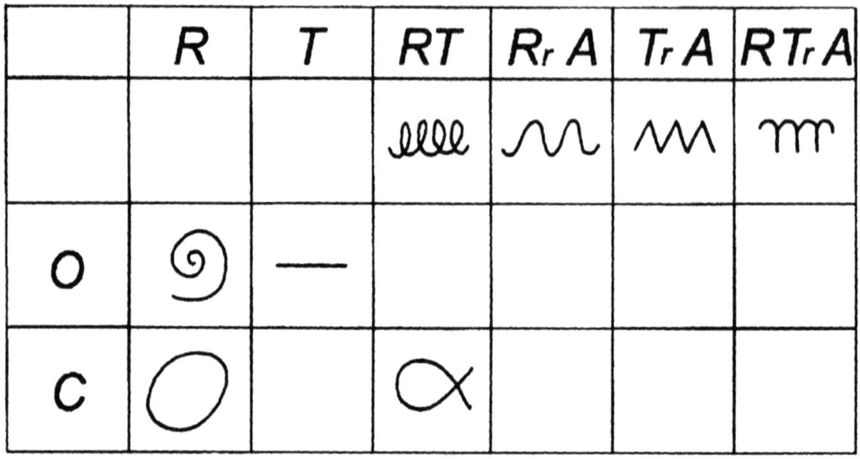

Figure 15. The eight graphemes of system G1

Figure 16. A zigzag drawn by Congo

The eight graphemes of the first and eight of the second system can now be further combined by the new mode of junction, together with intersection, inclusion, and repetition (Figure 17).

Third Grapheme System (G3)

The third set of graphemes is added to the previous stock by a new and distinctive feature resulting from the child's sensitivity to horizontal and vertical symmetry possibly related to the horizontal and vertical axes of

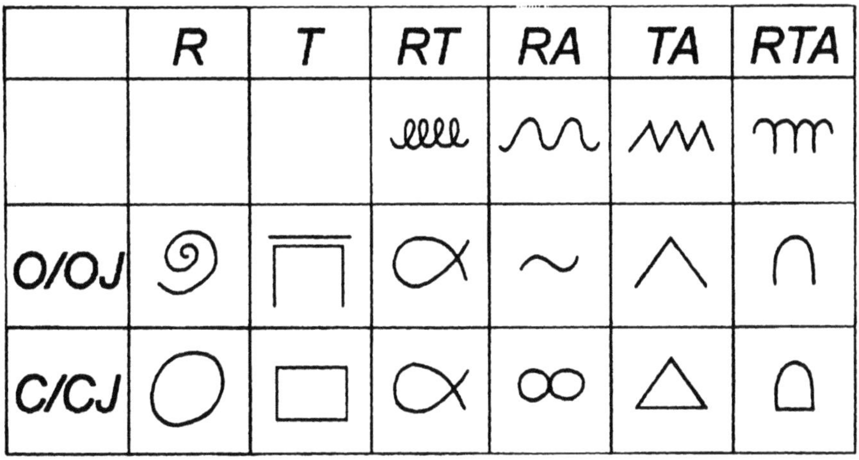

Figure 17. The 16 graphemes of system G2, including the eight of system G1 and eight new ones

the body. The forms thus acquire the feature of being symmetrical (S) around one, two, or three axes. Furthermore, another late distinctive feature appearing together with S seems to originate in a double alternation of an angle (aA), such as a diamond (TaAC). These two new distinctive features permit the differentiation of 14 more graphemes (Figure 18, overleaf).

All 30 single graphemes from the three grapheme systems can now be combined by symmetrical juxtaposition in addition to the previous modes of intersection, inclusion, repetition, and junction. Thus, at the end of grapheme development, the nine aforementioned distinctive features (R, T, O, C, rA, A, J, S, aA) differentiate 30 graphemes that may be composed in five modes.

The quantitative tests confirmed the results of the longitudinal study. The progressive disappearance of modulations with age shows that an "older" graphic behavior, modulation, is slowly replaced by a new one, the production of graphemes. This suggests that during the phase of modulation children have little or no production capacity at their disposal for expressing the configurations of mental images. The strongest drop in modulations occurs only at the end of the 5th year when the children reach their peak (93 %) in differentiating between four closed symmetrical graphemes. The strong increase (from 0 to 73%) in the symmetrical graphemes of the free drawings occurs at 4:0 to 4:6 years, when 43 % of the children differentiate correctly between the

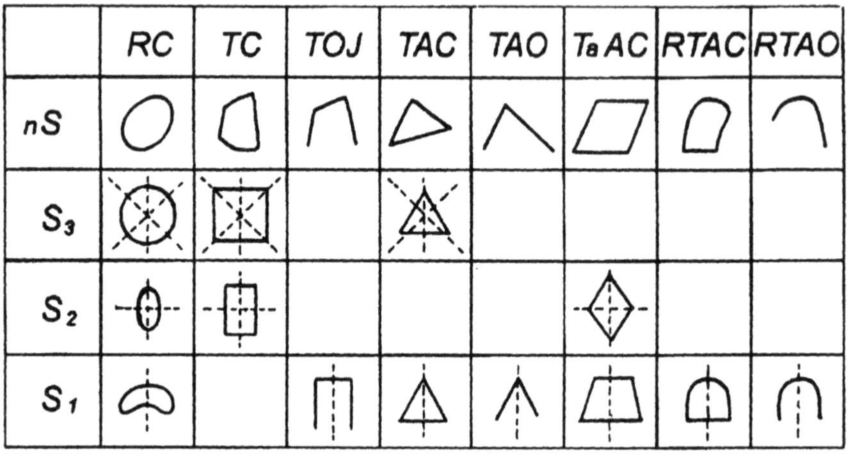

Figure 18. The 21 graphemes of system G3, including one of G1 and six of G2 (first row)

three closed forms. It is at this age, when all five modes of composition are present, that the children can begin to acquire grapheme system G3. With this production capacity, it can also be expected that interesting renderings of the configurations of mental images are now possible.

Up to now, the evidence on the increase in graphemes considered to be the formal building blocks of children's drawings has been reviewed. But so far, the question as to how formal units (signifiers) are connected to the content they transmit (signified) remains open. Olivier's hypothesis opens up the question of whether children's drawings might be doubly articulated in meaningful units (morphemes) composed in turn of merely differential but otherwise meaningless units (graphemes). An examination of children's drawings shows that in contrast to language, where phonemes remain mere differentiating units, graphemes may acquire the status of morphemes in the context of other graphemes. Thus, it is true that the round (RC) is in itself a minimal differentiating element, as is the line (TO). But in combination with other rounds and lines, they may make up meaningful units such as the head, eyes, body, legs, arms, mouth, and nose. But these meanings do not seem to be fixed to the graphemes as they are to the morphemes in language.

A commutation test can furnish additional proof. The trunk of a human figure can be portrayed in children's drawings by a round (RC), a triangle (TACJ), and a quadrangle (TCJ). A triangle may be used indiscriminately as the roof of a house, the sail of a boat, or the top of a pine tree.

It may not be substituted, however, for the wheels of a cart, which must always be presented as round. Graphemes, therefore, in contrast to phonemes, seem to have an ambiguous status, being minimal differentiating units in a grapheme system and becoming morphemes in the context of other graphemes.

This double function suggests that graphemes could be iconic, that is, structurally "similar" to (parts of) the objects they are used to represent. While Olivier does not wholly exclude a limited iconic grapheme function, he nevertheless insists that they are selectively acquired in the course of spontaneous exercises based on motoric and perceptual development. Only subsequently are they used to compose meaningful units stemming from the environment of the children. Since the environmental units are separate from the organism and permeated by culture, they must have existed prior to, and independently of, the children's ability to portray them, like words antedating the ability to speak them.

7. Nature and culture in children's drawings

The fact that children's ability to draw is based on the physical maturation of arm and wrist joints and the coordination of hand and eye movements shows the strong part "nature" plays in children's drawings. The part of "culture" can easily be demonstrated by the example of the divergent "developmental carriers" which drawings of religious buildings take in different cultures (Krampen 1991). If Turkish and German children are asked to draw a religious building (a mosque or a church), their drawings become increasingly different with age. In Figure 19 (overleaf), "phantom drawings" based on the drawings of 79 Turkish and 63 German children are shown (Krampen 1991).

Figure 19 demonstrates that at the age of 6-7 years, Turkish children begin to draw their religious building with a dome and a minaret, German children with a nave and a spire. This difference cannot stem from anything but their different cultural environments.

The phylogenetic roots of children's drawings become evident if we ask what children can do in their drawings that apes cannot do. As we have seen, apes can master single and double control and can, therefore, beyond scribbling, produce some graphemes. This means that they also could produce some letters (since writing and drawing stem from the same unitary graphic activity).

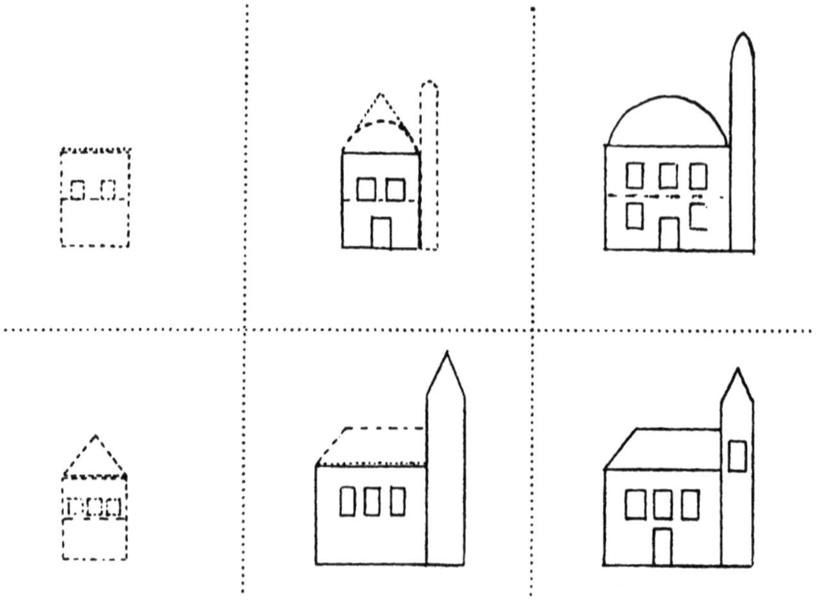

Figure 19. "Phantom drawing" representing the image of a religious building for 3-5, 6-7, and 8-12 year old children. Top row Turkish, bottom row German children.

It was probably in 1876 that Friedrich Engels wrote an article with the title "Anteil der Arbeit an der Menschwerdung des Affen" [The share of work in the humanization of the ape], later published in his *Dialectics of Nature* (Engels 1974). In this article, he speculates that the hands of apes always played a special part in climbing, but that walking upright, which became more and more a necessity for them, freed their hands more and more for other activities. While the number and position of the bones and muscles corresponded approximately in ape and man, the hands of the most primitive tribesman can apparently do innumerable things that no ape's hand can imitate - notably the production of tools. In the transitional species between ape and man the activities done by the hand became, therefore, more and more numerous. Engels (1974: 445-446) concludes:

> But the decisive step was: the hand became free and could acquire more and more skills, and its flexibility won thereby was bequeathed from generation to generation. Thus, the hand is not only the organ of work, but also its product. Only by work; by adapting ever new activities; by hereditary transmission of the special formation of muscles and

> ligaments produced by it; in the long run also of the bones; by always new application of inherited refinements to new, more and more developed activities, has the human hand acquired the high level of perfection, on which it could produce, as by magic, paintings of Raffael, statues of Thorvaldsen, and music of Paganini.

Apart from some corrections which might be necessary in detail, the speculative scenario of Engels makes still sense today, even if his political ideas have suffered a historical setback.

References

Engels, Friedrich
1876 "Anteil der Arbeit an der Menschwerdung des Affen".
[1974] [Reprinted in: "Dialektik der Natur", in: Karl Marx, Friedrich Engels, Werke. Vol. 20. Berlin: Dietz, 444-453.]

Jakobson, Roman
1942 *Kindersprache, Aphasie und allgemeine Lautgesetze*. Uppsala (= Universitets Arsskrift 9).
[1971] [Reprinted in: R. Jakobson, *Selected Writings*. Vol. 1. The Hague: Mouton, 328-401.]

Koch, Walter A. (ed.)
1989 *The nature of culture*. Bochum: Brockmeyer.

Krampen, Martin
1989 "Grapheme in Kinderzeichnungen", in: Walter A. Koch (ed.), 521-541.
1991 *Children's drawings: Iconic coding of the environment*. New York: Plenum.

Luquet, Georges H.
1913 *Les dessins d'un enfant*. Paris: Alcan.
1922 "Genèse de l'art figuré", *Journal de psychologie* 19: 695-719.
1927 *Le dessin enfantin*. Paris: Alcan.

Lurçat, Liliane
1979 *L'activité graphique à l'école maternelle*. Paris: Editions ESF.

Olivier, Fernand
1974 "Le dessin enfantin est-il une écriture?", *Enfance* 3-5: 183-216.

Piaget, Jean – Bärbel Inhelder
1948 *La représentation de l'espace chez l'enfant*. Paris: Presses Universitaires de France.

1966 *L'image mentale chez l'enfant.* Paris: Presses Universitaires de France.

1967 *The child's conception of space.* Translated by F. J. Langdon and J. L. Lunzer. New York: Norton.

1969 *The psychology of the child.* Translated by H. Weaver. New York: Basic Books.

Sebeok, Thomas A.

1977 "Ecumenicalism in semiotics", in: T.A. Sebeok (ed.), 180-206.

Sebeok, Thomas A. (ed.)

1977 *A perfusion of signs.* Bloomington: Indiana University Press.

Can a picture tell a thousand words? Interpreting sequential vs. holistic graphic messages

Richard W. Janney and Horst Arndt

What's in the braine that incke
may character?
Shakespeare (Sonnet 108)

1. Introduction

It is generally thought that attributions of significance depend, at least at the most basic level, on patterns of perception, and that patterns of perception involve scanning: the eye that is not allowed to scan ceases to see; the ear that is subjected to an unchanging tone ceases to hear; the finger that is not allowed to move across a surface ceases to register the texture, and so on. Such sensory perceptual phenomena are sometimes invoked in support of the notion that, strictly speaking, it is not perceptions of discrete, isolated things or events that give significance to human experience, but perceptions of patterned relationships between things (cf. Ruesch – Bateson 1951: 173). In their discussion of the pragmatics of human communication, Watzlawick, Beavin, and Jackson (1967: 36) use this notion to argue that only where there is perceived pattern is there potential significance in communication. Where there is no recognizable pattern, according to Watzlawick, Beavin, and Jackson, messages are uninterpretable.

The subject of this paper is the perception and interpretation of graphic messages. The interpretation of graphic messages is a complicated, learned, goal-directed, constructive activity that depends not only on global assumptions about what count as significant graphic marks in the interpreter's culture, but also on assumptions about how graphic messages are organized, what general types of information they convey, and how they should be scanned in different situations (cf. Arndt – Janney 1987: 55 ff.). As perceivers and interpreters of graphic messages, we do not simply see marks in isolation; rather, we see them in relation to each other, and, equally importantly, as Aicher (1991: 39) says, "we see [them] against the background of our knowledge." This knowledge includes not only what we literally know, but also what we guess, imagine, hypothesize, and/or presuppose about the conditions under which, and the reasons for which, graphic messages are generally produced.

The paper compares the interpretation of sequential word messages with the interpretation of holistic picture messages. A message in words usually appears as a linear sequence of abstract graphic marks (symbols, characters, letters) that stand for units of a language (phonemes, morphemes, syllables, words), and are arbitrarily related by linguistic convention to each other and to the conceptualized things or thoughts to be communicated (see Figure 1).

Dear son, please come home. I am sending you $53 for your travel expenses. Your father.

Figure 1. A sequential word message

A picture message usually appears as a holistic configuration of graphic marks in the form of a drawing (likeness, pictograph, pictogram, hieroglyph, cartouche), which bears some kind of iconic resemblance to the things or thoughts to be communicated (see Figure 2).

Figure 2. A holistic picture message (after Mallery 1893)

The paper points out some important features of these two forms of graphic communication, discussing the roles of information patterning and presupposition in the interpretation of written, as compared with pictorial, messages. It assumes that the significance, and hence the interpretability, of both types of graphic messages depends in some sense on the recognition of patterned relationships.

The discussion centers around the following questions: to what extent is the perception of a linear sequence of abstract symbols comparable with the perception of a holistic picture? What types of relationships are perceptually relevant, and how are "messages" identified in the two modes? Are the kinds and/or qualities of interpretations that can be attached to word messages and picture messages comparable?

According to conventional wisdom, a picture can tell a thousand words and a sequence of words can convey a picture. This amounts to imagining that sequential and holistic graphic messages communicate roughly similar kinds of information. In the case of the examples in Figures 1 and 2, however, the similarity of the information that is conveyed is not immediately evident. Figure 1 is an English translation of the Cheyenne Indian picture message in Figure 2 (cf. Coulmas 1989: 20). The two messages seem rather obviously to be based on different principles, and seem to convey different types of information. But on the other hand, from a perceptual point of view, similar problems seem to arise in attempting to interpret them. In both cases, the significance attributed to what is perceived depends less on the individual symbols or iconic markings than on the perceived patterned relationships between these, and on the perceived relationships between the graphic representations and the contexts in which the interpreter imagines them to have been produced. One of the purposes of this paper is to discuss these paradoxical differences and similarities, and to trace some of their implications for the interpretation of both types of graphic messages.

2. Configurations of abstract symbols and pictorial likenesses

Abstract symbols and pictorial likenesses have been complexly interrelated in graphic communication throughout history (cf. Harris 1986; Coulmas 1989). Although the present paper focuses specifically on contrasts between sequential symbolic messages and holistic pictorial messages, it should be noted that symbols and pictures appear in other configurations as well. For instance, symbols appear in holistic configurations in some Egyptian cartouches (Figure 3); pictures appear in sequential configurations in some ancient disk inscriptions (Figure 4); and symbols and pictures are combined in medieval illuminated manuscripts and stained glass windows, and in modern cartoons (Figure 5).

cartouche	*phonetic value*	*name*
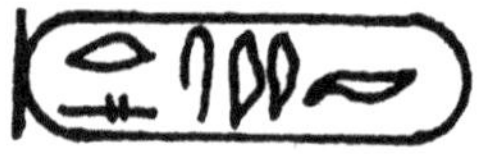	*Kjśrs*	Caesar

Figure 3. An Egyptian cartouche: abstract symbols in a holistic configuration (after Zauzich 1980)

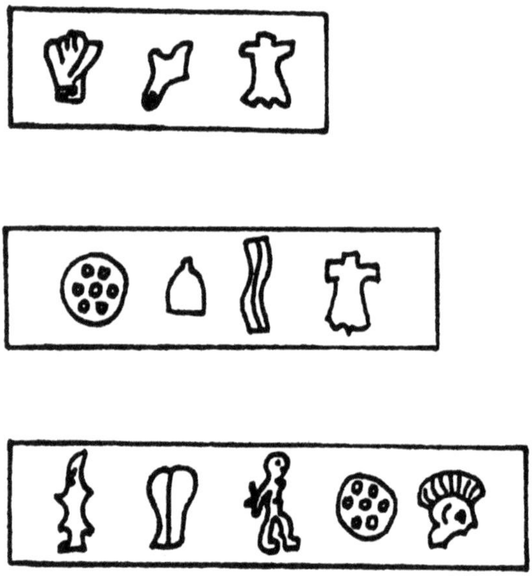

Figure 4. Details from the Phaistos disk: pictorial likenesses in a sequential configuration (after Keenan 1985)

The capacity to perform "cross-modal transfers" (cf. Davenport 1977), or to produce and interpret different types of messages simultaneously in different modes, is widely discussed in the literature on brain functions and the emergence of human communication (cf. Janney – Arndt [in press]). Not only do people habitually substitute nonverbal gestures for words, and words for nonverbal gestures in speech (cf. Arndt – Janney 1987, 1991); they also substitute abstract symbols for pictures, and pictures for abstract symbols in graphic communication. As said, however, for the sake of simplicity, here, the authors will discuss only instances in which symbols appear in sequences and pictures appear in holistic configurations.

Figure 5. Roy Lichtenstein's "M-Maybe" (1965): A mixture of the sequential symbolic and holistic pictorial graphic modes.

3. Sequential graphic messages

Sequential symbolic writing is often said to have begun with early accounting practices (cf. Bernal 1954; Schmandt-Besserat 1978, 1979; Coulmas 1989). There are striking similarities between the shapes of tokens used in Middle Eastern accounting systems at around 8,500 B.C. and the first written symbols (cf. Schmandt-Besserat 1978: 56). Token tally systems based on simple correspondences between symbols and objects or events are much older, reaching back to the early bone counting systems of ancient paleolithic people about 12,000-30,000 years ago (cf. Schmandt-Besserat 1979: 23).

The idea that sequential symbolic writing "emerged from accountability", as Bernal (1954 [1971]: 119) puts it, partially explains the digital, self-contained quality of this form of graphic communication. In the sequential symbolic mode, the message, as it were, is reduced to a series of discrete, digital, conceptual units, which are combined additively, according to the rules of linguistic syntax, yielding a sort of logical sum or product of the operations performed on the individual units within the message sequence. The result - a written sentence - is a version of what the writer wishes to communicate. It does not convey literally everything that is metaphorically on, or in, the writer's mind at the moment it is produced, but only those aspects of the writer's thoughts which are reducible to, and ultimately accountable for within, the terms and rules (and hence the expressive possibilities) of the sequential symbolic graphic mode.

What kinds of information are reducible to sequences of abstract linguistic symbols? The written message in Figure 1 seems mainly to convey ideas and facts: 'Son, use these $53 to come home.' The original pictorial message in Figure 2, on the other hand, contains a great deal of additional relational information that is not represented in the written version. Note, for example, the marked connection between the large figure on the left and the two turtles in the upper left corner, the connections between the figures in the foreground and the smaller figures in the background, the proxemic orientations of the three figures on the right, who are all facing the large figure on the left, etc.

Following Malinowski (1923), it can be said that a sequence of written words is better at conveying ideational information than at conveying information about feelings, relationships, and/or bonds of union. In Malinowski's terminology, this makes writing better suited for communication than for communion. Sequential symbolic messages are more complex, versatile, and abstract than holistic pictorial messages. They abound in syntactic and semantic qualifiers, and in linguistic indicators that help reduce ideational ambiguity. This is one reason for the impression of semantic clarity in written messages, as opposed to the impression of semantic fuzziness in pictorial messages. Scholars traditionally assume that semantic clarity is a prerequisite for the graphic transmission of knowledge across time and space. To the extent that writing possesses this clarity, and is capable of binding time, it is an important prerequisite for human civilization.

4. Holistic graphic messages

When we turn from written messages to messages in the form of holistic pictures, it becomes clear at once that the conception of meaning as something that is contained within the graphic artifact itself becomes problematical. A defining characteristic of picture messages is their weak connection to language (cf. Coulmas 1989: 20). Holistic messages lack anything comparable to the morphology, syntax, or semantics of sequential symbolic writing. This means, among other things, that there are generally no equivalents of concepts such as "either-or", "if-then", "yes-no", "certainly-maybe" in picture messages, and no qualifiers to suggest which of two or more possible readings is implied. Holistic graphic communication lacks the simple negative - an expression for *not* (modern European traffic signs require abstract symbols to indicate that the thing pictured is not allowed) - and has generally got only very restricted means of indicating distinctions between past, present, or future (cf. Watzlawick – Beavin – Jackson 1967: 65-66).

In order to interpret the "message" of a holistic picture letter, the interpreter thus must supply most of the conceptual content, the underlying motivation, and the surrounding qualifications and distinctions him- or herself. The situation resembles the interpretation of a dream, where "sense" must be somehow introduced into configurations of images. What is literally there to be interpreted is only a field of iconic likenesses, in which there are only implicit relationships between things. For this reason, picture messages, like object messages, knot messages, and other forms of quasigestural, nonlinguistic communication, tend to lend themselves to different, sometimes incompatible, interpretations.

Herodotus (ca. 490-430 B.C.) describes a fatal instance of incompatible interpretations in connection with an object message received by Darius the Great during the Persian war against the Scythians in the 5th century B.C. (cf. Herodotus 1955: 131-133). As Darius and his army were preparing to attack the Scythians, a message was received, consisting of a bird, a mouse, a frog, and five arrows. Assuming that the Scythians were about to surrender, Darius interpreted the message as a capitulation: take our earth (mouse), air (bird), water (frog), and weapons (arrows). He then ordered his army to march into the Scythian city to claim their victory. Upon entering the city, Darius's soldiers were massacred by the Scythians, whose message had been intended as a threat: if you Persians do not rise into heaven like birds, creep into the earth like mice, or dive into the swamps like frogs, you will be struck by these arrows, and you will never see your homeland again. The interpretation of holistic pictorial messages, like the interpretation of object messages, depends, to a much greater extent than the interpretation of sequential symbolic messages (although in an analogous fashion, finally), on assumptions about the possible contexts in which, and reasons for which, they are produced. Interpretation depends on reasoning by analogy, which contrasts sharply with the digital reasoning processes involved in the interpretation of sequential symbolic messages.

Figure 6 is a petition from seven North American Indian tribes to the US Congress for fishing rights in several lakes. The tribes are represented analogically by animal icons; the hearts of the animals are connected by lines to the heart of the crane (the tribe which sends the petition); the eyes of the animals are connected by lines to the lakes; the crane and the other animals are turned eastward, toward the US Congress.

Figure 6. A petition from seven North American Indian tribes to the US Congress for fishing rights in several lakes

Although the intended function of the message in Figure 6 may indeed, like that of a sequential symbolic graphic message, be to express thoughts or facts, the holistic graphic message mode itself is much more "a mode of action", in Malinowski's (1923: 310) terms, than a mode of reflected or conceptualized communication. The picture in Figure 6 is actually a type of enactment, or a graphic gestural performance, in which the referential function is subordinate to the Indians' attempt to express the nature of their relationships to each other, their lakes, and the US Congress. The message is not essentially a statement about concrete facts, but an evocation of a particular constellation of relationships (cf. Watzlawick – Beavin – Jackson 1967: 63); its relational message is: 'Connected in our hearts, our eyes on the lakes, we turn toward you.'

Generally speaking, the relational content of a picture letter is never fully detached from the situation in which the letter is produced. Figure 7 illustrates how difficult it is to perceive relevant patterns, reconstruct relations, and assign significance or meaning to holistic messages if we can imagine no context for their production. It is a Siberian love-letter. But this bit of information (in contrast to the case of the American Indian petition for fishing rights) does not help us interpret the message. In spite of the drawing's highly patterned appearance, there is little means of assigning significance to the individual markings without knowing what the various lines and feather-like structures refer to (or are "about") in the shared assumptive context of the writer and recipient of the message.

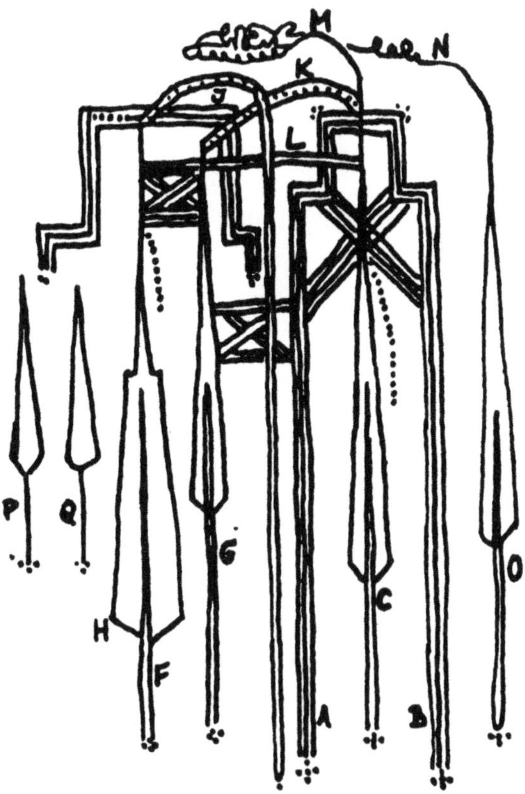

Figure 7. A holistic graphic message out of context

To interpret the Siberian love letter, we must know, or guess, among other things, that: (1) the form bounded by A/B is a house; (2) in this house lives C, a girl (with a braid, symbolized by the falling dotted lines); (3) C is unhappy (symbolized by the thick, x-formed lines over C inside the house); (4) F (a woman) and G (a man) live together in another house; (5) F is a Russian woman (symbolized by the wider skirt, H); (6) the house is incomplete (symbolized by the vertical lines, which are not fully extended downward); (7) F and G love each other (symbolized by the thick, Roman numeral ten-shaped figure over F and G); (8) C also loves G (symbolized by the thin, Roman numeral ten-shaped figure between C's house and G); (9) C's feelings go out to G (symbolized by lines K and L); (10) but they are cut through by F (symbolized by line J, which divides everything in C's world from G); (11) in spite of this, C is thinking of G (symbolized by line M); (12) another man, O, is thinking of C; (13) P and Q are children. The original translation of this message is as follows: 'You are gone; you love a Russian woman who keeps us apart; the two of you will have children; but I am still thinking of you, even if there is another man who thinks of me.'

4. What gets lost in the translation

It is interesting to consider what happens to holistic message material when it is translated into the sequential symbolic mode. The result, as is noticeable in this paper in the written translations of Figures 2, 6, and 7, is almost inevitably a loss of affective relational information that is comparable to the loss of cognitive ideational information that arises when sequential symbolic messages are translated into the holistic pictorial mode. Naturally, it is possible in principle to translate pictures into words, and words into pictures, but in each case there is a loss of information.

Written accounts of events often remain curiously unconvincing at the affective level, just as pictorial accounts often remain conceptually unclear. We get a sense of this ambiguity if we compare the description of the bombing of Guernica in Figure 8 with Picasso's painting "Guernica" in Figure 9.

Picasso's pictorial version of the bombing of Guernica lacks the conceptualized "facts" cited in the written version. It provides little clear cognitive information with which one could rationally reconstruct what literally happened in Guernica on April 27, 1936. Questions related to the who, what, when, where, and why aspects of the bombing of Guernica cannot be answered adequately on the basis of such a holistic account. The written account, on the other hand, which does provide

Guernica Spanish (gairneeka) 43 19N, 2 40W, pop (1981) 17836. Basque town in Vizcaya province, NE Spain, 25km/15 miles ENE of Bilbao, on an inlet of the Bay of Biscay. During the Spanish Civil War, on 27 April 1937, planes of the German Legion Condor, in aid of insurgent forces, bombed and strafed the defenseless town. The attack on Guernica, which was a nonmilitary objective, provoked widespread indignation, and is recalled in the famous painting by Picasso (now in Madrid).

Figure 8. A written account of the bombing of Guernica

Figure 9. Picasso's "Guernica" (1937)

answers to these questions, does not convey enough affective, relational information for an adequate emotional reconstruction of the bombing of Guernica. It explains what literally happened but not how it happened or how it was experienced by the inhabitants of the town.

Metaphorically speaking, the sequential symbolic mode possesses a powerful cognitive syntax, but lacks an adequate semantics in the realm of emotions and affective relationships. The holistic pictorial mode, on the other hand, seems to possess the emotional semantics, but lacks the necessary cognitive syntax for unambiguously defining the nature of the affective relationships it represents (cf. Watzlawick – Beavin – Jackson 1967: 66-67). Thus, sequential graphic messages convey facts more clearly than feelings, and holistic graphic messages convey feelings more clearly than facts. This makes both types of messages ambiguous, to a certain extent, in each other's terms, and complicates the translation of information from one mode into the other (see Table 1 overleaf).

An unambiguous representation of the emotional impact of the bombing of Guernica is perhaps not fully possible in a sequence of words, and an unambiguous representation of what literally was done by whom, where, when, and why in Guernica is perhaps not fully possible in a holistic picture. What is left unexpressed in each mode must thus be actively reconstructed by the interpreter, on the basis of his or her knowledge, assumptions, feelings, and/or intuitions about the reasons

Table 1. Graphic messages, expressive handicaps, and interpretive consequences

Graphic Message	holistic pictorial	sequential symbolic
	↓	↓
Expressive Handicap	ideational ambiguity	affective ambiguity
	↓	↓
Interpretive Consequence	Interpreter tends to focus on affective, relational content of the message.	Interpreter tends to focus on cognitive, ideational content of the message.
	↙ (from right column)	↘ (from left column)
	Pictures cannot convey sequences of words unless the interpreter supplies the missing cognitive ideational background.	Words cannot convey holistic pictures unless the interpreter supplies the missing affective, relational background.

for, and the contexts in which the graphic messages were produced (see Table 1). We interpret Picasso's painting, that is, against the background of what we know about the bombing of Guernica, and we interpret what we read about the bombing of Guernica against the background of what we know about Picasso's painting. To paraphrase Dürrenmatt (1990: 12), the graphic artifact itself, whether holistic or sequential, whether iconic or symbolic, is finally something "aimed at", from which our thoughts, assumptions, and expectations rebound, or are reflected - as interpretations.

If we return to the questions raised at the beginning of the paper (Can a picture tell a thousand words? Can a thousand words convey a picture?), it is perhaps possible now to suggest that these questions might be reformulated as follows: Can a holistic affective message be fully interpreted in the sequential symbolic mode? Can a sequential ideational message be fully interpreted in the holistic pictorial mode? The answer, from the standpoint of the information carrying capabilities of these two graphic modes themselves, is perhaps *no*. Nevertheless, a picture *can* potentially tell a sequence of a thousand words with the active, constructive participation of an interpreter; and a thousand words *can* potentially convey a holistic picture with the active, constructive participation of an interpreter (see Table 1). That is, as interpreters of graphic messages, we take over the task, in a sense, of filling in the missing information. In the opening quotation of the paper, Shakespeare asks, what is in the brain that ink can characterize? The answer is, everything that we can imagine.

References

Aicher, Otl
1991 *Analog und digital.* Berlin: Ernst.

Arndt, Horst – Richard W. Janney
1987 *InterGrammar: Toward an integrative model of verbal, prosodic and kinesic choices in speech.* Berlin: Mouton de Gruyter.
1991 "Verbal, prosodic, and kinesic emotive contrasts in speech", *Journal of Pragmatics* 15: 521-549.

Bernal, J.D.
1954 *Science in history.* 4 vols. London: Watts.
[1971] [Reprinted Cambridge, MA: MIT Press.]

Coulmas, Florian
1989 *The writing systems of the world.* London: Blackwell.
Davenport, Richard K.
1977 "Cross-modal perception: A basis for language?" in: Duane M. Rumbaugh (ed.), 73-83.
Dürrenmatt, Friedrich
1990 *Turmbau. Stoffe IV-IX.* Zürich: Diogenes.
Harris, Roy
1986 *The origin of writing.* Manchester: Manchester University Press.
Herodotus
1955 *Geschichtsbuch.* Vol. 4. [Trans. A. Horneffer.] Stuttgart: Kröner.
Janney, Richard W. – Horst Arndt
in press "The 'central organizer' concept: A neurological contribution to linguistics?" in: *Proceedings of the 13th Scandinavian Conference of Linguistics.* Roskilde: Roskilde University Press.
Keenan, Victor J.
1985 *The disk from Phaistos.* Athens: Efstathiadis.
Malinowski, Bronislaw
1923 "The problem of meaning in primitive languages", in: C.K. Ogden – I.A. Richards, 296-336.
Mallery, Garrick
1893 *Picture-writing of the American Indians* [= U.S. Bureau of American Ethnology. Tenth annual report (1888-1889)] Washington: U.S. Government printing office.
[1972] [Reprinted New York: Dover, 2 vols.]
Ogden, Charles Kay – Ivor Armstrong Richards
1923 *The meaning of meaning.* London: Routledge.
Ruesch, Jurgen – Gregory Bateson
1951 *Communication: The social matrix of psychiatry.* New York: Norton.
Rumbaugh, Duane M. (ed.)
1977 *Language learning by a chimpanzee: The Lana project.* New York: Academic Press.
Schmandt-Besserat, Denise
1978 "The earliest precursors of writing", *Scientific American* 238: 50-59.
1979 "Reckoning before writing", *Archeology* 32: 23-31.
Vargha, D. – J. Botos
1979 *An attempt at homophony: Free deciphering of the Phaistos disk.* Budapest: Hungarian Acadamy of Science.

Watzlawick, Paul – Janet Helmick Beavin – Don D. Jackson
1967 *Pragmatics of communication: A study of interactional patterns, pathologies, and paradoxes.* New York: Norton.
Zauzich, Karl Th.
1980 *Hieroglyphen ohne Geheimnis.* Mainz: Zabern.

Writing, inscription, and text

Götz Wienold

1. Inscriptions: A step in the development of language use

Writing, inscription, and fixed text are born together. In the course of histories and cultures, they develop in different directions. Yet, even today, something of the original unity is present.

With the development of a general fit between writing and speech, written textuality also develops. This is textuality founded on cohesion in contiguous utterances, a textuality which is not entirely alike for speech and written language, but which shares many traits in the two modalities. Before and besides this development, we have text types which are limited to inscriptions. They represent a use of writing which does not presuppose that the whole of language can be written. Such inscriptional text types are: indications, lists, packs, and orientations in a system (cf. Wienold in print).

Indication is a basic category which can become a part of the other three categories of texts, that is lists, packs, and orientations. An instance of the category "indication" can make a language participant aware that something of a certain nature - expressed as a noun or a partially incomplete noun phrase in some languages - can be found either at the place of the inscription or in its vicinity (e.g., a street sign with the name of the street, a name plate at or in front of a house, etc.) or that something can be found following a direction - in many cases, inscriptions of this kind are associated with pointers (e.g., traffic signs) - or that something is of a certain class (*milk* on a package containing milk), is meant to be used for something (*tea* on a cup out of which people are supposed to drink tea), or that something has a certain value (price, height, distance, and so on).

I have said "before and besides" the development of fit in textuality between speech and writing until now. Inscriptions are written language in a sense which is strange to the notion of language and textuality prevalent today. And it is important to get a feeling for this strangeness. Inscriptions are strange because they are not written in the sense that one writes language according to the textual cohesion of spoken language (as used in conversation or narration). Inscriptions are language which ordinarily does not occur in language as it is spoken. A name for some object or situation can occur as an inscriptional indication. But in many languages as they are spoken, a name for an object cannot function by itself except when citing a word, otherwise one has to have a noun phrase. As inscriptional indication, however, *entrance* (on or close to a door or the like) can be used by itself. There is a corollary: inscriptional indications do not allow - except within quite restricted limitations - for definiteness (e.g., *The entrance*) and quantification (e.g., *All entrances*) which are so important to the referential functions of noun phrases in conversation. (One example of what I believe to be a restricted use of quantification in inscriptional indications are traffic signs like *Alle Richtungen* or *Toutes directions*.) This means: indications as the primary type of inscriptions from the beginnings up to now are a special category of language use not met with in conversation. Wittgenstein (1970: 151) writes in his notes *On Gewißheit,* published only many years after his death: "In einem zoologischen Garten könnte die Aufschrift stehen 'Das ist ein Zebra'; aber doch nicht 'Ich weiß, daß das ein Zebra ist'." Wittgenstein, in this remark, is concerned with the use of *wissen*, 'to know', particularly with the use of *Ich weiß, daß* . . ., 'I know that . . .' To use *Ich weiß, daß* meaningfully, one has to have a reason for using it, e.g., in upholding a disputed argument. Such utterances are made in conversational exchange. Signs in a zoo bearing a name of an animal or in a botanical garden with the name of a plant do not occur in such exchanges. There are no people who question or answer, doubt or support a point within the realm of the inscription. People reading the sign may question, call into doubt, reaffirm and so on, but this is a different sphere of language use, or in Wittgenstein's terminology, a different language game. I think even Wittgenstein's phrasing *Das ist ein Zebra* is quite unusual for such a sign. The common format is *Zebra* followed, as the case may be, by information on its (Latin) scientific name, on regions of the globe where zebras live and so on. The meaning of a sign "zebra" in a zoo is, I submit, not equivalent to 'This is a zebra'. The point is subtle, but I think important. A sign like the sign "zebra" in a zoo makes do without having to refer to an object by a demonstrative, makes do without classifying *is*, makes do without making a statement. It is a simpler way of using language in which the need for making explicit reference and the need for using and distinguishing between illocutionary forces does not arise.[1] This is what I propose to call an indication. As one can see from a selection of examples, *zebra*, *out*, *3m* (height), *3km* (distance), *$20*, such indications are of quite variable use. They are there to be used if one needs information of a certain kind and in their simplest form do not go beyond this.

In the beginnings of written running text, all writing is, so to speak, "inscriptional". Letters, narratives, treatises, instructions: everything is written on tablets (cf., e.g., Kienast 1969: 51ff.) of varying shape (cf. Avrin 1991: 70f.). This, however, is not what I mean by "inscription". I use "inscription" as a term in linguistic pragmatics to characterize a type of language use. Inscriptions are written uses of language which do not have a recognizable producer (or emitter) and are not meant for specific individual receivers. They can be read/received by anyone coming into an appropriate distance. This distance, normally, is much larger than the normal reading distance for writing (e.g., of letters or notes, not of graffiti) and print (e.g., of books, not of posters). They do not arise out of, establish, or promote personal relationships and are not interpreted that way (as conversation is) (cf. Wienold 1991, in print). Inscriptions introduce the new category of language use with the beginning of writing. Indication is already present in isolated uses of single graphic signs, e.g., an arrow (cf. Krampen 1988: 40; Stankowski – Stankowski – Gomringer 1972; Mukooda 1985: 114ff; cf. also Harweg 1979: 17ff.). When such signs acquire conventional representativeness for linguistic symbols, one fundamental step in fitting a graphic representation to language as spoken has been taken. This is what is at the origin of cuneiform writing as reconstructed by Schmandt-Besserat (1977, 1979, 1982a, b, 1986a, b, 1992) and surmised also for the origins of Egyptian hieroglyphs by Schenkel (1983).

According to Schmandt-Besserat, cuneiform writing was not invented all of a sudden, but was a new step in the development of a system of recording in the ancient Middle East which can be traced back as far as the eighth or ninth millennium before our current era. Record-keeping in households, business, and in administration was done by means of clay or stone tokens of various shapes, some of which apparently had numerical value, some had incisions or punched markings upon them, some had iconic shapes of animal heads, vessels and utensils. "Various shapes of tokens are strikingly similar to some signs of the archaic Sumerian writing system" (Schmandt-Besserat 1979: 24). A decisive step towards writing was taken when tokens were stored in archives by means of sealed envelopes, "really, hollow globular cases" (Schmandt-Besserat 1986a: 34). Besides the seals, these envelopes had markings on their exterior corresponding to the shape and number of tokens included inside. "The innovation of using the images of the tokens - i.e., signs - rather than the tokens themselves proved much more efficient. Hollow clay balls filled with tokens were in time replaced by full clay balls - i.e., tablets - marked with signs" (Schmandt-Besserat 1986a: 25).

At these beginnings of writing, we find isolated single graphic marks on objects with a representational function referring to other objects. When one interprets them as "writing", that is as a conventional graphic representation for specific expressions of a human language, i.e., words or names, these bits and pieces of language have come to be "written". Only very gradually does such writing acquire the capacity of representing language to a larger extent. Inscriptions - particularly those of a single sign instance which are also observed in the pre-stages of Chinese writing, e.g., signs on pot shells of pre-Shang cultures (Toodoo 1982: 44ff.; Chueng 1983, cf. also Kaizuka – Ogawa 1981: 26ff.; Chang 1986: 122ff., 150ff., 172f., 295ff., 392ff.), do not necessarily "write" a language in the sense that potentially everything expressible in that language could also be written. Originally, only rudimentary texts could be written. Invention of a script does not mean just invention of an inventory of signs. It also means inventing a practice of writing texts, and this practice develops only gradually (cf. also Wienold 1991: 171, fn. 9). Civil and Biggs (1966: 12ff.) have shown that in early cuneiform texts, parts are missing, and they have suggested that these texts represent a stage where writing has not yet achieved the status of a general representational medium for language. Similar observations have been made for early so-called hieroglyphic Hittite or Luwian: "Les vieilles inscriptions, réduites à des 'cartouches' négligent l'expression des rapports grammaticaux; la syntaxe apparaît à peine, ou plutôt elle réside dans l'ordre des mots. La phrase écrite consiste en une succession de groupes pictographiques qui ne réalisent que l'essentiel du sens; le lecteur reconstitue la suite de la pensée sans effort" (Laroche 1960: 252; cf. also Hawkins – Morpurgo-Davies – Neumann 1973: 146).

The first point, then, I want to make here is: early writing has a specific rudimentary textuality, so to speak the extreme limiting case for textuality, namely, one sign by itself. Such single signs, though, have a specific function. They have a pragmatic force functioning upon a sign. The pragmatic force and the sign together function in some way like an utterance in speech.[2] But I also pointed out differences which go down into the structure of noun phrases. Indications of the nature of the pre-stages of cuneiform and Chinese writing have been preserved as a category of inscriptional writing up to now. They participate as a basic category of two more complex classes of inscriptional texts, lists (e.g., price lists, lists of players participating in a football or hockey match and so on) and orientations in a system (traffic signs guiding one from place to place, sign boards in buildings, stations, airports and the like). These text classes are basic to inscriptional writing, and they are text classes which do not belong to the spoken language of conversation.

The second point I want to make is that inscriptions make a specific use of language which differs from other uses of language. I call it a specific type of language use (next to the types of conversation, narration, recitation, teaching/learning of language/soliloquy; cf. Wienold 1985, 1986, 1991, in print). There is some overlap with other types, but there are basic pragmatic differences. When script is invented and develops into writing, it maintains a type of language use different from the other uses of the language it is applied to. Writing does develop in a way that conversation[3] can be written in some form (e.g., in letters), that narration can be written, or that language to be recited can be written. Nevertheless, writing maintains a basic pragmatic difference from these other uses of language, and this difference becomes apparent in inscriptions as a separate type of language use.[4]

The oracle is an interesting example of inscriptions adding a new dimension to a cultural practice. I will insist on saying that inscriptions add this dimension rather than just writing, because inscriptions are a specific cultural practice of writing, and in the beginnings of writing in various cultures may have been the most preponderant, if not the only one. Oracles, of course, are not dependent on writing. But writing adds to oracles a use of language of its own. It is highly interesting to note that the first use of Chinese writing in anything like sentences or texts are the by now famous turtle backs and bovine bones carrying oracular inscriptions (cf. Ngo 1976: 9ff.; Serruys 1986; Chang 1986: 122, 128, 150f.). According to current interpretation, these oracular inscriptions are records of oracular questions after the event to keep a record of the way the question was posed and of the outcome. The outcome can be read from the cracks fire produces when put into small grooves in the bones. That is, it can be read if one knows the art of divination with oracular bones. The question was written on the bone close to the place where the divinatory cracks were obtained. Many bones were used more than once and, thus, carry several unrelated inscriptions incised at different points in time. (As often with inscriptions, contiguity in space does not lead to textual cohesion; cf. Wienold in print). Also the textual format of the recording of an oracular "question" is interesting. It often takes the form of a sequence of two sentences, the second being the negation of the first. The divinatory crack next to the inscription determined which of the two sentences - the unnegated or the negated one - was the correct oracular prediction (Serruys 1986: 205f.). Such a sequence of sentences is a very rare instance of a cohesive device. It will be quite hard to find similar instances outside of oracular practice. This is what I mean by the claim that inscription adds quite a specific dimension to oracle. This specific use comes out of this specific practice of oracles and of recording oracles.

Toodoo rightly maintains that writing is only present when at least two conventional graphic signs fitted to conventional spoken linguistic signs are conjoined, that is when we find an inscription which clearly corresponds to a spoken sentence. Hereby, Toodoo (1982: 44ff.) considers all isolated occurences of single graphic signs in China as pre-writing (cf. Chang 1986: 295f.).[5] I agree with Toodoo insofar as writing is defined as a transcription of language fitting somehow the way it is used when spoken. But it seems to me a point well worth considering that single sign inscriptions which represent indications - that is, uses of language applying a pragmatic force to signs - also are writing of language, but not in the sense of writing language as it existed without writing. Writing from the beginning adds a new type of use: inscription.

Textuality in its graphic presentation develops only gradually. Evenness of direction of writing, line grouping, sentence and word spacing, punctuation, as well as modes of displaying a text graphically (columns, paragraphs and other ways of signalling sections, e.g., by lines or end signs) come into regular use only gradually in the course of histories and cultures. Particularly, indication of sections, spacing, and punctuation are handled quite differently in various writing systems. For example, Japanese and Thai writing do not use word spacing. Thai writing has a specific way of spacing groups smaller than sentences, but not necessarily equal to clauses. However, as Caflisch (1977: 14-20) has observed, quite early scribes set out to achieve a "logical arrangement" of signs:

> Hittite clay tablets of the 15th or 13th centuries B.C.. . .show a two-column arrangement of cuneiform characters, in between which lines are drawn not only to mark the space between the columns but also to separate the different paragraphs. . . On a monument with an inscription by King Kilamuwa of the Aramaeans in North Syria, the last lines of the paragraphs to be read from right to left, end with two adjacent floral ornaments. . . and the different paragraphs are separated by three horizontal lines, and the rows separated from each other by lines.

Crystal (1981: 69ff.) pointed out that the line is something specific to writing and has no equivalent in speech at all. Writing thus produces textual arrangement and structuring unlike what is found in language as spoken. As is well known, our notion of text up to now is deeply influenced by written texts.

Writing also develops characteristic fits with other types of language use. Up to the present time we find inscriptions consisting of sayings, proverbs, and the like. A well-known use of inscriptional sayings are inscriptions on houses (cf., e.g., Widera 1990). We also find inscriptional sayings on grave stones. Grave stones or grave cross inscriptions show quite a complicated text structure mixing indications as described earlier (e.g., names and dates which may take a sentential format but do not necessarily do so) with sentential sayings, a kind of inscriptional recitation (for examples, cf. Baur-Heinhold 1984; Dickenberger 1990). Ritual uses of inscriptions are quite common: name tablets of deceased persons in ancestor rites in Japan or Korea (Smith 1976; Janelli – Janelli 1982), names written on altars as representation of the deceased and as memorial aid for intercessions during mass for people alive or dead,[6] inscriptions on amulets which people wear around their necks. (For Islamic examples, where such inscriptions have verses[7] which are also recited in rituals, see Welch 1980: 167ff.) Inscriptions of texts which are also recited in religious rituals are found in India as well (cf. Renou – Filliozat 1985: 322f.) and are indeed common enough.

A further indication of the specific status of inscriptions is that they align themselves in specific ways in multimedial texts. Posters are a multimedial text form which combines linguistic expression (mostly in non-sentential forms, often of the category indication) and other kinds of visual expressions (cf., e.g., Metzl 1963; Barnicoat 1975; Sahlström 1990). Moreover, certain inscriptions also combine in specific ways with the objects on which they are written. Inscriptions which report historical events (a victory, ascension to the throne, the conversion of a people to a new religion[8]) are celebrations made permanent. As inscriptions on monuments are sometimes found to be factually incorrect and, indeed, can sometimes be shown to have intentionally falsified history (cf. Mittig 1987: 468ff.), the ensemble of inscription and monument or memorial stone is something which goes quite beyond other types of language use. Textuality of inscription, then, shows special facets not to be found in conversation, narration, recitation and so on (which, in turn, of course, have their own special facets not shared by inscriptions).

Coins, medals, and seals, yet again, are objects carrying inscriptions which show a combination of objects and inscriptional texts as representatives of authority and power. Coins in the possession of and spent by individuals, are founded, nevertheless, on power and authority far beyond individual existence. The inscriptions on coins express this power and authority (reign of a king, the issuing state, invocation of a god etc.). At the foundation of inscriptions, we find superindividual power and authority. As a script comes into existence with specific structures of superindividual authorities - and since its maintenance requires such structures - inscriptions carry this link from the beginning of writing to this very day. Messick (1993) makes for highly instructive reading on the relationship between writing and authority.

2. Inscriptions vs. literature

Inscriptions form one root of literature, a root which has had a good deal less chances of developing in most cultures than the other two roots, narration and recitation. Whereas narration and recitation have flourished generally, inscription as a channel of literary life has been flowing meagerly. Inscriptions do not belong to the mainstream of language and the (linguistic) textual culture which has gained the attention of a large group of practitioners, connoisseurs, and theoreticians. Consider the development of narration and recitation in more silent individualized reception practices since about the thirteenth century, particularly since the advent of print (cf. Saenger 1982) and since literacy became more and more a common cultural asset. (Notice that in Western Europe the latter seems to have developed on the basis of the former, whereas in Japan, for instance, such a connection seems to have been much less obvious if it existed at all.) This development has made narrative and recitation even more diverse to the point that people may no longer be aware that narrative literature has its foundation in a specific type of language use practiced orally, namely narration. People seem to have become almost equally oblivious to the fact that poetry has its roots - not only historically, but structurally up to this very day - in recitation. They are no longer aware of the fact that poetry is or was at least at some time related to inscriptions in epigraphic poetry, epitaphs and the like, which is conserved as a specific genre in the epigram. Epic has died away in many cultures anyhow. Only drama still practices recitation widely. Drama has successfully withstood becoming primarily a matter of reading and of printed books. Movies turned into video may perhaps in the future become the graveyard of recited drama. Inscriptions were a way of publishing and practicing literature in the early written cultures of the Near East. In China, inscriptions as early as the famous Shang and Zhou bronzes with dedicatory pronouncements of veneration for the ancestor have to some degree been part of a literary practice. In India, at least the Asoka inscriptions seem to be a case of interest in this connection.

But inscriptions in the West do play a role in literary sedimentation. Culling pieces of literary expression from their original context and putting them up, cast in iron, cut into stone, or written on wood with brush or pen in public places has been practiced in various ways in several cultures. To this very day we find them around in our daily lives; and they do have an effect. When, in 1957, the University of Munich, upon completion of the restoration of its *Lichthof*, decided to also have the erstwhile iron casts of Latin literary phrases restored, students launched public protests against having Horace's *Dulce et decorum est pro patria mori* up again. Inscriptions of this kind are fragmented literature broken away from context and made even more permanent, seemingly unquestionable, self-evident.

There have been inscriptions of poetry in China for many centuries (e.g., in the mausoleum for Confucius) and in Japan (cf. Mori 1988: 233ff. on the Basho House in Sekiguchi, Tokyo). In Latin literature, "the literary epitaph was an accepted verse-form, the elegists worked funerary and dedicatory half-lines and couplets into their poems . . ." (Ireland 1983: 233). Today, the practice of putting up a stone with a poem (or part of a poem) in commemoration of a poet is flourishing in Japan. Takamura Kootaroo, sculptor and poet, incised a large part of his fellow poet Miyazawa Kenji's famous poem *Amenimo makezu* ('Not yielding to the rain nor . . .') for a large commemorative slate displayed publicly in Hanamaki, Iwate prefecture, a town that played a role in both poets' lives. As the text of such poems is usually inscribed entirely or at least to a large extent, this is a different practice than the quotational inscription discussed earlier.

For a third instance of inscription as a form of publishing and practicing literature we may briefly look at graffiti. Graffiti are a form of unofficial literature, probably rarely admitted to a candidacy for even that. Yet, graffiti display some traits of literature, if we use literature as a notion to describe cultural practice rather than as a normative notion resting on sanctioned cultural codifications (cf. Neumann 1986; Hashiguchi 1991; Roettig 1992).

Genuine literary practices in some cultures (of greater or smaller extent), literary sedimentation, and unofficial literature ("counterliterature"), hence, show inscriptions to have a place in the evolution of language as literature when it started to embark upon the passage ways of writing.

There may exist or may have existed a culture somewhere at some time which assigned inscriptions a larger segment of its literary life. We may imagine a "modern" culture with such a practice. There are parts and environments in (almost) every human settlement of this imaginary culture displaying texts incised, inscribed, impressed, penned, or maybe just scribbled on/in stone, iron, wood, plastic. Texts that win attention and distinction will be displayed here as well as texts which the curators of the inscription parks and inscription halls deem well deserving of attention and acclaim. Artists of the inscriptional school will also publish new works here. People will gather and draw each other's critical sense and inclination to adore these texts. If that culture be a ("post")modern one, computer/light display equipment will have a store of texts as well as active publishing facilities. Let us at least speculate about such a culture to get an idea whether it is really incompatible with what we know literature to be and whether it may really be incompatible with what literature may possibly develop to be in some culture, a culture oriented towards literature, permanency of media, and communal sharing of art all at the same time.

3. Writing, typography and legibility

In the history of alphabetic writing and its development in print, media legibility has been a major issue. This is a conclusion one may draw from the literature on both strains of development, the development of alphabetic script in handwritten use and its development in the print media.[9] Again, in the development of types for computer use, legibility has been a major issue. The typographical reforms in the nineteen twenties even set themselves the goal of maximum legibility which could not be improved on (cf. Gerstner 1985: 138).[10] Standards of legibility, on the other hand, are sometimes encroached upon or weakened when aesthetic principles come into play or when writing, particularly inscriptional writing, turns into ornament.[11]

Legibility of written language not only refers to the legibility of the typefaces when combined into words and running text (body type or main type) but also to all the aspects of displaying and organizing a written text on surfaces by giving it a "macropunctuation" (as Waller 1980 calls it; cf. also Morison 1967; Tschichold 1975: 24ff.). In an earlier section we quoted Caflisch's notes on forms of macropunctuation in inscriptions as early as in the second millennium before our current era.

In modern treatments of the history of the alphabet in the West, i.e., the history of writing based on the Roman alphabet, time and again, the importance of the Caroline minuscule is stressed. In the Caroline minuscule, earlier steps towards ascenders and descenders in uncial, younger Roman cursive writing and semiuncial reach their fully developed form. Ascenders and descenders not only achieve greater distinctiveness of letter shapes, but give definite contours to letter combinations in words. The combination of letter shapes with and without ascenders and descenders in different places articulates and distinguishes graphic word shapes. Distinctiveness of letter shapes and delineation of word shapes through contours is also aided by a mixture of straight and round forms (Tschichold 1942: 10ff.).[12] Also, spaces between (graphic) words have only been introduced since Irish writing, which is somewhat earlier than the Caroline minuscule, and Caroline writing. "Word separation was the singular contribution of the early Middle Ages to the evolution of Western written communication" (Saenger 1982: 377). With Caroline minuscule, the form of alphabetic writing, as we know it also from print, has been achieved almost completely. Only the dot on the letter *i* and the top of the *t* were added later. The Antiqua typefaces since the early Italian masters of the 15th century (cf. Morison 1925) and the French type cutters, foremost Claude Garamond, are a combination of Caroline minuscule for lowercase and Roman capital letters. The full-fledged form of what have become ordinary letters today was established. When the Renaissance designers of print mistook the inherited minuscule - in their time, the humanistic minuscule of Arrighi and others - for the Roman way of writing, added to them capital letters as derived from the Roman Capitalis,[13] and implemented the i-dot, the basic form of all print not derived from the fractured Gothic writing had been established, namely, the Antiqua type in all its versions, which has also been the basis of the grotesque and all sanserif typefaces since the 19th century.[14] Humanist scribes also improved on textual legibility by inventing modern quotation marks, introducing the parenthesis and evolving "full syntactical sentence punctuation with the characteristically modern usage of the comma and period" (Saenger 1982: 410; following Roncaglia 1941).

However, there are some interesting further steps worth noting. The first is the development of grotesque typography, doing away with serifs and thereby deleting an important link between printed type and handwritten letters. Renner (1925), the creator of Futura, one of the most successful 20th century sanserif types, notes how persistent over almost five centuries the inheritance from writing in print has been, that serifs were preserved in print although they were only a part of letter shapes resulting from the use of the pen. The reformers or revolutionizers of typography in the twenties were very much influenced by artistic movements of the time, particularly Russian constructivism and Dutch De Stijl. Lissitzky's work is strongly represented in Tschichold's manifesto type presentation of the new typography of 1925 (Tschichold 1925; cf. also Lissitzky 1925; Tschichold 1928). De Stijl was directly involved with Bauhaus typography once Theo van Doesburg assumed the directorship of the Bauhaus. Tschichold's typography of that time as well as all Bauhaus inspired typography, inspired as they were by constructivism (cf. Tschichold 1988), derived the type shapes from geometric forms, e.g., the letter *o* of such typefaces, was based on the circle.[15] Tschichold, when threatened by Nazism, moved to Switzerland, gradually distanced himself from the new typography and more or less from sanserif typography in general at least for the typography of books, and also criticized its dependency on pure geometric forms for disregarding the ways letters are perceived by human eyes (cf. Tschichold 1942, 1975). Human eyes tend to "magnify all horizontal values and to diminish vertical ones" (Ruder 1967: 94). But what Tschichold criticized in his own work as well as in sanserifs in general is due at least in part to the application of constructivist geometric principles. More recent creations of sanserif types, foremost Frutiger's *univers* and Aicher's *rotis* deviate quite deliberately from pure geometrical shapes (cf. Ruder 1967: 96f.; Frutiger 1980: 16ff.; Ruder 1980; Aicher 1988).[16] Interestingly enough, both Tschichold and Frutiger praise Antiqua types, Frutiger particularly following Nikolaus Jenson when designing his *meridien* type, Tschichold the Hungarian Niklaus Kis, a type cutter of the 17th century (Tschichold 1960: 45ff.).[17] Thus, sanserif types have seen a new future after Tschichold's propagation and renunciation of them. And thus, a class of typefaces which has severed some links with handwriting, as it persisted from Renaissance typography onwards, may not only survive but gain ascendancy.[18]

In the last paragraph, I have followed some details of typographical developments in the 20th century. In all of these discussions, legibility has been a main issue, although Tschichold as well as Frutiger or Aicher, of course, also stress the artistic and aesthetic qualities of typography and writing in general. Legibility as a measure for the quality of writing shows that in the history of material and form of writing standards of conversation have become preponderant. Writing has taken on the function of representing language on a grand scale. Although, of course, it cannot present spoken language, it can write language to a large extent as it is spoken. And in this it has opened up functions for language which had not been open to spoken language by itself. The standards that the developments of written language follow largely derive from the basic principles of one type of language use: conversation. Similarly, our notion of text is very much related to these developments in the history of writing.

Yet, writing in inscriptions has preserved very early features of inscriptional writings, pragmatic forces (as in the class of indication) and text types (as in lists) up to date. These pragmatic features of inscriptions, however, are not noticeable to a way of thinking about language which is mostly, if not solely, attending to one type of language use only, i.e., to conversation.

Conversation, narration, and recitation (including ritual) as types of language use probably antedate inscriptions by at least tens of thousands of years in the evolution of language. Writing aligns itself with some of the uses but before doing so has given birth to a new type. Writing has been used to develop forms of language extending the realms of conversation. It has also been used to take over narration, and the development of poetry shows that it has also nearly replaced recitation in one of its forms (not so yet in the form of drama and ritual, although it has also intruded into these areas). The importance of writing for history, language, cultural development has often been portrayed. But what is being neglected, I submit, is that the type of language use specific to writing, that is its use in inscriptions which came into existence only with writing and has been with it in its very early form of use when writing could be used only for segments of language, is still with us today.

Writing in a very modern use, writing in the design and application of trade marks (cf., e.g., Igarashi 1987), still follows these very early uses of inscriptional writing in indications. The design aspects of this inscriptional use, however, are new to writing and take it into a semiotic sphere which has only begun to be explored (cf. Bense 1971). Inscriptions, as I have argued elsewhere, have taken on the value of an ideational ornament, in which more than the visual form the ideational content is ornamental (Wienold 1993a). This step in the history of writing, however, is dependent on typography. Typography, particularly 20th century typography, seems to have carved out in writing an ideational value. The typographers of the twenties were all artists who did not shun at all from applying their talents to the propagation of new industrial goods and the industrial world. From that time onwards this ideational value of type has been linked to representing trades, brands and the corporate world.[19]

Notes

1. Similarly I would argue against Harweg (1979) who maintains that a street sign with the name of the street is to read: "Here is the . . . street." According to Harweg, the sign itself has the same function with regard to the inscription that a human speaker has with regard to an utterance he or she makes. Harweg even calls the sign "Äußerungsträger" and writes: "Der Stein spricht" (1979: 9; cf. 14ff.). In my view, even the term 'utterance' ("Äußerung") may be misleading when we apply it to inscriptions. Harweg's problem and view of the problem only arise when one construes inscriptions along the lines of utterances and acts in conversations. Harweg rightly rejects the view a sign with an inscription should be construed as the utterance of a magistrate or functionary. The sign and its inscription remain, and their authority remains when the individual functionary has left office, has died and so on. They are independent of such individuals. That is, behind inscriptions of this kind, there is a supraindividual authority which does not speak and behave the way individuals do in conversation. There are inscriptions which contain uses of deictics. But indicational inscriptions, which consist of nouns or rudimentary (with regard to conversational noun phrases: defective) noun phrases, do not. This as well as the consistent lack of full noun phrase structures and of a sentential format of such inscriptions give support to the claim that indications are a use of language in some way simpler than acts of speech in conversation. They are there for a "reader" seeking information in a situation. Harweg takes note of this to some degree when he recognizes that references to users of an inscription as well as cases of temporal references have to be interpreted differently for each occasion when an inscription is interpreted. Deictics have to be dealt with for each type of language use, which will be distinguished later on in this paper, separately.
2. "Pragmatic force functioning on a sign": cf. Searle and Vanderveken (1985: 15), who distinguish illocutionary force (or point) and propositional content and say that "the illocutionary point is achieved on propositional content". On the force of inscriptions cf. Wienold (1993b). - A more restricted and limited case of inscriptions are inscriptions of single letters only. These cannot be treated within a theory of language bound to referential semantics. One finds inscriptions consisting of single letters - sometimes assemblages of such single letter inscriptions - in Shingon Buddhism in Japan. The letters

used are letters of the so-called Siddham script, derived from the Sanskrit Devanagari script and introduced via China. The letters stand for the sounds used in meditative practice, but the latter shapes are also a part of meditative and ritual practice (cf., e.g., van Gulik 1956; Yamazaki 1974: 156ff.).

3. I use "conversation" in a technical sense as a term for a type of language use which is not to be identified with spoken language (which has also types like recitation, narration and so on) and is not confined to spoken language. This technical sense is taken from Grice (1975).
4. I do not distinguish between a linguistic expression incised in some material and a linguistic expression written on some material or between German *Inschrift* and *Aufschrift* (cf. also Wittgenstein's use of *Aufschrift* in the earlier quotation). I see no difference as to the basic pragmatic type of language use which I wish to characterize. Harweg (1979) uses the term *Beschriftung*.
5. Most of all, these isolated graphic signs appear on pottery (cf. Chueng 1983). Again, it is interesting to note that this use of graphic signs - debatable inscriptions, at least - is one of the oldest and found in several cultures in their developmental phases of writing (also, e.g., found in Egypt before hieroglyphic writing was extant in the form we know it, cf. Schenkel 1983; Petrie 1912) and is one continued over many centuries of full-fledged writing outside inscriptional writing up to today (cf., e.g., Chaffers 1979). A stamp indicating the origin or the place of production, in recent times, has mostly been placed on the outer bottom of ceramics or metal objects, more rarely on the outer visible rim of a vessel (examples in Gysling-Billeter 1975: 60f.) or its inner bottom (again see Gysling-Billeter 1975: 106). In the collection of objects presented in Gysling-Billeter (1975), one often finds a name and a figurative or abstract graphic mark (a signet) juxtaposed. When the indications of origin are in writing only, they tend to be decorative signatures. Thus, a relationship between craftsmen's marks and ornaments has formed.
6. A very interesting example can be found in the names written in ink or incised onto the altar plate of the church in Reichenau-Niederzell, Germany. Cf. Geuenich – Neumüllers-Klauser – Schmid (1983, particularly pp. 39f.). I have to thank Alexander Patschovsky (Konstanz) for introducing me to this most interesting specimen of inscriptions.
7. Metrical inscriptions are not uncommon at all. Cf. the references to related publications in the index to Koch (1987: 138).
8. E.g., von Gabain (1979: 50): "Im Jahre 832 errichtete ein uigurischer Steppenfürst in Kara-balgasun eine uigurisch-sogdisch-chinesische Inschrift und berichtete von der einstigen Bekehrung seines Volkes vom

Buddhismus zum Manichäismus". The most famous example may be the Asoka inscriptions in India (cf. Bloch 1950).

9. Cf. Tschichold 1942: 13f., 21f.; Ruder 1967: 6; Spencer 1969; Carter – Day – Meggs 1985: 81ff.; Gerstner 1985: 130ff.; Aicher 1989: 128ff.
10. Cf. also Tschichold's (1975: 18ff) criticism of his own contribution to the typographic reform movement of the twenties.
11. Cf. Gray 1976; Gray 1986: 9; Ruder 1967: 6; Ettinghausen 1974.
12. Tschichold (1942: 14): "Auch die Karolingische Minuskel wurde mit einer Kielfeder geringer Breite geschrieben. Ausläufer der Buchstaben nach oben und unten, die von den spätrömischen Verkehrsschriften vorbereitet worden waren, dienen dazu, den Wortbildern charakteristische Umrisse zu erteilen, die die Lesbarkeit außerordentlich verbessern. Die römischen Buchstaben, darunter die Rustika, hatten den Wörtern keinen anderen Umriß gegeben als den eines verschieden langen Rechtecks. Die Schreibschnelligkeit wird durch die arkadenförmigen Hauptzeichen n, m, u und durch die Einführung vieler runder Zeichen wie e, t, l bedeutend erhöht."
13. Only the lower case *v*, *w*, *x*, *y*, *z* types are not derived via the development of cursive, uncial and semi-uncial writing into Caroline minuscule and the scripts descending from it. Thus, there has been, up to now, no distinction between the lower case type and the capitals in these five letters. Hence, even today, our letter shapes are still on the level of development achieved in Caroline writing, a rather remarkable historical fact to note. Tschichold stressed this fact time and again.
14. On the 19th century grotesque types, see Gray 1976: 37ff.; Bartram 1975: 7ff., 46ff., 72ff., 84ff. On sanserif lettering in architecture and art of the 18th century, see Mosley 1965.
15. Renaissance artists had also been much preoccupied with the geometrical forms of the Roman capital letters (cf. Avrin 1991: 195).
16. Also, e.g., the o's of 19th century grotesques deviated from pure circles. Alan Bartram refers to "the slight angularity of the grotesque". "The o's for instance are not at all circular, but resemble rectangles with rounded corners. This angularity is a common feature of grotesques, and also of much of the most successful architectural lettering from 1820 onwards. The explanation, presumably, is that it relates to the angularity of the architectural form" (Bartram 1975: 48). Entirely round o's, on the other hand, are found in sanserif inscriptions of the 20th century, e.g., of fascist inscriptions in Italy (cf. examples in Bartram 1975: 144ff.).
17. On the life and work of Jan, or as he called himself for a while, Iwan, Tschichold, with Stanley Morison and Giovanni Mardersteig, one of the most influential typographers of the 20th century, see McLean (1975) and Klemke (1988).

18. Still, it is noteworthy that Frutiger in developing his new types (univers, meridien and so on) often refers to very specific features of the history of handwritten alphabetical writing (cf., e.g., Frutiger 1980: 18).
19. For a presentation of the relationship between the artistic principles of the Bauhaus and applications in advertising, cf. Gottschall – Lubalin (1964: 15ff.).

References

Aicher, Otl
1989 *typographie*. (2nd edition.) Lüdenscheid: ernst & sohn.

Assmann, Aleida (ed.)
1991 *Weisheit: Archäologie der literarischen Kommunikation III*. München: Fink.

Assmann, Jan – Aleida Assmann – Christoph Hardmeier (eds.)
1983 *Schrift und Gedächtnis: Beiträge zur Archäologie der literarischen Kommunikation*. München: Fink.

Avrin, Leila
1991 *Scribes, script and books: The book arts from antiquity to the Renaissance*. Chicago: American Library Association.

Barnicoat, John
1975 *A concise history of posters*. London: Thames & Hudson.

Bartram, Alan
1975 *Lettering in architecture*. London: Lund Humphries.

Baur-Heinhold, Margarete
1984 *Schmiedeeisen - Grabkreuze*. München: Callwey.

Bense, Max
1971 *Zeichen und Design: Semiotische Ästhetik*. Baden-Baden: Agis.

Bloch, Jules
1950. *Les inscriptions d'Asoka*. Paris: Les Belles Lettres.

Busch, Werner – Peter Schmoock (eds.)
1987 *Kunst: Die Geschichte ihrer Funktionen*. Weinheim and Berlin: Quadriga, Beltz.

Caflisch, Max
1977 *Typographie braucht Schrift / Typography needs type / La typographie a besoin des caractères*. Kiel: Rudolf Hell.

Carter, Rob – Ben Day – Philip Meggs
1985 *Typographic design: Form and communication*. New York: Van Nostrand Reinhold.

Chaffers, William
1979 *Marks and monograms on European and Oriental pottery and porcelain.* (15th rev. edition.) 2 vols. British section ed. by Geoffrey A. Godden, European and Oriental sections ed. by Frederick Litchfield – R.L. Hobson. London: W. Reeves.

Chang, Kwang-Chih
1986 *The archaeology of ancient China.* (4th edition.) New Haven: Yale University Press.

Chueng, Kwong-Yue
1983 "Recent archaeological evidence relating to the origin of Chinese characters", in: David N. Keightley (ed.), 323-391.

Civil, M.–. R.D. Biggs
1966 "Notes sur des textes sumériens archaiques", *Revue d'assyriologie et d'archéologie orientale* 60: 1-16.

Cole, Peter – Jerry L. Morgan (eds.)
1975 *Syntax and semantics, vol. 3: Speech acts.* New York: Academic Press.

Crystal, David
1981 *Directions in applied linguistics.* London: Academic Press.

Deydier, Christian
1981 *Chinesische Bronzen.* München: Hirmer.

Dickenberger, Udo
1990 *Liebe, Geist, Unendlichkeit: Die Inschriften des Stuttgarter Hoppenlau-Friedhofs und die poetische Kultur um 1800.* Hildesheim: Olms.

Enzweiler, Jo – Heinrich Popp (eds.)
1993 *Der Zaun: Visign & Totentanz* (= Studien. Schriftenreihe des Instituts für aktuelle Kunst im Saarland, an der Hochschule für Bildende Künste Saar 1). Saarbrücken: St. Johann.

Ettinghausen, Richard
1974 "Arabic epigraphy: Communication or symbolic affirmation", in: Dickran K. Kouymjian (ed.), 297-317.

Friedl, Friedrich
1986 "Lernen von Jan Tschichold: Echo und Reaktionen auf das Sonderheft 'Elementare Typographie'", in: Tschichold (ed.) 1925 [reedition 1986], 1-12.

Frutiger, Adrian
1980 *Type sign symbol.* Zürich: ABC.

Gerstner, Karl
1985 *Kompendium für Alphabeten: Systematik der Schrift.* Teufen AR: Arthur Niggli. [First published 1972].

Geuenich, Dieter – Renate Neumüllers-Klauser – Karl Schmid (eds.)
1983 *Die Altarplatte von Reichenau-Niederzell.* (Monumenta Germaniae Historica: Libri memoriales et necrologia n.s. 1.) Hannover: Hahnsche Buchhandlung.
Gottschall, Edward M. (ed.) – Herbert Lubalin (Designer)
1964 *Typographic directions: Advertising directions 4 (Trends in visual advertising).* New York: Art Directions.
Gray, Nicolete
1976 *Nineteenth century ornamented typefaces.* Berkeley and Los Angeles: University of California Press.
1986 *A history of lettering: Creative experiment and letter identity.* Oxford: Phaidon.
Grice, H. Paul
1975 "Logic and Conversation", in: Peter Cole – Jerry L. Morgan (eds.), 41-58.
Gysling-Billeter, Erika (ed.)
1975 *Objekte des Jugendstils aus der Sammlung des Kunstgewerbemuseums Zürich.* Bern: Benteli.
Harweg, Roland
1979 "Dauer-Deixis oder Wie sind Beschriftungen zu lesen?" *Orbis* 28: 5-26.
Hashiguchi, George
1991 *Die Mauer - The Berlin Wall 1981-1991 - Berurin-no kabe 1981-1991.* Tokyo: Jotto Center.
Hawkins, D.J. – Anna Morpurgo-Davies – Günter Neumann
1973 *Hittite hieroglyphs and Luwian: New evidence for the connection* (Nachrichten der Akademie der Wissenschaften in Göttingen, I. Phil.-hist. Kl. 1973, No. 6), 145-197.
Henig, Martin (ed.)
1983 *A handbook of Roman art: A comprehensive survey of all the arts of the Roman world.* Ithaca, N.Y.: Cornell University Press.
Igarashi, Takenobu (ed.)
1987 *World trademarks and logotypes II.* Tokyo: Graphic-Sha.
Ireland, Robert
1983 "Epigraphy", in: Martin Henig (ed.), 220-233.
Janelli, Roger L. – Dawnhee Yim Janelli
1982 *Ancestor worship and Korean society.* Stanford, Cal.: Stanford University Press.
Kaizuka, Shigeki – Tamaki, Ogawa
1981 *Chuugoku-no kanji* ['The Chinese characters of China']. Tokyo: Chuuookooronsha.

Keightley, David N. (ed.)
1983 *The origin of Chinese civilization*. Berkeley and Los Angeles: University of California Press.

Kienast, Burkhart
1969 "Keilschrift und Keilschriftliteratur", in: *Frühe Schriftzeugnissse der Menschheit*, ed. Joachim-Jungius-Gesellschaft der Wissenschaften (Hamburg). Göttingen: Vandenhoeck & Ruprecht, 39-55.

Klemke, Werner (ed.)
1988 *Leben und Werk des Typographen Jan Tschichold*. [Mit einer Einleitung von Werner Klemke, der Bibliographie aller Schriften und fünf großen Aufsätzen von Jan Tschichold.] München: K.G. Saur. [First Dresden, 1977].

Koch, Walter
1987 *Literaturbericht zur mittelalterlichen und neuzeitlichen Epigraphik (1976-1984)*. München: Monumenta Germaniae Historica.

Kolers, Paul A. – Merald E. Wrolstad – Herman Bouma (eds.)
1980 *Processing of visible language 2*. New York: Plenum.

Kouymjian, Dickran K. (ed.)
1974 *Near Eastern numismatics, iconography, epigraphy and history: Studies in honor of George C. Miles*. Beirut: American University of Beirut.

Krampen, Martin
1988 *Geschichte der Straßenverkehrszeichen: Diachronische Analyse eines Zeichensystems*. Tübingen: Stauffenburg.

Laroche, Emmanuel
1960 *Les hiéroglyphes hittites*, Part I: *L'écriture*. Paris: Centre national de la recherche scientifique.

Lissitzky, El
1925 "Typographische Tatsachen", in: Aloys Ruppel (ed.), 152-154.
1988 *Werke und Aufsätze*. Berlin: Gerhardt.

McCoy, J. – T. Light (eds.)
1986 *Contributions to Sino-Tibetan Studies*. Leiden: Brill.

McLean, Ruari
1975 *Jan Tschichold: Typographer*. London: Lund Humphries.

Messick, Brinkley
1993 *The calligraphic state: Textual domination and history in a Muslim society*. Berkeley: University of California Press.

Metzl, Ervine
1963 *The poster: Its history and art*. New York: Watson & Guptill.

Mittig, Hans-Ernst
1987 "Das Denkmal", in: Werner Busch – Peter Schmoock (eds.), 457-489, 755-756, 782-783.
Mori, Shooji
1988 *Hikokutechoo: mikata-to gijutsu* ['A guide to the incision of inscriptions: what they look like and how they are made']. Tokyo: Mokujisha.
Morison, Stanley
1925 "The type of the Hypnerotomachia Poliphili", in: Aloys Ruppel (ed.), 254-258.
1967 *First principles of typography.* (2nd edition.) (=Cambridge author's and printer's guides 1.) Cambridge: University Press.
Mosley, James
1965 "The nymph and the grot: The revival of the sanserif letter", *Typographica* 12: 2-20.
Mukooda, Naoki
1985 *Sekai-no kazami: The weather vanes of the world.* Tokyo: Bijutsushuppansha.
Neumann, Renate
1986 *Das wilde Schreiben: Graffiti. Sprüche und Zeichen am Rande der Straßen.* Essen: Blaue Eule.
Newell, William H. (ed.)
Ancestors. The Hague: Mouton.
Ngo, Van Xuyet
1976 *Divination, magie et politique dans la Chine ancienne.* Paris: Presses Universitaires de France.
Petrie, W.M. Flinders
1912 *The formation of the alphabet.* London: Macmillan.
Renner, Paul
1925 "Revolution der Buchschrift", in: Aloys Ruppel (ed.), 279-282.
Renou, Louis – Jean Filliozat
1985 *L'Inde classique: Manuel des études indiennes.* Tome II. Paris: École Française d'Extrême Orient.
Roettig, Petra
1992 "'Proletarier der Welt, vergebt mir': Von der Inschrift zu den Graffiti-Formen der Denkmalkommentare", *Neue Zürcher Zeitung*, Fernausgabe 210, 11.9.1992: 43-44.
Roncaglia, Aurelio
1941 "Note sulle punteggiatura medievale e il segno di parentesi", *Lingua nostra* 3: 6-9.

Ruder, Emil
1967 *Typographie: Ein Gestaltungslehrbuch*. Teufen AR: Arthur Niggli.
1980 "Das Phänomen Grotesk und die Bedeutung der Univers", in: Adrian Frutiger (ed.), 13-14.

Ruppel, Aloys (ed.)
1925 *Gutenberg Festschrift zur Feier des 25jährigen Bestehens des Gutenbergmuseums in Mainz*. Mainz: Verlag der Gutenberg-Gesellschaft.

Sabban, Annette – Christian Schmitt (eds.)
in print *Der sprachliche Alltag: Festschrift für Wolf Dieter Stempel*.

Saenger, Paul
1982 "Silent reading: Its impact on late medieval script and society", *Viator: Medieval and Renaissance Studies* 13: 367-414.

Sahlström, Berit
1990 *Political posters in Ethiopia and Mozambique: Visual imagery in a revolutionary context*. Stockholm: Almqvist & Wiksell International.

Schenkel, Wolfgang
1983 "Wozu die Ägypter eine Schrift brauchten", in: Jan Assmann – Aleida Assmann – Christoph Hardmeier (eds.), 45-63.

Schmandt-Besserat, Denise
1977 "An archaic recording system and the origin of writing", *Syro-Mesopotamian Studies* 1/2: 1-33.
1979 "An archaic recording system in the Uruk-Jemdet Nasr period", *American Journal of Archaeology* 83: 19-48.
1982a. "The emergence of recording", *American Anthropologist* 84: 871-878.
1982b "How writing came about", *Zeitschrift für Papyrologie und Epigraphik* 47: 1-5.
1986a "An ancient token system: The precursor to numerals and writing", *Archaeology* 39, 6: 32-39.
1986b "The origins of writing: An archaeologist's perspective", *Written Communication* 3, 1: 31-45.
1992 *Before writing I: From counting to cuneiform*. Austin: University of Texas Press.

Searle, John R. – David Vanderveken
1985 *Foundations of illocutionary logic*. Cambridge: University Press.

Serruys, P.L.-M.
1986 "Notes on the grammar of the oracular inscriptions of Shan", in: J.McCoy – T. Light (eds.), 205-257.

Smith, Robert J.
1976 "Who are the 'ancestors' in Japan? A 1963 census of memorial tablets", in: William H. Newell (ed.), 33-60.

Spencer, Herbert
1969 *The visible word*. London: Lund Humphries.

Stankowski, Anton – Joachim Stankowski – Eugen Gomringer
1972 *Der Pfeil: Spiel - Gleichnis - Kommunikation*. Starnberg: Josef Keller.

Toodoo, Akiyasu
1982 *Kanji-no kako-to mirai* ['Past and future of Chinese characters']. Tokyo: Iwanami.

Tschichold, Iwan (ed.)
1925 *typographische mitteilungen, sonderheft (=oktoberheft 1925): elementare typographie*. Leipzig.
[1986] [Reprinted Mainz: H. Schmidt.]

Tschichold, Jan
1928 *Die neue typographie. Ein Handbuch für zeitgemäss Schaffende*. Berlin: Verlag des Bildungsverbandes der deutschen Buchdrucker.
[1987] [Reprinted Berlin: Brinkmann & Bose.]
1942 *Schriftkunde, Schreibübungen und Skizzieren für Setzer*. Basel: Schwabe.
1960 *Erfreuliche Drucksachen durch gute Typographie: Eine Fibel für Jedermann*. Ravensburg: Otto Maier.
[1988] [Reedition Augsburg: Maro.]
1975 *Ausgewählte Aufsätze über Fragen der Gestalt des Buches und der Typographie*. Basel and Stuttgart: Birkhäuser
1988 "El Lissitzky (1890-1941)", in: El Lissitzky, 3-8.

Van Gulik, R.H.
1956 *Siddham: An essay on the history of Sanskrit studies in China and Japan*. Nagpur.

Von Gabain, Annemarie
1979 *Einführung in die Zentralasienkunde*. Darmstadt: Wissenschaftliche Buchgesellschaft.

Waller, Robert H.W.
1980 "Graphic aspects of complex texts: Typography as macropunctuation", in: Paul A. Kolers– Merald E. Wrolstad – Herman Bouma (eds.), 241-253.

Welch, Anthony
1980 "Islamic calligraphy: Meaning and symbol", in: Paul A. Kolers – Merald E. Wrolstad – Herman Bouma (eds.), 157-167.

Widera, Joachim
1990 *Möglichkeiten und Grenzen volkskundlicher Interpretationen von Hausinschriften*. Frankfurt a. M.: Lang.

Wienold, Götz
1985 "Types of language use: A notion relevant to the writing of literary history?" *Poetics* 14: 345-363.
1986 "Rezitation und Konversation: Linguistische und semiotische Aspekte dramatischer Texte und des Theaters", *Zeitschrift für deutsche Philologie* 105: 257-282.
1991 "Inschriften - die Weisheit an der Wand", in: Aleida Assmann (ed.), 155-175.
1993a "Inschrift, Design, Objekt: Über eine Tendenz zur inneren Sprache im japanischen Design", in: Jo Enzweiler – Heinrich Popp (eds.).
1993b "Inschriftliche Illokutionen", *Dokkyo Daigaku Doitsugakukenkyuu - Dokkyo-Universität Germanistische Forschungsbeiträge* 30: 1-26.
in print "Inscriptions in daily life", in: Annette Sabban – Christian Schmitt (eds.).

Wittgenstein, Ludwig
1970 *Über Gewißheit*. Ed. by G.E.M. Anscombe – George Henrik von Wright. Frankfurt a.M.: Suhrkamp.

Yamazaki, Taikoo
1974 *Mikkyoomeisoohoo* ['The way of meditation in the secret teaching'.] Kyoto: Nagata Bunshodoo.

PART V

Appendix

List of contributors

Robin Allott
Generalist and linguist, independent scholar. Fitzgerald Park, Seaford, East Sussex, England BN25 1AX. Recent work includes *The Motor Theory of Language Origin* (1989) and "Evolutionary aspects of love and empathy" in *Journal of Social and Evolutionary Systems* (1992).

Emmanuel Anati
Archaeologist. Department of Paleo-Ethnology, University of Lecce, Italy. Director Centro Camuno di Studi Preistorici, I-25044 Capo di Ponte, Brescia, Italy. Author of numerous articles and books on prehistoric art and religions in several languages.

Myrdene Anderson
Anthropologist. Department of Sociology and Anthropology, Purdue University, West Lafayette, Indiana 47 907, USA. Her most recent major work, *On Semiotic Modeling*, coedited with Floyd Merrell, was published in 1991.

Horst Arndt
Applied linguist. Department of English, EZW, University of Cologne, D-50931 Cologne, Germany. Co-author, with Richard W. Janney, of *InterGrammar: Toward an Integrative Model of Verbal, Prosodic, and Kinesic Choices in Speech* (1987).

Bernard Bichakjian
Evolutionary linguist. Department of French, University of Nijmegen, Holland. Editor of the *LOS* (Language Origins Society) *Forum* and (co-) editor of books on Romance linguistics and on the development of speech and languages. Major publication: *Evolution in Language* (1988).

Paul Bouissac
Semiotician. Departments of French and Linguistics, University of Toronto, Canada. Editor of *The Semiotic Review of Books*. Author of books and articles on the ethnosemiotics of circus and performance in general, nonverbal communication, iconicity, and evolutionary semiosis.

Udo L. Figge
Linguist. Department of Romance Languages, Ruhr-Universität Bochum, D-44780 Bochum, Germany. Publications on cognitive linguistics, computational semiotics, language origins, e.g., in *Geneses of Language* (1990).

Peter Grzybek
Slavist, semiotician. Department of Slavic Languages and Literatures, University of Graz (KFUG), A-8010 Graz, Austria. Co-editor of *Znakolog - An International Yearbook of Slavic Semiotics* (1989ff.), editor and author of numerous publications on Slavic semiotics, semiotics of folklore, and literature.

Roland Harweg
Linguist. German Department, Ruhr-Universität Bochum, D-44780 Bochum, Germany. Author of *Kompositum und Katalysationstext, vornehmlich im späten Sanskrit* (1964), *Pronomina und Textkonstitution* ([1]1968, [2]1979), *Studien zur Deixis* (1990), and *Studien über Zeitstufen und ihre Aspektualität* (1994) as well as of numerous articles in the fields of text linguistics, German and general grammar, contrastive linguistics, and semiotics.

Gordon W. Hewes
Anthropologist. Department of Anthropology, University of Colorado, Boulder, Colorado 80309, USA. *Language Origins: A Bibliography* (1975), "History of Glottogonic Theories", in *Language Origin: A Multidisciplinary Approach* (NATO ASI Series 1992).

Yoshihiko Ikegami
Linguist with special interest in semantics, semiotics, and poetics. College of Arts and Sciences, University of Tokyo, 3-8-1 Komaba, Meguro-ku, Tokyo 153, Japan. Among the numerous publications are *Grammar of English Poetry* (1967), *Semantics* (1975), *Poetics and Cultural Semiotics* (1983), and (Ed.) *The Empire of Signs: Semiotic Essays on Japanese Culture* (1991).

Richard W. Janney
Applied linguist. Department of English, EZW, University of Cologne, D-50931 Cologne, Germany. Co-author, with Horst Arndt, of *Inter-Grammar: Toward an Integrative Model of Verbal, Prosodic, and*

Kinesic Choices in Speech (1987), and Executive Editor of the *Journal of Pragmatics*.

Martin Krampen
Psychologist. Professor emeritus. Am Hochstaess 18, D-89081 Ulm, Germany. Major publications include: *Meaning in the Urban Environment* (1979), *Icons of the Road* (1983), *Children's Drawings* (1991) and numerous contributions to the *Encyclopedic Dictionary of Semiotics* (1986) and to the handbook *Semiotik* (forthcoming).

Paul Leyhausen
Ethologist. Former Head of Wuppertal Study Group, Max Planck Institute of Behavioral Physiology, Wuppertal, Germany, now in active retirement. Auf'm Driesch 22, D-51570 Windeck, Germany. Major publications include *Cat Behavior* (1979) and *Motivation of Human and Animal Behavior* (1973; co-author Konrad Lorenz).

Peter Meyer
Sociologist. Department of Economics and Social Sciences, University of Augsburg, Memminger Str. 16, D-86159 Augsburg, Germany. Major works include *Evolution und Gewalt* (1981), *Soziobiologie und Soziologie* (1982), *Human Nature and the Function of War in Social Evolution* (1990).

Winfried Nöth
Semiotician and linguist. Department of English, University of Kassel, D-34109 Kassel, Germany. His major publications include *Strukturen des Happenings* (1972), *Dynamik semiotischer Systeme* (1977), *Literatursemiotische Analysen* (1980), *Handbuch der Semiotik* (1985), and *Handbook of Semiotics* (1990).

Jörg Pfleiderer
Astronomer. Institut für Astronomie, Leopold-Franzens-Universität, A-6020 Innsbruck, Austria. Radio astronomy, photometry, space research, image restoration. Numerous articles in scientific journals. Books: *Astronomie für Jedermann* (1983), *Ursprung und Zukunft des Weltalls* (1987).

Posner, Roland
Professor of Linguistics and Semiotics, Director of the Research Center for Semiotics. Technische Universität, Ernst-Reuter-Platz 7, D-10587

Berlin, Germany. Vice President of the International Association for Semiotic Studies, Editor of *Zeitschrift für Semiotik* and three semiotic book series. Recently published *Warnungen an die ferne Zukunft: Atommüll als Kommunikationsproblem* (1990) and *Signs, Search, and Communication: Semiotic Aspects of Artificial Intelligence* (1993).

Holger Preuschoft
Anatomist, anthropologist. Abteilung Funktionelle Morphologie, Ruhr-Universität Bochum, D-44780 Bochum, Germany. Major field of research: human evolution and the question to what extent characteristics of humans can be found in nonhuman primates. Author of numerous articles and editor of six collective volumes.

Signe Preuschoft
Psychologist, primatologist. Ethologie en Socio-oecologie, Rijksuniversiteit Utrecht, Padualaan 14, NL-3508 TB Utrecht/Netherlands. Research on primate communication and social relationships, emphasis on the evolution of facial expressions.

Helmut Schnelle
Theoretical linguist. Sprachwissenschaftliches Institut, Ruhr-Universität Bochum, D-44780 Bochum, Germany. Recently authored *Die Natur der Sprache* (1991).

Vitaly Shevoroshkin
Historical linguist. Department of Slavic Languages and Literatures, University of Michigan, Ann Arbor, Mi. 48109-1275, USA. Editor of *Explorations in Language Macrofamilies* (1989), *Proto-Languages and Proto-Cultures* (1990), and *Nostratic, Dene-Caucasian, Austric, and Amerind* (1992).

Colwyn Trevarthen
Researches brain/mind growth and communication in infancy. Department of Psychology, University of Edinburgh, 7 George Square, Edinburgh EH8 9JZ, Scotland, UK. Works include *Brain Circuits and Functions of the Mind* (1990) and "Signs before speech" (1990), in *Semiotic Web 1989*.

Susan Vogel
Linguist and translator. Englisches Seminar, Ruhr-Universität Bochum, D-44780 Bochum, Germany. Author and translator of works in Semiot-

ics, English, and History. Author of *Humor: A Semiogenetic Approach* (1989). Most recent project: *News from the Land of Freedom* (1991).

Götz Wienold
Linguist, semiotician. Faculty of Foreign Languages, Dokkyo University, 1-1 Gakuencho, Soka-shi, 340, Japan. Linguistic Typology, Semantics, Language Acquisition. Recent publications include "Genus und Semantik im Indoeuropäischen", in: *Aufgaben, Rollen und Räume von Frau und Mann* (1989) and "Natur oder Naturdarstellung?" in *Zeitschrift für Semiotik* 13 (1991).

Index of names

Index of subjects

syntactic features 278
syntagmatic 38
syntax 278

Bei Fragen zur Produktsicherheit wenden Sie sich bitte an:
If you have any questions regarding product safety,
please contact:

Walter de Gruyter GmbH
Genthiner Straße 13
10785 Berlin
productsafety@degruyterbrill.com